# Communication & the Culture of Technology

# Communication
# &theCulture
## of
# Technology

edited by

## Martin J. Medhurst
## Alberto Gonzalez

and

## Tarla Rai Peterson

Washington State University Press
Pullman, Washington
1990

Washington State University Press, Pullman, Washington, 99164-5910

Printed and bound in the United States of America

00 99 98 97 96 95 94 93 92 91   1 2 3 4 5 6 7 8 9 10

*Library of Congress Cataloging-in-Publication Data*

Communication and the culture of technology / edited by Martin J.
   Medhurst, Alberto Gonzalez & Tarla Rai Peterson.
       p.  cm.
       ISBN 0-87422-069-6. -- ISBN 0-87422-068-8 (pbk.)
        1. Technology--Social aspects.  2. Communication and culture.
   I. Medhurst, Martin J.  II. Gonzalez, Alberto, 1954- .
   III. Peterson, Tarla Rai.
   T14.5.C6523 1990
   303.48'3--dc20                        90-12257
                                                     CIP

*Communication and the Culture of Technology* was underwritten in part by the Program for Excellence in the Humanities and Social Sciences, College of Liberal Arts, Texas A&M University.

This book is printed on pH neutral, acid-free paper.

# Contents

ACKNOWLEDGEMENTS ............................................................... vii

INTRODUCTION

Human Values and the Culture of Technology ............................................. ix
    Martin J. Medhurst

PART ONE: CONSCIOUSNESS AND CULTURE

Communication Technology, Consciousness, and Culture:
Supplementing FM-2030's View of Transhumanity ..................................... 3
    Bruce E. Gronbeck

The Language of Technology: Talk, Text, and Template as
Metaphors for Communication ....................................................... 19
    James W. Carey

PART TWO: CULTURE AND IDEOLOGY

Representation of Interests and the New Communication Technologies:
Issues in Democracy and Policy .................................................... 43
    Stanley Deetz

Breaking into Silence: Technology Transfer and Mythical
Knowledge Among the Acomas of *Nuevo Mexico* ..................................... 63
    Alberto Gonzalez and Charmaine Bradley

Structuring Closure through Technological Discourse:
The Mormon Priesthood Correlation Program ............................................77
　　Tarla Rai Peterson

Modern Discourse on American Home Technologies ................................. 95
　　Karen E. Altman

## PART THREE: IDEOLOGY AND LANGUAGE

Experts, Rhetoric, and the Dilemmas of Medical Technology:
Investigating a Problem of Progressive Ideology ........................................115
　　Michael J. Hyde

Rhetorical Maintenance of Technological Society:
Commercial Nuclear Power and Social Orthodoxy ....................................137
　　Micheal R. Vickery

The Technological Priesthood: A Case Study of
Scientists, Engineers, and Physicians for Johnson-Humphrey ....................157
　　David Henry

## PART FOUR: LANGUAGE AND MEDIA

Technical Heterogeneity, Specialization, and Differing Motives:
An Examination of the Influences of Technology on Group and
Organizational Decision Making ...............................................................183
　　Roger C. Pace and Steven Hartwell

Language As and In Technology: Facilitating Topic Organization in a
Videotex Focus Group Meeting .................................................................197
　　Wayne A. Beach

Telephone Speaking and the Rediscovery of Conversation .........................221
　　Robert Hopper

## PART FIVE: MEDIA AND CONSCIOUSNESS

The Technological Shadow in *The Manchurian Candidate* .........................239
　　Thomas S. Frentz and Janice Hocker Rushing

The Ultimate Technology: Frederick Wiseman's *Missile* ...........................257
　　Thomas W. Benson and Carolyn Anderson

# Acknowledgements

The essays in this volume were originally presented at the 1989 Conference on Communication and the Culture of Technology, hosted by Texas A&M University. The Conference was underwritten, in part, by the Program for Excellence in the Humanities and Social Sciences and by the College of Liberal Arts at Texas A&M. We are especially grateful to Daniel Fallon, Dean of the College of Liberal Arts, and Associate Deans Charles Johnson and Paul Parrish for their encouragement and support.

A special word of appreciation is due Professor Peter McIntyre of the Texas A&M Physics Department. Professor McIntyre, one of the originators of the idea of a Superconducting Supercollider, made a particularly stimulating presentation and proved to the satisfaction of all in attendance that superior communication skills are not the exclusive province of those in the liberal arts.

Finally, we would like to thank all those who participated in the Conference, both those who presented papers and those who listened, questioned, and responded. The essays that compose this volume are much better because of the give-and-take that followed the initial presentations.

# Human Values and the Culture of Technology

*Martin J. Medhurst*

It is commonly said that we live in a technological age. If this is so, then one might reasonably inquire as to the moment at which this "age" began. Did it commence with the introduction of the personal computer or the explosion of the first atomic bomb? Was it the steam engine or the Model-T that ushered in the age of technology, or perhaps telegraphy or the discovery of electricity? Some point to the introduction of moveable type and printing as the dawn of technological man. But already we have retreated to the fifteenth century and there is no end or, rather, no beginning in sight. From the moment ancient hominoids became *homo faber*, users of tools, ours has been a technological age.

Indeed, one might well make a case that the dawn of the technological age is co-terminus with the birth of humanity's most advanced technological instrument: language. If mankind invented language, then just as surely language invented mankind—or so says Kenneth Burke.[1] It is certainly true that without the capacity for symbolic action in the form of language use we would not, today, find ourselves in a technological dilemma of our own making. Konrad Lorenz states the case succinctly when he observes that "all the great dangers threatening humanity with extinction are direct consequences of conceptual thought and verbal speech."[2] In short, "without symbols there could be no technologies,"[3] neither those that threaten humanity nor those that liberate it.

Yet even the terms of debate are ambiguous for that which threatens at one moment (nuclear weapons) liberates at the next (nuclear medicine). The opposite is also true. That which is thought to be a form of "liberation" when conceived and introduced is sometimes found, upon reflection, to be a threat to self, others, or society. Medical advances save and prolong lives, but at what cost to self-dignity, interpersonal relations, and larger social goods? Interactive cable television brings sports, banking, and shopping into our living rooms, but with what consequences for self-concept, interpersonal communication skills, or social knowledge?

To study "Communication and the Culture of Technology" is to be interested in the matrix that is formed by human curiosity, making, use, and reaction, for the culture of technology is more than the sum of its parts. It is more than the discovery of means to accomplish an end; more than the engineering of a specific design; more than the application of that design to solve a particular problem; and more than the subsequent actions taken to correct for unexpected consequences of that application.

John Kenneth Galbraith was correct when he defined technology as "the systematic application of scientific or other organized knowledge to practical tasks,"[4] but such a definition tells us almost nothing about the specifically human dimensions that accompany the creation of a culture of technology. It tells us nothing of the political motives that call forth technological means; nothing of the subtle shaping of consciousness that the design and utilization of technological means fosters; nothing of the social and psychological effects of living in a technological culture; nothing of the reflexive relationships among technology, economics, ideology, and power; and nothing of what it means to be a human being engaged in communication, be it through face-to-face conversation, telephone, television, newspaper, or film.

## Technological Faith

The problematics of the culture of technology have very little to do with applying organized knowledge to accomplish practical ends. Humans have been extraordinarily successful in such endeavors. So much so that until quite recently faith in humanity's abilities to solve its problems with a technological fix was "the implicit 'religion' of the West."[5] Technology would be our salvation if only we would put our faith in its omnipotent designs. That such faith parallels and fulfills the same functions as traditional religion has been noted by several scholars. Stanley Jaki explains:

> Those mindful of the name *Yahweh*, or HE WHO IS—that is, the very foundation of existence—will easily notice a theological undertone in the declaration according to which technology enables man to reshape his very existence. Those mindful of that omnipotence that has been traditionally ascribed to God, will not fail to think of theology on hearing some engineers declare than man, by virtue of technology, can do almost anything.[6]

Not so long ago, such a faith in humanity's ability to engineer its fate was commonplace. But something happened on the way to paradise: automation/unemployment; industrialization/air pollution; fluorocarbons/ozone depletion; synthetics/Love Canal; lasers/Star Wars; nuclear power/Three Mile Island; atomic fission/Hiroshima. The problem is not that our technologies fail to work, but that they work too well, with consequences that outpace even the most farsighted among us. Langdon Gilkey poses the crucial question: "If a valid science and a reliable technology can really compound our problems rather than dissolve them, what does *that* mean about man and about the history he helps to create?"[7] Are we the masters of our own fate, the captains of our souls; or, are we merely the servants of our own devices? And if servants, must we be forever indentured or does an emancipation await us, an emancipation not so much from the tools that would destroy us but from the modes of thought, values, and desires that played a central role in the creation and use of those tools?

Gilkey locates the source of the technological dilemma when he writes: "For what clearly is amiss here, what reintroduces the ills we thought almost banished, is not our intelligence, inquiry and technology per se; our creativity in itself is not at fault. Rather it is the demonic use to which it is put. At fault, as our religious traditions have emphasized, are the infinite desire and concupiscence, the greed and selfishness which motivate our use of scientific intelligence and technological power."[8] Such greed and selfishness, says Gilkey, are the seedbeds of technological problems. Human nature, not human tools, is in need of re-examination for it is out of the desires of the heart that humanity's technological "miracles" come into being.

Put differently, it is not the tools that humans have created that constitute the problem. Instead, it is the way humans have conceptualized, communicated, and created various cultures with those tools; cultures that body forth values, attitudes, and incipient belief systems; cultures that privilege some and disenfranchise others; cultures that revolutionize concepts of selfhood and understandings of who, in fact, is one's neighbor; cultures that often take on a life of their own, apart from conscious, human decision making; cultures that are in need of examination, analysis, and criticism.

Too often the debate about technology and its effects on human society has been dominated by the extremes: those who hold technology to be the way of humanity's salvation from all manner of evil and those who hold technology, itself, to be an evil; something to be exorcised from the human spirit. Neither extreme is helpful, for as Michael Hyde has noted, "the danger is not technology; rather, the danger is us—we who do not question, we who do not understand, we who do not communicate beyond the rhetoric of either/or."[9]

This volume is an attempt to move beyond either/or reasoning. Between technology as miracle and technology as mirage lies the gray area of human choice making, human valuing. This is the intellectual arena that the contributors to this volume inhabit. Like Stephen Cutcliffe, the authors represented in this book realize that "neither science nor technology are autonomous juggernauts with lives of their own, but neither are they simply mental tools subject to ready redirection and utilization by just any passing need or interest. Rather, they are complex enterprises taking place in specific contexts shaped by, and in turn shaping, human values as reflected and refracted in cultural, political, and economic institutions."[10]

## TECHNOLOGY AS CONSCIOUSNESS AND CULTURE

In section one, Bruce E. Gronbeck and James W. Carey explore the role of communication in the shaping of consciousness and culture. Gronbeck reviews the theories of McLuhan, Ong, Ellul, and others, and warns about the twin dangers of technological determinism and what he calls "cultural transformation theory."

Gronbeck posits that "cultural evolution is, not a biological process that we can visualize literally in our archaeological records, but a human construct bounded by rhetorical brackets." Such cultural transformations, he holds, are "human products engineered by that special group of technicians we call rhetoricians."

Carey, too, is interested in social and cultural transformation. He writes about "the story of social evolution told as the evolution of communications." In this story, "to speak or to write or to program . . . is not merely to pick up a tool or to exercise a skill. It is to constitute a world, to bring a world into existence, and to simultaneously constitute a self. The artifacts of communications differ, as do the social practices they engender, but they are linked in a chain of transformation: a process whereby the world and the self is reconstituted." Carey explores the metaphors of talk, text, and template through the writings of Goody, Eisenstein, and Innis, concluding that "artifacts have a deep moral significance precisely because they can be made to speak to larger issues—our liberties and our enslavements."

## TECHNOLOGY AS CULTURE AND IDEOLOGY

In section two, specific examples of the "liberties" and "enslavements" of human technology are analyzed. Stanley Deetz sets the tone for the case studies that follow by meditating on the Politics of technological development. Deetz argues that "each technology structures sensory advantages, provides ways of knowing the world, privileges certain notions of what is real and posits personal identities. A technology posits a subject, has an epistemology, and structures value choices." Most importantly, "technology is never Politically neutral." Deetz searches for the building blocks from which to construct "a moral foundation for communication technology policy and collective self-formation."

It is the problem of self-formation in a particular collectivity that motivates Gonzalez and Bradley to investigate technology transfer among the Acomas of New Mexico. "At Acoma," argue the authors, "the state has unilaterally prescribed and exported to the reservation a particular communication technology without reference to the specific cultural relevancies of the recipient." By taking into consideration the roles of cultural "significance, translation, and integration" in technology transfer, Gonzalez and Bradley hope to forestall future instances of "ethnocentric and repressive" telecommunication initiatives.

No less repressive than cultural exploitation is the use of language as a technological means of marginalizing people. Tarla Rai Peterson, focusing on the "correlation program" of the Mormon Church, shows how technological discourse—*"language used to structure human action according to rules of closed systems"*—is used by the Church to define what it means to be a man or a woman. "By formally removing women from the direct line of institutional authority," Peterson argues, "correlation potentially undermines the ability it claims to promote."

While Peterson examines a type of discourse, Karen E. Altman excavates a particular technological site: the American home. Altman demonstrates how "discourses and images that circulate widely in American culture . . . construct ways of seeing, imagining, or valuing technology in the home." In so doing, she shows that "repeated themes in the modern formation constituted everyday knowledges about who owned technologies, who operated them, for which purposes, and in what interests."

## TECHNOLOGY AS IDEOLOGY AND LANGUAGE

Michael Hyde opens section three by investigating the "problem of progressive ideology" as it is manifest in the "dilemmas of medical technology." Hyde offers "an examination of a professional community whose sovereign status in the twentieth century was made possible by its appropriation of the ways and means of Progressive ideology and, consequently, by its ability to use its scientific and technological expertise for the purpose of meeting the expectations of its clientele." By examining the public moral argument that swirls around the use (and misuse) of medical technology, Hyde provides a striking case study of the interface among expertise, rhetoric, and ideology.

Micheal R. Vickery is also interested in this interface, but as it applies to the commercial nuclear power industry. He argues that "comparatively little has been done to reveal the ways in which technology functions as a premise in the communicative process through which a social order is constituted and maintained," and proceeds to detail the rhetorical implications of such a premise. One implication is that the rhetoric of nuclear power constitutes "a secular faith in the absolute potency of human will-to-control and the legitimacy of technological agencies through which that will is exercised."

It is this same secular faith bodied forth in a "technological priesthood" that motivates David Henry's case study of "Scientists, Engineers, and Physicians for Johnson-Humphrey." Henry holds that "the activism of Scientists, Engineers, and Physicians for Johnson-Humphrey evidences the working in tandem of the bardic and priestly voices to create a scientific ethos unique to the technological priesthood." Like Hyde and Vickery, Henry is interested in the ideological implications of such a creation of character. He warns us "to guard against unwitting acceptance of apparent expertise and presumed authority" and to remember that "a technological world is too important to be left to the technologists alone."

## TECHNOLOGY AS LANGUAGE AND MEDIA

While section three deals with the ideological and linguistic fallout of technological expertise, section four identifies both language and logic as, themselves, technologies—technologies that are made manifest in and through various media. Too often, we tend to conceptualize technology in terms of things—medical devices, power plants, cable television—when, in reality, all uses of language and logic are technological acts: acts that proceed by their own rules, norms, and conventions; acts that are purposive, task-oriented, and productive of real-world consequences. The essays by Roger C. Pace and Steven Hartwell, Wayne A. Beach, and Robert Hopper all focus on language or logic as technologies for getting things done.

Pace and Hartwell explore the logic of causal reasoning as it applies to the American legal system. "Law-trained people," they argue, "tend to understand causation as deductive and operating in specific instances. The legal model of causality draws from a deductive, Newtonian world perspective." Courts of law tend to reject arguments predicated on induction, especially those based on statistical models of probability. Pace and Hartwell investigate the potential consequences of privileging one type of logic or technology over another and speculate that the problems of "technical heterogeneity" account for many of the "divisions that impede decision making" in groups and organizations.

Beach is also interested in group communication, especially in how language is used as a technology for accomplishing a task. Beach argues that "conversation and technology are reflexively coupled: Conversational activities are technological achievements in and through co-participants' methods for getting tasks done, just as descriptions of the impact of specific technologies on everyday life are possible *only* through the language employed to produce such descriptions." Conversation, for Beach, is a "technological resource, the organization of which is ultimately rooted in practical circumstances of everyday choice and action."

While Beach is primarily interested in the technology of talk, Robert Hopper explores the potential connections between conversation and the development of a specific technology: the telephone. Hopper claims that "we have rediscovered the human conversation through our telephone conversations." He offers four lessons that telephone conversations teach about the nature of language and conversation, concluding that "telephone conversation is pure speech communication." As such, telephone conversations "help us pay attention to the importance of audition to speech communication and to the centrality of turn taking in interaction."

## TECHNOLOGY AS MEDIA AND CONSCIOUSNESS

Section five continues the emphasis on a particular medium, but this time the subject is film, perhaps the most powerful medium for presenting a vision of technology's relationship to humanity. Thomas S. Frentz and Janice Hocker Rushing explore this relationship by conducting a Jungian analysis of *The Manchurian Candidate*, a film that "presents a portrait of the psyche as losing its identity with the eruption of the technological shadow." The authors find the film to be "uncannily prophetic in its depiction of the peculiarly American form of technological dehumanization."

Thomas W. Benson and Carolyn Anderson analyze Frederick Wiseman's documentary film, *Missile*, as "the ultimate technology." They study "how Wiseman's text presents itself as rhetoric, how he engages us as participating spectators, and how he draws us into his own peculiar contemplation of the apocalypse." Will we control our technologies or will our technologies control—and perhaps ultimately destroy—their makers? And what, precisely, does it mean to be in control? These are questions that inform both analyses in this section and that point us, once again, to the realization that each particular medium—film, telephone, television, newspaper, computer—eventually leads us back to questions of consciousness, culture, and communication.

## TECHNOLOGY AS RHETORICAL TRADITION

Though one can make somewhat arbitrary divisions between and among consciousness, culture, ideology, language, and media, each remains inextricably linked to the others. And the glue that binds—the common denominator—is human communication. Indeed, the very term "technology" comes from two Greek words—*techne* and *logos*—that were central terms for classical rhetorical theory. Classical rhetoric can be read as the ongoing battle between these two concepts. Was the purposive use of human language to be a craft (*techne*) somewhat akin to cooking or woodworking or basketry? Or, was rhetoric to be a creative art, an art that required invention, reason, and prudential judgment, an art that could not be reduced to a handbook or a set of rules or techniques because it involved the power to call ideas into being, the power to create the world through human speech (*logos*), thereby endowing it with purpose and meaning.

Just as the ancients failed to solve all the ambiguities inherent in the theory/art/craft of rhetoric, so have the contributors to this volume failed to "solve" all of the problems associated with the culture of technology. But by drawing upon that ancient rhetorical tradition, modern scholars of communication have employed one of the most important inheritances from that tradition: a focus on prudential judgment. Such judgment can never be absolute, such conclusions never final, never closed to debate, revision, or rethinking. And this is as it should be, for technology, like rhetoric, never operates in a vacuum, divorced from the contingencies of time and place. If this volume adds but a little wisdom to the ongoing debate over technology and the culture it has created, if it raises questions, challenges assumptions, provides insights, motivates change, or provokes a responding voice, then it has fulfilled its role as a rhetoric worthy of serious contemplation.

## NOTES

1 Kenneth Burke, "Definition of Man," in his *Language as Symbolic Action* (Berkeley: University of California Press, 1966), 9.
2 Konrad Lorenz, *On Aggression* (New York: Harcourt Brace & World, 1966), 238.
3 Leo J. Moser, *The Technology Trap: Survival in a Man-Made Environment* (Chicago: Nelson-Hall, 1979), 97.
4 John Kenneth Galbraith, *The New Industrial State* (Boston: Houghton Mifflin, 1985), 12.
5 Langdon Gilkey, "The Religious Dilemmas of Scientific Culture: The Interface of Technology, History, and Religion," in *Being Human in a Technological Age*, ed. Donald M. Borchert and David Stewart (Athens: Ohio University Press, 1979), 78.
6 Stanley L. Jaki, "The Three Faces of Technology: Idol, Nemesis, Marvel, " *The Intercollegiate Review* 23 (Spring 1988): 40.
7 Gilkey, 78.
8 Gilkey, 84.
9 Michael J. Hyde, "Introduction: The Debate Concerning Technology," in *Communication Philosophy and the Technological Age*, ed. Michael J. Hyde (University: University of Alabama Press, 1982), 4.
10 Stephen H. Cutcliffe, "Science, Technology and Society: An Interdisciplinary Academic Field," *Phi Kappa Phi Journal* (Spring 1989): 25.

# Part One

# Consciousness and Culture

# Communication Technology, Consciousness, and Culture: Supplementing FM-2030's View of Transhumanity

## Bruce E. Gronbeck

Do you use your intelligence adequately to monitor the world around you? Do you think things through?

Do you manage your emotions intelligently?

Does the quality of your everyday life need improving?

Does your leisure/fun/work ratio need balancing?

Are you behind in your use of new technology?

Are you telefficient—making effective use of new telecom to access information and services?

How far along are you in shifting from a high-stress low-yield industrial-age track to a low-stress high-yield telespheral life?

Do your values need updating? For example does your competitiveness depreciate the quality of your life and your potentials for growth in all areas?

Does your lifestyle need realigning? How aware and open are you to new methods of procreation—new options for shared parenting—new networks of intimacy?

Are you sufficiently fluid in an increasingly fluid world?

Does your appreciation of art and culture need updating?

Does your ideological orientation need adjusting?

Does your level of humanity need refining?

How mobile and telecommunitized are you?

Are your loyalties and commitments keeping up with an ever-expanding global environment?

How involved are you in our new extraterrestrial environment?

Are your attitudes to[ward] life and death keeping up with all the gains we are making in the immortality movement?

In the Age of Information how information rich are you? How updated are you on the accelerating pace of advances in all areas of life?[1]

These are the questions a futurist named "FM-2030" asks of those who wish to accelerate their RPGs, or rates of personal growth. FM-2030 has devised a series of twenty-five tests people can use to assess their preparation for a new age. In that age, which the author assumes will be on us in the year 2030, humans will be evolving past and through "the premises of biological terrestrial life that have always defined the human."[2] In that time, thanks to shifts in consciousness and culture rooted in technological transformations, we will pass out of the age of humanity into one more state of post—this time, the age of posthumanity. FM-2030's twenty-five monitoring tests cover changes in one's vocabulary and ideology, understanding of time in both daily and epochal measures, powers of mind as well as expressions of emotion, both family and cosmic orientations, attachments to rituals and power—in all, changes in technology that alter the constituents of both consciousness and culture. In the sweep of FM-2030's queries we find a totalizing vision that befits our subject matter. It is on that vision, a bifocal vision looking inward to consciousness and outward to culture, that I will concentrate.

My title is suggested by FM-2030's breadth of vision. His analyses carry with them the notion of technological determinism—the assumption that new technologies act as "causes" that produce discrete individual and social "effects." I want to challenge that assumption with much the same spirit but in ways different from those with which Raymond Williams challenged it in 1975:

> The most precise and discriminating local study of 'effects' can remain superficial if we have not looked into the notions of cause and effect, as between a technology and a society, a technology and a culture, a technology and a psychology, which underlie our questions and may often determine our answers. . . . Until we have begun to answer them, we really do not know, in any particular case, whether, for example, we are talking about a technology or about the uses of a technology; about necessary institutions or particular and changeable institutions; about a content or about a form.[3]

Now, Williams' challenge to technological determinism focused on human instrumentalities—on human *intentions*—that bring technologies into being so that they can serve human social and political *needs*. His is an important argument, one that has shaped British Cultural Studies.[4] Too often, however, British Cultural Studies's examinations of communication technology move facilely from an intention/needs analysis to an evil intention/artificial needs analysis, in studies of how a class-based society allows media oligarchists to construct ideologies able to lock the masses into hegemonic structures of domination. A view this politicized has seldom done well in American scholarly circles, and hence we need to reframe Williams' attack upon technological determinism to make it fit American intellectual predilections.

FM-2030's vision suggests a second task. FM-2030 sees humanity breaking with its "biological terrestrial life,"[5] as though we somehow could escape history when the right technology frees us. In his view, we live in a post-industrial age where the most advanced among us are "transhumans," waiting for technologies to make us "posthumans." He says, "Many of the breakthroughs embodied in transhumans are nothing less than the beginnings of the eventual transformation of the human species."[6] In such transformations, can humans thus slough off their history, as so much cocooned containment, and fly away as so many mechanized monarchs? I think not, and wish to make some points about cultural transformation theory, especially the versions being circulated in communication studies.

Overall, I hope to come out of this brief look at technological determinism and cultural transformation theory better able to identify some research projects awaiting execution. To help, we should review the work of two scholars who have shaped much American thinking about communication technology, consciousness, and culture: Marshall McLuhan and, more deeply, Walter Ong. I want to use the so-called orality-literacy theorems as means to frame the two principal issues facing us and as prods that push research in particular directions.

## THE ORALITY-LITERACY THEOREMS

First, then, how can we construe the orality-literacy theorems in ways useful for our foci? I would start us where the classicist Eric Havelock liked to start, within a twelve-month period in 1962 and 1963. In that period, he noted, five important works were published: Claude Lévi-Strauss's *The Savage Mind* (1962), Goody and Watts's "The Consequences of Literacy" (1968), Marshall McLuhan's *The Gutenberg Galaxy* (1962), Ernst Mayr's *Animal Species and Evolution* (1963), and Havelock's own *Preface to Plato* (1963). These works produced what Havelock talked about as the "oral-literate equation,"[7] or what Ong more generally discusses as "orality-literacy theorems."[8] The mathematical metaphors used here are appropriate, for both men advance basic axioms as givens from which to deduce explanations for characteristics of oral, written, electronic, and other sorts of discourse. Lévi-Strauss's *The Savage Mind* traversed the relationships between tribal myth and contemporary language, especially nomination; Mayr found language the great link between biology and society; Goody and Watt attended to the particular collective functions of orality; McLuhan dealt with the sociopolitical and to a lesser extent the psychological effects of the literacy revolution following the printing press; and, in the Greek move to writing, Havelock discovered a key to fifth-century B.C. Greece's cultural revolution.

## MARSHALL MCLUHAN ON MEDIA

For these investigators, the terms "orality" and "literacy" connoted more than mere media, more than "a pipeline transfer of units of material called 'information' from one place to another."[9] McLuhan got most of us thinking about media with his aphorisms "The medium is the message," "The medium is the massage," and "The medium is the mess-age."[10] In those three terms "message," "massage," and "mess-age," we have reference to communication systems, psychological perception, and cultural orientation—the three primary concepts we are confronting.

Whatever one might think about his scholarship and pronouncements as the Oracle of the Electronic Age, McLuhan's role in articulating and populariz-ing new views of the media cannot be overemphasized. In the 1960s, he played the fool, slipping from conference to television show to a $100,000 chair at Fordham, offering comic and ironic commentaries on the foibles of academic inquiry and illustrated cultural critiques of America's insensitivity to change. He taught us new concepts: "hot" and "cool" media, "synesthetic" experience, the "outering" of knowledge by media, culture as a "rear view mirror," and media technologies as "extensions of our physical being."[11] In all of this intellectual activity a quarter of a century ago, Marshall McLuhan was raising the profile of communication technology, examining its effects upon mind and society.

In the 1970s, however, he began to challenge his own technological determinism. "To dismiss McLuhan as a technological determinist," as Kroker notes, "is to miss entirely the point of his intellectual contribution."[12] While he never totally abandoned his theories of being-extension and technological determinism, he did change emphases. He turned in his clown makeup for a seer's rod. As he formulated what he termed "McLuhan's Laws of the Media"—which were to culminate in a book of exemplars—he downplayed the linearity of his earlier paradigm and emphasized instead a structuralist view of cultural history. His laws were based on what he variously called "a four-part analogy" (that is, a proportion, A:B::C:D) or "tetrad" or an expansion of the usual figure-ground dichotomy into a four-member metaphorical structure with two figures and two grounds or "the four irreducible relations in technology."[13] The four psychosocial changes that McLuhan found repeatedly locked into tetrads were amplification of some situation, obsoletising of earlier psychosocial balances, recreation of older contents of communication, and a perceptual flipping from one psychosocial state into another. To quote from his 1977 article:

> For example—THE LAWS OF EQUILIBRIUM: (A) Any input [into a social system] amplifies or intensifies some situation (inflates); (B) Obsolesces existing homeostasis or balance; (C) Recreates an older mode of equilibrium (e.g. Eliot—Auditory Imagination); (D) When pushed to its limits, the system reverses its modali-ties . . .
>
> PRINTED WORD: (A) Amplifies private authorship, the competitive, goal-oriented individual; (letters are an extension of the teeth, the only lineal and repetitive part of the body; as Harold Innis explains, writing on paper leads to military bureaucracies); (B) Obsolesces slang, dialects, and group identity; separates composition and perfor-mance, divorces eye and ear; (C) Retrieves tribal elitism, charmed circles, the "neck verse"; (print makes everyone a reader, and Xerox makes everyone a publisher); (D) With flip from manuscript into mass production via print, there comes the corporate reading public and the "historical sense."[14]

Such passages highlight McLuhan's structuralist thought, illustrate the operation of the "laws" that transcend a linear model of technological determin-ism, and show what Curtis has termed the "polyphonic" character of the theory.[15]

Locked into historical relationships with each new technology were the ampli-
fying of some situation and an obsolescing of a past social practice, and a
remaking of an old medium simultaneously with a major change in some
communication technology. Thus, in his example of print technology, private
authorship is amplified even as oralisms are made obsolete, and tribal oligarchies
reemerge even as mass literacy is made possible, leading to the power struggle
of nationalism that McLuhan discusses at length in *The Gutenberg Galaxy*. The
element of simultaneity visible in such laws all but destroyed causal analysis.
That McLuhan believed he had firm control of a central principle of human
history can be seen in the issue of *Technology and Society* where he introduced
the laws:

> You will note that, although these are called Laws of the Media, only a few of them
> deal with communications media narrowly conceived. Instead, I am talking about
> "media" in terms of a larger entity of information and perception which forms our
> thoughts, structures our experience, and determines our view of the world about us. . . . I
> call them "laws" because they represent, as do scientific "laws," an ordering of thought
> and experience which has not yet been disproved; I call them "laws of the media" because
> the channels and impact of today's electronic communication systems provide the infor-
> mational foundation upon which we order, or structure, these experiential perceptions.[16]

In that article McLuhan then went on to illustrate the laws by presenting
outlined analyses of housing, the elevator, clothing, number, steamboats, rail-
ways, the Copernican revolution, photoduplication, microphones, money, the
wheel, printing, instant replay technology, satellites, and electric media. The
structuralistic bases for this work are clear, as he took cross-sections of technological developments in the psychosocial world and riveted them into synchro-
nous relationships to each other. These relationships are not really causal, for
McLuhan termed the space between any two of the analogical elements "a
resonating interval."[17]

Perhaps because words tumbled out of his mouth as either epigrams or
preachments, perhaps because his style was too public for the academy, and
perhaps because he talked about media at the height of social science's ascen-
dency in mass communication research, Marshall McLuhan did not strike deeply
enough to be taken seriously by most scholars. The same has not been true for
Walter Ong, dubbed by some the thinking man's McLuhan.[18]

## WALTER ONG'S THEOREMS

Walter Ong, who had McLuhan as a teacher before World War II, is best
known internationally for his 1982 book, *Orality and Literacy*, already translated
into multiple languages. His version of the orality-literacy theorems is sophisticated enough to avoid technological determinism, though he writes on the brink
of cultural transformation theory and hence is a fitting subject to study such
predicates. I first want to review some of the primary features of Ong's thinking
about orality, literacy, and electronic media, beginning with his older developmental model of communication media.

To Ong, in "primary oral cultures"—societies that have no literate modes of communication—thought and expression have a series of identifiable features. They are (a) additive rather than subordinate, with details or items piled one upon the other; (b) aggregative rather than analytic, with ideas clustered in clichés and maxims that aid memory; (c) redundant or "copious," with much repetition of the "just said," which keeps hearers and speakers on the same track; (d) tied to the experiential lifeworld for communication and images, with knowledge concretized rather than abstracted; (e) agonistically toned, with knowledge claims tested in combat rather than contemplated in individualized essay; (f) conservative or traditionalist, with the culture's primary commitments frozen in narratives and aphorisms that can be memorized and repeated easily to the next generation, since there is no other way of keeping social history; (g) empathetic and participatory rather than objectively distanced, featuring involving, personalized formulaic expressions of thought rather than the objectifications of the world possible in print cultures; (h) homeostatic, with societies living in a kind of permanent present, shedding the old that does not serve the here-and-now, yet retaining what is useful; and (i) situational rather than abstract, for oral language users cannot keep in mind the abstractions that can be recorded on paper because memory is largely concrete.[19]

In a second stage of evolution, writing restructures consciousness. Writing is a technology, a means of exteriorizing thought, that alienates the self from nature and even, by allowing for individuation, from other selves. Writing allows for the development of lists, "facts," science, and other marks of exteriorization of knowledge. It distances people by interposing texts between them, and permits the development of the feminine in culture. And, writing even produces totally reorganized societies.[20]

Electronic media—Ong includes the telegraph, telephone, radio, sound motion pictures, television, and computers—comprise a third stage of human evolution. In the 1960s, he tended to discuss them along McLuhanesque lines, as in "Transformations of the Word" (1967); that led him to think of electronic media in evolutionary terms, in terms of both the Freudian psychosexual developmental model and a sociobiological model.[21] "Transformations of the Word," first developed as a Terry Lecture for Yale in 1964, is perhaps Ong's most historicist conceptualization of humanity's three-stage, linear evolution, his most blatant statement of cultural transformation theory. In this period, he is breaking the back of cyclicist thinking and reconciling Darwinist evolutionary thought with Christianity.

Ong struggles to free himself from linearity, however, because he is convinced that change always brings with it "residues" from older cultural dispensations. Thus, the idea of phylogeny, of cultural developments in which stores of experience and knowledge are accumulated and passed on genetically from one generation to another, becomes appealing to him in the 1970s.[22] Orations are seen as "oral residues," written speeches containing traces of preliterate oral culture, and the key phrase moving Ong out of the 1970s and into the 1980s is "secondary orality." He notes that "the basic orality of language is permanent,"[23] by which he means both that the great majority of the world's languages are not written and that traces of orality are present in all languages. As early as his essay "The Literate Orality of Popular Culture" (1971), Ong says: "Secondary orality is founded on—though it departs from—the individualized introversion of the age of writing, print, and rationalism which intervened between it and primary orality and which remains as part of us. History is deposited permanently, but not unalterably, as personality structure."[24]

Here is a kind of hermeneutic understanding of mediation, one wherein the past is re-presented albeit in altered forms in current practice. So, for example, Ong argues: "Secondary orality has generated a strong group sense, for listening to spoken words forms hearers into a group, a true audience . . . But secondary orality generates a sense of groups immeasurably larger than those of primary oral culture—McLuhan's 'global village.'"[25] Secondary orality also carries forward the agonistic practices of older eras, not whole, but in forms altered by succeeding communication technologies. Hence, Ong says of televised political debate:

> On television contending presidential candidates do not stomp about a platform flailing their arms or even stand out in the open, like earlier orators metonymically claiming possession of the field, but install themselves behind protective lecterns for genteel exchanges of words projecting images of their self-contained selves instead of pacing up and down a rostrum flailing verbally at one another. They have texts in front of them— a state of affairs unknown to orators from antiquity through the Renaissance and beyond.
>
> Writing governs our oral delivery as never before, and since, as has been seen, writing is interiorizing and nonforensic, the agonistic edge of oratory is dulled.[26]

Ong thus complicates his early linear model of cultural transformation, producing a kind of hermeneutic figure. In two of his latest books, *Fighting for Life* (1981) and *Hopkins, the Self, and God* (1986), Ong completes his portrait of the present age by showing how the past is tied to the present in productive tension. While humankind always has valued historicity, following the spread of writing it could invent history proper, recorded chronicles and their interpretations.[27] In that way, both histories and historicity, that is, both outered and inner knowledge of the past, combine to produce a noetic state in which even technologically sophisticated media function with characteristics of ancient media, and hence take on the status of secondary orality; and thus, our cultures exist as the global villages McLuhan depicted, diversified, fragmented, and yet united in a re-presented past that is part of our present.

Ong's orality-literacy theorems, therefore, in good structuralist practice are founded on oppositions. The two terms "orality" and "literacy" can be combined in different ways to produce a sophisticated analysis. They can be opposed, used with one modifying the other ("literate orality" and "oral literature"), employed adjectivally ("literate popular culture" and "oral noetics"), and even examined derivatively, with both orality and literacy seen as "transformations of the word." Out of such differentiations is born a playful, critical scholarship. Ong returns time and again to orality-literacy theorems because they widen his vision and enlarge his generalizations so he can subsume under them conceptualizations of both self and society—both consciousness and culture.

With this much background, let us move to a discussion of two ideas that often accompany writing about technology—technological determinism and cultural transformation theory.

## BLUNTING TECHNOLOGICAL DETERMINISM

Ever since he wrote *The Technological Society* in 1964, Jacques Ellul has been a most consistent advocate of technological determinism. His concept of la technique is sweeping and all-pervasive. He depicts technology, not simply as a body of technical knowledge or rules, not as engineering practices, not as physical tools or instruments, or, for that matter, not even as the organization of people and resources into large-scale social systems and institutions—common enough definitions. Rather, he speaks of a technological condition, an essential character and quality of life that results from the development and unavoidable use of technological knowledge. To Ellul and others like him, autonomous technology is self-perpetuating and destructive of human values—even humanity itself. In the words of Hans Jonas, "the triumph of *homo faber* over his external object means also his triumph in the internal constitution of *homo sapiens*, of which he used to be a subsidiary part," and hence "man himself has been added to the objects of technology."[28]

As Emmanuel Mesthene has noted, technological determinism is a position articulated and attacked by two sustained ideologies—the "back to nature" philosophies of Rousseau, Thoreau, and some branches of 20th-century ecologists, and variations on socialism and Marxism that associate technology with capitalist values. For various reasons, neither of these philosophies provides a satisfying understanding of technological determinism. Insofar as life is irreversible, it is impossible for 240 million Americans to return to states of berry-eating and wood-burning; technology is not so easily shed as the naturists think, nor, for that matter, ought it to be. And, as Turkle argues, we have gotten beyond "the romanticism of the 1960s, when people set themselves in opposition to technology and rationalism."[29] Alternatively, the 19th-century socialist-Marxist critique of technology and capitalism does not get us much farther. Insofar as the view asks us to see technology employed in the service of capitalist ideology—or, now, technology in service of the perverse new sciences—it asks us to judge the ends to which technologies are put.[30] In both of those Marxist views, we are forced to separate technology from the rest of existence, to judge its instrumentalities rather than its relationships to consciousness and culture.

I mean to suggest that some of the attacks on technology merely posit a determinist position so as to ring alarm bells, to cry out a warning against mechanical wolves. Students of communication studies face much stronger arguments favoring technological determinism. Certainly one of Marshall McLuhan's mentors, economic historian Harold Adams Innis, offered a dazzling transnational (and determinist) narrative, relating communication technologies to internal and external governmental systems across the centuries of Western existence.[31] More recently, Alvin Toffler's argument in *The Third Wave* was built on a technological determinist view of the Industrial Revolution and a causal analysis of new social arrangements possible in the wake of new energy, agricultural, technological, and communication systems. And, too, most developers of contemporary electronic communication systems shiver when thinking about the social impacts they can have. As the editors of a recent anthology on communication technology and information tell us, "the new devices for gathering, storing, transmitting—and even initiating—information present the human mind with exhilarating possibilities. It is hardly to be wondered that some, including even scientists working in the field of communication, stand in awe of what is happening, and refer to the results of the new technology as having implications equaling those that followed the invention of printing or the industrial revolution."[32]

Hence, technological determinism does not go away, and paeans are sung to it in our own studies. That is one reason why we ought to study the gospel according to McLuhan and Ong—stories of two great thinkers about communication media who dallied with the determinist thesis, attractive as it was as a way of explaining the evolution of media, consciousness, and culture, but who then fought it off in favor of other strategies.

McLuhan's cross-sectional approach, even if one wishes to discard the particular elements he arrays in his tetrads, has much to recommend it. He asks researchers to examine multiple variables as voices in dialogue: some voices, characteristics of communication media, and other voices, from both the perceptual and social realms of daily living. McLuhan's laws are based on a kind of paradigmatic understanding of relationships among key features of media, consciousness, and culture at any given time of innovation.

Ong's historical approach is quite different. While based on multiple relationships between "orality" and "literacy" as modes of thought and social organization, it never stresses the innovative aspects of new communication technologies at the expense of the older traditions that are incorporated into the new. His concept of residues, whether articulated in phylogenetic or in cultural terms, leaves the present always in conversation with the past. And thus, new technologies are depicted as building upon and with old technologies, and by implication upon and with previous psychological and cultural states. Cultural evolution always contains feedback loops, and multiple communication media coexist—a state of being that rings true when we examine our own lives. Oral group discussions, written messages in letters and books, and electronic radio, television, and satellite communications all are parts of the communication environment of our time.

The communication determinism thesis, thus, while powerfully attractive, must die if we are to better understand not only ourselves but also our forebearers and our progeny. This is one of the great challenges of our era.

## AVOIDING THE TRAPS OF CULTURAL TRANSFORMATION THEORY

When history-writing in some distant past became professionalized, when the singers of tales began writing down chronicles marked with concrete references to time and place, humanity set itself up for the traps of cultural transformation theory. The mythic "once upon a time" was recited over the ages as means of defining peoples and socializing them into the expectations of collectivities. But, once time and place were concretized, the mythic became the historical. Interestingly, however, I am not at all sure that the social functions served by the old myths disappeared. The stories had less whimsy about them, and the recitations of historical facts gave them an intellectual aridity that still makes children shudder when asked to read a history book. But, we use historical narratives, yes, to define peoples and socialize them into the expectations of collectivities. Historical narratives can have every bit as much ideological potency as mythic narratives.

I wish to suggest, more specifically, that cultural transformation theory is a minefield of ideological traps. By "cultural transformation theory," I refer to historical accounts of how some collectivity, or even all people in all places, undergo complete reconstitution as the result of some cataclysmic change in their situation. Academicians are fond of positing cultural transformations: the agricultural revolution that made permanent towns and hence the beginnings of stable society possible;[33] the Kurgan raids from the east that swept through southern and central Europe 6,000 years ago and shifted the existing societies from partnership cultures to patriarchal cultures;[34] or, the discovery of the stirrup and mounted shock combat, which feudalized society.[35]

Communication scholars are equally fascinated by cultural transformations. At least since Homer talked about Achilles as a speaker of words and doer of deeds, Western commentators have explored the power of communication processes. At least since Plato asserted that writing destroys memory, students of rhetorical and communication studies have investigated the transformative force of media. Walter Ong made his scholarly reputation writing about the coming of print and the transformative power of its visualist and monological possibilities in *Ramus, Method, and the Decay of Eloquence* (1958); and while McLuhan mined that same thesis four years later in *The Gutenberg Galaxy*, as we have noted he took us by storm when probing the electronic revolution in *Understanding Media*. These and others have depicted the transformative power of communication technologies.

What is wrong with that? The traps are several. Most obvious is the sense of entelechial force accompanying the major shifts. A guiding *telos* often seems to be built into discussions of cultural transformations. Now, that entelechy can be viewed positively, as it was in Hugh Blair's 18th-century review of the progress of civilization produced by increasingly sophisticated rhetorical prowess and as it is today in Toffler's moves from the first to the second to the third wave. Or, entelechy can be viewed negatively, as increasingly enslaving, as it was to Plato and as it is today in Ellul's vision of life in *The Technological System* (1980). The trap of historical inevitability is its ideological power to stop thought or resistance to its projection.

A second trap in the transformation thesis is its sense of irreversibility. Not all communication technologies carry us "forward." The women's liberation movement in the early 1970s, for example, employed a "backward" technology, the consciousness-raising session, often even with tokens that had to be spent when one spoke so as to equalize participation;[36] or again, today the small press, the basement operation where art books are handprinted for limited distribution, flourishes in greater numbers than ever in the history of the United States, in direct contrast to mass-printed paperbacks. Neither of these loops back in communication time is anachronistic. Both were ventured into at points in the twentieth century when a subculture felt the need for communication technologies with particular characteristics: the one, when women wanted the self-defining power of face-to-face community building, the other, when book lovers wanted the printer to be recognized as an *auteur* every bit as important to the pleasures of reading as the writer. Our uses of communication technologies vary by circumstance, not simply by our position on an historical timeline.

A third difficulty with the idea of cultural transformation is that it silently brackets human experience. As historians know, periodization is a prime interpretive strategy, one that can be used for sociopolitical purposes. When Riane Eisler in The *Chalice and the Blade* sets the Kurgan First Wave of Indo-European migrations in about 4,200 B.C.E. as *the* set of transformational events that ensconced dominator patriarchies in the Western world, she posits a different starting point for modern civilization than do others who discuss, say, the rise of Greek civilization between 2,000 and 1,000 B.C., the development of the Roman Empire hundreds of years later, and so on. Starting and ending points often control the stories we tell.

We all employ historical brackets to mark off now from the good old days, then from the brave new world of now. Cultural transformation theorists like to think of the junctures in human history as natural, akin to the punctuated equilibria or chaos-producing events Stephen Jay Gould and others see accompanying biological evolution. On the contrary, it is clear enough that cultural evolution is, not a biological process that we can visualize literally in our archaeological records, but a human construct bounded by rhetorical brackets.

This last point suggests a perspective we ought to be taking on the questions raised in this essay. Cultural transformations are human products engineered by that special group of technicians we call rhetoricians. People dwell within a world marked by the succession of the sun and the moon, but they

see the progression of time—and of everything else—through rhetorical constructions. In the final analysis, therefore, both the arguments favoring technological determinism and the traps populating cultural transformation theory are
interpretations of life set out purposively by rhetors who want to control the way
we envisage ourselves and our world.

## THE RHETORIC OF COMMUNICATION TECHNOLOGY

We ought to consider what a rhetoric of communication technology might
be and do. My fellow rhetoricians have offered some interesting studies of
communication technologies in recent years. Most have concentrated upon
television, including this volume's lead editor.[37] Among other examples, Brown
has explored primetime television's powers to construct sociopolitical visions of
life, as have Slater and Elliott.[38] Breen and Corcoran approach television
discourse as myth-making, and Corcoran by himself, as ideological apparatus.[39]
Many rhetoricians have focused on the mass-mediated political campaign: for
example, Swanson studied television's presentation of the 1976 presidential
campaign; Tiemens won an award for his study of camera shots in the 1976
presidential debates; Berquist and Golden examined ways in which television
affected public perception of the 1980 presidential debates.[40]

One could reference many other studies, though most would drive us to the
same conclusions these do: as yet, rhetoricians have paid far too little attention
to communication technology per se. Consider a topic close to rhetorical studies'
heart: televised politics. While Postman has argued pointedly for television's
destruction of political dialogue in modern America, while Meyrowitz has
demonstrated that television puts us not backstage or frontstage but sidestage to
political events, ultimately demythifying and hence weakening leadership, and
while Jamieson has examined the kind of political style that works best on
American television, we have barely scratched the surface when it comes to
understanding politics and technology.[41]

We need two sorts of studies of this relationship. First, we need to go much
further in studying technology's interaction with our expectations for political
communication. In overexposing us to political messages, especially in the
sixties and seventies—from campaigns to assassinations to riots to white papers/
documentaries and back to political ads—television inured, or "numbed" as
McLuhan said, the American conscience to traditional political messages. If in
fact it is true that the average presidential candidate gets only seventeen seconds
a crack on network nightly news and prefers the 15- and 30-second polispot to
longer ads, then we have witnessed a fundamental change in American political
campaigning. We need to study that change.

Likewise, we must understand that not only have millions been poured into
television campaigns, but Benjamin Franklin's gift to America, the postal
service, has allowed a parallel mini-technological system to interact with politics
in equally significant ways. Richard Armstrong's new book, *The Next Hurrah;*

*The Communications Revolution in American Politics*, opens with the sentence, "I'm a junk-mail writer."[42] While Armstrong's is not the first book on direct-mail politicking, it is probably the best since Sabato's 1981 book, for it interconnects the technology with both campaigning as a cultural practice and shifts he sees in the thought processes of the electorate. For good measure, Armstrong urges us to look at politics' use of other electronic technologies—telemarketing, cable, satellites, and computers.

The point here is not that electronic media have caused major changes in campaigning. Rather, what must be studied via close examinations of politics and technology in particular situations are the alignments or interrelations between habitual uses of technology, the communication requirements of some part of culture, and the audience's mindsets as they interpret messages. So, for example, while Castro's Cuba apparently can tolerate three- to four-hour political speeches, Reagan's America could not. Our television heritage trained us to expect mass-mediated messages chopped into small units. As political processes were remade by the television ad that paid for our media programming, so were voter expectations. The idea of riding two days to hear politicians each speak for one-and-a-half hours—which is what Illinoisians did when Lincoln and Douglas sought a Senate seat in 1858—is an idea whose time has come and gone.

Somehow, we must be able to see resonances among characteristics of a cultural arena, available technology, and mindsets of people who populate that arena. In the case of politics, the size and complexity of America have made face-to-face campaigning for national office an impossibility, creating cultural stress points and audience needs. Appropriate media—both mass media and such mini-media as mail, billboards, and bumperstickers—are used in configurations that reduce the stress and meet the needs. The use of some technology should resonate with features of the situation, the culture, and the audience members. The laws of resonance need to be postulated and tested. So, one sort of study needing to be done is the study of structural relationships between and among media, culture, and mind.

A second sort of study is better represented here: the study of rhetoric about technology. If we are to understand how technology enters into the public sphere, and how (if at all) we construct technology rhetorically, then we must pursue case studies of debates over technology. As well, while we might find Ellul an alarmist, his central point about technology as background to living cannot be ignored. Ellul argues: "Man can choose, but in a system of options established by the technological process. He can direct, but in terms of the technological given. He can never get out of it at any time, and the intellectual systems he constructs are ultimately expressions or justifications of technology."[43] He is right, and scholars with rhetorical sensitivities need to investigate technology-as-background: technology-as-background in American movies, from iceboxes in the 1930s to cars in the 1990s, that naturalized convenience technology; technology-as-background visible in the ways computer talk has worked itself into everyday language; technology-as-background shaped like the mushroom-shaped cloud that overshadows talk about technology itself.

As Barthes has noted, any human institution or phenomenon that becomes wholly naturalized into our environments takes on undeniable ideological force.[44] Or, as McLuhan said it more simply in his film "This is Marshall McLuhan," "We shape our world, and our world shapes us." The shaping being referenced here is a fully conditioned set of relationships between and among mind, self, and society—between and among socialized citizens, modes of communication, and cultural environment. Our everyday world must become naturalized, mundane, before we can function in it competently and effectively. The technological system, as Ellul calls it, by now is fully naturalized. How it got that way and what its implications are demand study by scholars sensitive to the power of verbal and behavioral practices—by rhetoricians, in the broad meaning of that word.[45]

## A Farewell to FM-2030

In drawing this essay to a close, I would like to leave a message for FM-2030. No, we are not transhuman, and no matter how many spare parts become buried in our joints, hearts, teeth, or even brains, we never will become posthuman. Technology undoubtedly will extend our lives and even have a central place in our psychocultural worlds. But, one of the reasons Ong is so significant is that he always concludes his journeys into past or present culture with a return to self, to the "I" of self-definition. Of this "I" he says:

> [T]he "I" identifies itself by its self-awareness and by this alone. The "I" has no name. Others know me as Walter, but of myself and to myself I am no more Walter than I am Tom or Dick or Harry, however much I may have accustomed myself to referring to myself or being referred to as Walter. To myself, I am simply "I." The "I" I alone can find. . . . All others besides myself, even father and mother, can contact this "I" only indirectly.[46]

"When in the series of anthropoids and prehominids," continues Ong, "some beings appeared who were capable of the reflective self-possession expressed in the saying of 'I,' at whatever point they did so, the leap into human existence had clearly taken place."[47]

Once anthropoids made that leap into human existence, they took a step that could never be retreaded; they took a step toward self-knowledge that we never will surrender to any technology, no matter what FM-2030 and other futurists promise us down the road. The move to self-awareness was determinative, in an absolute sense, for humanity. This book explores the implications of that fact of existence.

# NOTES

1 FM-2030, *Are You a Transhuman? Monitoring and Stimulating Your Personal Rate of Growth in a Rapidly Changing World* (New York: Warner Books, 1989), 209-210.

2 FM-2030, 205.

3 Raymond Williams, *Television: Technology and Cultural Form* (New York: Schocken Books, 1975), 9, 10.

4 For example, see John Fiske, *Television Culture* (London: Methuen, 1987), and Michael Gurevitch et al., eds., *Culture, Society and Media* (London: Methuen, 1982).

5 FM-2030, 205.

6 FM-2030, 205.

7 Eric A. Havelock, *The Muse Learns to Write; Reflections on Orality and Literacy From Antiquity to the Present* (New Haven: Yale University Press, 1986).

8 Walter J. Ong, *Orality and Literacy; The Technologizing of the Word*, New Accents Series (London: Methuen, 1982), chap. 7.

9 Ong, 176.

10 Respectively, Marshall McLuhan, *Understanding Media: The Extensions of Man* (New York: McGraw-Hill, 1965); Marshall McLuhan and Quentin Fiore, *The Medium is the Massage: An Inventory of Effects* (New York: Bantam Books, 1967); and Marshall McLuhan, Quentin Fiore, and Jerome Angel, *The Medium is the Massage*, record (New York: Columbia Records/CBS, 1967).

11 McLuhan, *Understanding*, 182.

12 Arthur Kroker, "Processed World: Technology and Culture in the Thought of Marshall McLuhan," *Philosophy of the Social Sciences* 14 (1984): 548.

13 Marshall McLuhan, "Laws of the Media," *Et cetera* 34 (1977): 177.

14 McLuhan, "Laws," 178.

15 James M. Curtis, *Culture as Polyphony: An Essay on the Nature of Paradigms* (Columbia: University of Missouri Press, 1978).

16 Marshall McLuhan, "Communication: McLuhan's Laws of the Media," *Technology and Society* 16 (1975): 75.

17 McLuhan, "Laws," 176.

18 Michael Heim, *Electric Language: A Philosophical Study of Word Processing* (New Haven: Yale University Press, 1987), 57-58.

19 Ong, 36-57.

20 These ideas are developed across a series of Ong's books: *The Barbarian Within; And Other Fugitive Essays and Studies* (New York: Macmillan, 1960); *Darwin's Vision and Christian Perspective* (New York: Macmillan, 1962); *Interfaces of the Word; Studies in the Evolution of Consciousness and Culture* (Ithaca: Cornell University Press, 1977); *Fighting for Life; Contest, Sexuality, and Consciousness* (Ithaca: Cornell University Press, 1981); *Orality and Literacy* (1982); and *Hopkins, the Self, and God* (Toronto: University of Toronto Press, 1986).

21 See his essays in *In the Human Grain: Further Explorations of Contemporary Culture* (New York: Macmillan, 1967) and *The Presence of the Word; Some Prolegomena for Cultural and Religious History* (New Haven: Yale University Press, 1967).

22 See "Oral Residue in Tudor Prose Style" (pp. 23-47) in *Rhetoric, Romance, and Technology; Studies in the Interaction of Expression and Culture* (Ithaca: Cornell University Press, 1971) as well as essays in *Interfaces*.

23 Ong, *Orality*, 7.

24 Ong, *Rhetoric*, 285.

25 Ong, *Orality*, 136.

26 Ong, *Orality*, 142.

27 See Ong, "The Writer's Audience Is Always a Fiction" (pp. 53-81) in *Interfaces*, and *Orality*, 96-101.

28 Quoted in Norman J. Vig, "Technology, Philosophy, and the State: An Overview," in *Technology and Politics*, ed. Michael E. Kraft and Norman J. Vig (Durham, NC: Duke University Press, 1988), p. 11.

29 Emmanuel G. Mesthene, *Technological Change; Its Impact on Man and Society* (New York: New American Library, 1970), and Turkle, 325-326.

30 Alvin W. Gouldner, *The Dialectic of Ideology and Technology; The Origins, Grammar, and Future of Ideology* (New York: Oxford University Press, 1976), chap. 1.

31 Harold Adams Innis, *Empire & Communications*, rev. Mary Q. Innis (1950; Toronto: University of Toronto Press, 1972).

32 Thomas J. M. Burke and Maxwell Lehman, eds., *Communication Technologies and Information Flow* (New York: Pergamon Press, 1981), p. ix.

33 Alvin Toffler, *The Third Wave* (New York: William Morrow and Co., 1980), passim., for discussions of the so-called First Wave (agriculture).

34 Riane Eisler, *The Chalice and the Blade; Our History, Our Future* (San Francisco: Harper & Row, 1987).

35 Lynn White, Jr., *Medieval Technology and Social Change* (New York: Galaxy Books, 1966).

36 Karen A. Foss, "Ideological Manifestations in the Discourse of Contemporary Feminism" (Ph.D. diss., University of Iowa, 1976).

37 Martin J. Medhurst and Thomas W. Benson, eds., *Rhetorical Dimensions in Media; A Critical Casebook* (Dubuque, Iowa: Kendall/Hunt, 1984).

38 William R. Brown, "The Prime-Time Television Environment and Emerging Rhetorical Visions," *Quarterly Journal of Speech* 62 (1976): 389-399; Dan Slater and William R. Elliott, "Television's Influence on Social Reality," *Quarterly Journal of Speech* 68 (1982): 69-79.

39 Myles Breen and Farrel Corcoran, "Myth in the Television Discourse," *Communication Monographs* 49 (1982): 127-136; Farrel Corcoran, "Television as Ideological Apparatus: The Power and the Pleasure," *Critical Studies in Mass Communication* 1 (1984): 131-145.

40 David L. Swanson, "And That's the Way it Was? Television Covers the 1976 Presidential Campaign," *Quarterly Journal of Speech* (1977): 239-248; Robert K. Tiemens, "Television's Portrayal of the 1976 Presidential Debates: An Analysis of Visual Content," *Communication Monographs* 45 (1978): 362-370; Goodwin F. Berquist and James L. Golden, "Media Rhetoric, Criticism and the Public Perception of the 1980 Presidential Debates," *Quarterly Journal of Speech* 67 (1981): 125-137.

41 Neil Postman, *Amusing Ourselves to Death: Public Discourse in the Age of Show Business* (New York: Penguin, 1985); Joshua Meyrowitz, *No Sense of Place: The Impact of Electronic Media on Social Behavior* (New York: Oxford University Press, 1985); Kathleen Hall Jamieson, *Eloquence in an Electronic Age: The Transformation of Political Speechmaking* (New York: Oxford University Press, 1988).

42 Richard Armstrong, *The Next Hurrah; The Communications Revolution in American Politics* (New York: Beech Tree Books, 1988), 7.

43 Jacques Ellul, *The Technological System*, trans. Joachim Neugroschel (New York: Continuum, 1980), 325.

44 Roland Barthes, *Mythologies*, trans. Annette Lavers (New York: Hill & Wang, 1972), esp. "Myth Today," 109-159.

45 See the analysis of David Descutner and Delysa Burnier, "Toward a Justification of Rhetoric as Technique," in *The Underside of High-Tech: Technology and the Deformation of Human Sensibilities*, ed. John W. Murphy, Algis Mickunas, and Joseph J. Pilotta (New York: Greenwood Press, 1986), 147-158.

46 Ong, *Fighting*, 194, 195.

47 Ong, *Fighting*, 199.

# The Language of Technology: Talk, Text, and Template as Metaphors for Communication

*James W. Carey*

This is an essay on the use of metaphor in our understanding of communications history and technology. The alliteration in the title disguises in a conceit the meaning that is sought. Other words, alliterative still, would do: Performance, Print, and Program, for example. Better yet, would be to turn towards practice: Speaking, Printing, and Programming as human activities. Or, one can grasp at the social role behind the artifact and the activity: Speaker, Typographer, and Programmer. The terminology hovers indecisively for I am trying to compact into a phrase two different, though parallel, developments in intellectual life. The first is the spreading convention in historical writing about communications technology to partition time into three distinct phases each governed by a defining technology and master symbol: the oral tradition, the printing press, and the computer. This is the story of social evolution told as the evolution of communications. The narrative is organized around a series of decisive breaks or revolutions: from the voice to the printing press to the computer; from speech to print to electronics; from the performer or orator to the printer or typographer to the programmer or computer engineer; from the forming of sounds to the casting of letters to the "writing" of programs. This is a history with a telos. While much of the work of recent years has concentrated on the transition from speech to print, from a society in which speaking and performing are primary to a society in which reading and writing are primary, the larger objective is to understand the presumed communications revolution of our own time: the movement beyond literacy, beyond the printed word, to something quite new and problematic—visual literacy, computer literacy, the information society—a world in which the computer is the master trope. This is not the only way of writing history, and speech, print, and electronics are not the only metaphors about: the story of class struggle has hardly disappeared. While those great nineteenth-century historical actors—the proletariat and bourgeoisie—have something of the antique to them, the widely celebrated "new class," whose command and control of social power is based on the literacies of the information society, is very much available. Similarly, the evolution of social systems from feudal to capitalist to socialist continues to provide an architecture for considering the distinctive economic forms that the production of communications and culture takes: from patronage to the market to the plan. But the artifacts, roles, and practices encased in my title possess an allure to students of communications; they suggest laws of technological succession central to historical understanding.

But, there is a second intellectual movement parallel to this narrative of communications technology that needs to be considered. Three of the master tropes of contemporary philosophy and social theory are also organized around and privilege these same moments in the history of communications: talk, text, and template. It is, of course, a radical simplification to reduce social theory to a contest among metaphors of conversation, textuality, and structure but it is for our purposes a useful, heuristic simplification. The construction, reconstruction, or deconstruction of philosophical questions around the image of a conversation, text, or program is one way of looking at hermeneutics, deconstructionism, and structuralism. While they do not exactly exhaust current philosophical possibilities, they do constitute a significant part of contemporary debates among pragmatism, hermeneutics, structuralism, and post-structuralism. That is the hypothesis I want to pursue in this essay.

Such an hypothesis seeks to link, however indirectly and implicitly, these parallel developments in history and philosophy. It also is an attempt to review, critique, and take stock of the effort to write history as communications history: to clarify the enterprise, to highlight some of its dangers and to applaud, however tentatively, some of its successes.

# I

I want to begin by summarizing an argument presented elsewhere wherein I attempted to shift the ground of discussion within communications away from the contrast between administrative and critical research, a contrast by now tired, worn, and unpromising, and onto more promising terrain, namely the distinction between expressivist and objectivist conceptions of reality.[1] The movement from an expressivist to an objectivist conception of reality recapitulates part of the history I later want to exhume and is part, as well, of an episode in the history of the printing press.

In *Historical Consciousness*, John Lukacs argues that the vocabulary of history and the vocabulary of science were twin-born, both within the broad historical moment when the printing press began its diffusion throughout Europe.[2] The idea of history as a formal record makes its appearance, according to the *Oxford English Dictionary*, around 1482. Historian appears in English in 1500. The distinctions ancient and modern, retrograde and progressive, appear around 1600. Primitive, a sixteenth-century adjective was, according to Logan Persall Smith, "probably the first word in which our modern historical sense finds expression."[3] At the same moment the words original, novel, modern, and revolution undergo an inversion and take on their characteristic meanings: from referring to the cyclical, the beginnings, the point of origin, they take on the meaning of the latest, newest, most progressive. "As the seventeenth century progresses we find the sudden and at times tumultuous appearance of new words, one scientific (for example: acid, cohesion, elasticity, electric, equilibrium, fluid, gas, pressure, static, temperature, tension, volatile), the other historical (for example: antiquated, century, contemporary, decade, epoch, historic, out of date, primeval)."[4] The modern scientific, theoretical outlook emerges with the modern historical outlook, then, in seventeenth-century Europe and England.

In *The Printing Press as an Agent of Change*, Elizabeth Eisenstein relates, albeit implicitly, both these movements to the ways in which the printing press reorganized the Commonwealth of Learning and redeployed the activity of scholars.[5] The fixing of migratory manuscripts and the feedback and self-correction permitted by the wide dispersal of identical printed texts permitted, for the first time, something like an agreed upon historical record: a chronology of authors, texts, and events that revealed succession, distance, closeness. But the fixing of the old permitted a tradition of the new. Once energies were released from the tasks of copying manuscripts, and once the manuscript tradition was secured and straightened, intellectual energies could flow into new channels: the production of fresh deviation from the printed record. More importantly, the possibilities of travel and movement, of cataloguing in the largest sense, ushered in a new inventory of the phenomena of nature of which an accounting might be undertaken. Further, it was in the printer's workshop that the two traditions that formed modern science were brought together: the scholar's tradition of speculative philosophy (the tradition that later came to be known as arm-chair philosophy) with the lens grinders—the artisans and craftsmen who supplied the technical and experimental habits. It was this union of text and craft, thought and action, technology and logic which when merged produced the scientific method. Finally, I read out of Professor Eisenstein the conclusion that the creation of personal libraries, the ability for a scholar to have his own books, on his own shelf, in his own study, made possible the withdrawal into private thought, the production of unique combinatorials, the individuality and distance, that permitted the emergence of the social role of observer and spectator, the subjective consciousness of "*cogito ergo sum*," and the social practices through which received tradition could be deconstructed and scientific composition undertaken.

The scientific and historical, the theoretical and the chronological, emerge together. They are dialectically related practices. But they emerge, as well, as contradictory impulses. Science emerges as a means of deconstructing nature into its primary qualities, qualities that testify to its temporal continuity and spatial uniformity. Beneath the phenomena of appearance exists a transhistorical, transcultural world of nature: a language in which nature universally and unwaveringly speaks itself. This is the language Descartes discovers. The historical turn of consciousness apprehends, in the first instance, just the opposite. The world revealed to Vico and later Herder by the accumulation of manuscripts turned to printed texts, by texts placed in geographic, temporal, and linguistic order, was a world of discontinuity, rupture, break—a world of incommensurability. Here the metaphor of geological strata or archaeological levels, metaphors once again much in vogue, suggest themselves as against the smooth rising curve of continuity and progress. The historical impulse breaks forth in terms of culture not nature, particularity not universality, a language of humankind not a language of nature. Vico anticipates phenomenology and hermeneutics where the intellectual task is to seek the intelligibilities of culture rather than laws of nature. Isaiah Berlin is among the most sensitive interpreters of this dialectical turn and here he speaks in long paragraphs rather than pithy apothegms about the discoveries made by Vico:

> . . . there is a pervasive pattern which characterizes all activities of any given society: a common style reflected in the thought, the arts, the social institutions, the language, the ways of life and action of an entire society. The idea is tantamount to the concept of a culture; not necessarily of one culture but many; with the corollary that true understanding of human history cannot be achieved without the recognition of a succession of the phases of the culture of a given society or people. This further entails that this succession is intelligible, and not merely causal . . . is intelligible to those who possess a sufficient degree of self-awareness, and occurs in an order which is neither fortuitous nor mechanically determined, but flows from elements in and forms of, life, explicable solely in terms of human goal-directed activity.[6]

Later, Berlin turns his interpretation of Vico toward communication:

> . . . the creations of man—laws, institutions, religious rituals, works of art, language, song, rules of conduct and the like—are not artificial products created to please, or exalt, or teach wisdom, nor weapons deliberately invented to manipulate or dominate men, or promote social stability or security, but are natural forms of self-expression, of communication with other human beings or with God. The myths and fables, the ceremonies and monuments of early man, according to the view present in Vico's day, were absurd fantasies of helpless primitives, or deliberate inventions designed to delude the masses and secure their obedience to cunning and unscrupulous masters. This he regarded as a fundamental fallacy. Like the anthropomorphic metaphors of early speech, myths and fables and ritual are for Vico so many natural ways of conveying a coherent view of the world as it was seen and interpreted by primitive men. From which it follows that the way to understand such men and their worlds is by trying to enter their minds, by finding out what they are at, by learning the rules and significance of their methods of expression. . . .[7]

Berlin's rendering of Vico clearly reveals the origins of the distinction between *naturwissenschaften* and *geisteswissenschaften*. But the interest here is slightly different: namely, how contradictory impulses within scholarship were permitted or freed by the practice of typography. Cultural history and cultural anthropology—the discontinuity of time and the discontinuity of space—were twin-born as hermeneutic and phenomenological reactions to the natural sciences and to a printed record revealing historical discontinuities registered as cultural differences. It is not that simple of course—there was, after all, a canonical history as well as a hermeneutic one—but my simplification has a purpose. The critical move is the constitution of an expressivist tradition through the discontinuities revealed in a historical record. The distinction, indeed the difference and rupture, between a discovered and a constituted world, an expressed and an objective one, a received versus an achieved one—a world given in experience through language and a world achieved through activity in language—demands, of course, a division of loyalties, however provisional. In my earlier essay I cast that loyalty with the expressivist argument, particularly as formulated in American pragmatism. But the casting of loyalties is but a prelude to dissolving the distinction altogether, transcending it through a form of argument that makes the entire division between a science of nature and a science of culture moot.

My view of the expressionist tradition can be reduced to a series of propositions which I shall here repeat rather than explain. An adequate rendering of communications history requires embracing in some form the following propositions. First, that language is not merely an instrument that can be picked up and set down at will, a device that merely reveals nature to us. Rather, language is a vehicle of a certain form of consciousness that is characteristically human. In this sense we live in language before we live through it or with it. Language is a precipitate of an activity in which human consciousness comes to be; it is the characteristic form in which humans are in the world. Language, therefore, is not merely a vehicle of communication in the narrowed sense of a transmission system; it is more than a device of description. Language realizes a mode of consciousness and being.

Second, technical extensions of language—writing, printing, programming—are not merely instruments for the achievement of narrowly conceived purposes; they are not simply artificial products: utilities, or sources of power. As extensions of the primordial and defining power of language, they must be thought about first of all as precipitates of characteristic forms of human consciousness, ways of being in the world, in Berlin's words, "natural forms of self-expression, of communication." To speak or write or to program then is not merely to pick up a tool or to exercise a skill. It is to constitute a world, to bring a world into existence, and to simultaneously constitute a self. The artifacts of communications differ, as do the social practices they engender, but they are linked in a chain of transformation: a process whereby the world and the self is reconstituted.

Third, this argument is no mere idealist gloss on language. Rather, the mind is an instrument of production and a world is its most valued product.

Because human action depends upon a sedimented world, one not given in experience, the construction, maintenance, and transformation of the parameters of that world are the task of mind. The mind is not ancillary to technology but technology is one material form in which the power of the mind to render the world symbolically is realized.

Fourth, the mind operates as a means of production not merely by the symbolic transformation of nature into culture but also by the penetration of culture into the bosom of the natural order. The first process works whenever landscapes are clearly featured. The second is inevitable and indispensable on trackless deserts, open seas, and dark continents. If the parts of nature cannot be sufficiently disengaged so that they can represent themselves, then a grid will be laid across the natural order. Marking trees works in the forest; latitude and longitude are necessary on the open seas. But the movement from one process to the other is historical not merely logical: analog models precede digital ones, inclined planes precede split atoms.

Fifth, reality is a scarce resource and is, therefore, the site of a constant struggle. If the world is constituted in one way in the service of one set of purposes for one group of people, it is thereby preempted. This preemption, which ontologizes meaninglessness, means that conflict is not the property of this or that social system but is rather bound into the nature and conditions of human action.

These propositions merely attempt to reiterate, in my own idiosyncratic language, arguments made by many others. A part of it is captured by Pierre Bourdieu's suggestion that communication or symbolic action or cultural activity and our reflection on it involves three necessary and inevitably complementary aspects: structure, action, and power.[8] Communication, and the artifacts that manifest and organize it, is simultaneously a thing to think with, a thing to form social relations with, and a thing with which to exercise power. It is an inevitable tendency to reduce these three moments to one or, better, to choose one of the moments to represent the process as a whole, as if, for example, structure and action could be excluded and communication could be revealed as merely a form of power and domination. The writing of communications history in terms of technology is, then, more than a means of telling a story; it is also a form in which theory is privileged. Voice, text, and program become ways in which action, power, and structure are selected out to represent the history and essence of communications. There is a rough and homemade relation between parallel sets of terms: action, power, structure; voice, text, program; speech, press, computer; hermeneutics, deconstruction, structuralism; and the dimensions along which all communication artifacts effect their consequences—the creation, dissemination, and preservation of culture. In any given theoretical system or form of historical writing one of these dimensions is naturalized, rendered synecdochal, but it is the three dimensions—action, power, structure; creation, preservation, dissemination—that permit the production and reproduction, maintenance and transformation of a real world.

From the standpoint of action, communication is the instrument by which we construct reality. It refers to those interactions necessary to reach consensus on the structure of reality. This is, in Bourdieu's terms, its gnostic function and is the ground for a universal hermeneutics. Symbols are devices for social integration, for the agreement of subjectivities in the production of an objectively consensual world. It is the moment at which subjective experience is objectified and the objective world made part of subjective experience. Communication in this dimension is a structuring medium through which social relations of all types are formed and maintained. But communication also has a logical function; indeed, it is this logical function that in structuralism, systems theory, and cybernetics becomes the very meaning of communication. It displays a structured world that is only then achieved in interaction. Here the interest is not in agency—on the process of reality construction—but on the code, the underlying structure presupposed by any given activity. Agents on this view, are merely instruments for the reproduction of structure. Structure is a resource continuously, if unconsciously, drawn upon for the maintenance of any act. As a structured system, culture is a condition of the *intelligibility* of action; the structured medium that has to be constructed in order to account for the constant relationship between symbol and sense.

The third dimension of communications is power. The term is ambiguous, of course: power refers simultaneously to the potential and work necessary to effect a transformation of the materials of the world and also to the ability to exercise dominion over nature and other humans. But it is this political function or dimension of culture, symbol systems, and communications that dominates certain forms of analysis and subsumes, where it doesn't eradicate, the logical and gnostic aspects of communication. The analysis of power explains technology and symbols in relation to the interests of dominant classes—in relation to systems of stratification and inequality. In contrast to myth, a predominant feature of the oral tradition, a collective product collectively appropriated and consumed, ideologies, in Bourdieu's rendition, serve sectional interests which they represent as interests common to the society as a whole. Ideology contributes then to the integration of the dominant class by making communication possible among them. But it also provides a bogus integration of society as a whole and the demobilization of dominated classes. It establishes distinctions— including the distinction between literate and illiterate—and legitimates these distinctions as standards of judgment and self-worth. Communication that unites also divides. It legitimates all distinctions by defining all cultures by their distance from and by their deficiencies relative to the dominant culture. This ideological effect is disguised beneath the function of and within the very terms of communications.

This view of how power and ideology work as aspects of communications is illustrative only. The larger point is that each of these three general positions on communications corrects and completes the others; as each provides a figure it assumes the others as ground.

I want to illustrate the foregoing by briefly examining and radically simplifying three traditions of technological analysis roughly identified with Jack Goody, Elizabeth Eisenstein, and Harold Innis.

## II

The tradition of historical or crypto-historical writing (there are disputants) on communications technology can be decomposed into three separate problematics or subtraditions. The first is anthropological. It focuses upon literacy or the difference that literacy makes in the transit between, to use Levi-Strauss's and McLuhan's terminology, the cool world of the savage, oral, and preliterate and the hot world of the modern, industrial, and literate. The anthropological work was partly inspired by Harold Innis's colleague at the University of Toronto, Eric Havelock. Havelock's inventive untangling of questions of literacy from questions of philosophy in the Platonic and pre-Platonic texts set out the dimensions of the problem and the major terms of its solution. The best single summary of the argument of the anthropological tradition is Jack Goody's *The Domestication of the Savage Mind.*[9] The central question of the book is this: what are the effects of writing, of skills at reading and writing, on modes of thought and, secondarily, on major social institutions such as the family and education. The primary historical divide, the moment of social transit and transformation which accounts not only for the birth of writing but the birth of history, is the invention of literacy.

Goody attempts to describe how literacy mediates between those binary pairs that have preoccupied Western thought and which the computer and structuralism have made the *lingua franca* of social analysis: primitive *vs.* modern, myth *vs.* science, tradition *vs.* modernity, *gemeinschaft vs. gesellschaft*, expression *vs.* reflection, a science of the concrete *vs.* a science of the abstract. Goody argues that certain of the characteristics that Levi-Strauss and others have regarded as marking the distinction between "primitive and advanced . . . can be related to changes in the mode of communication, especially the introduction of various forms of writing."[10] He continues: "the advantage of this approach lies in the fact that it does not simply describe the differences but relates them to a third set of facts, and thus provides some kind of explanation, some kind of mechanism for the changes that are assumed to occur."[11]

This notion was not exactly lost on Levi-Strauss, for he once proposed that the term primitive should be replaced with "without writing," though, as Walter Ong notes, this is also a negative characterization, the identification of a group by what it is they lack rather than what they were and are.

Primitive or wild forms of thought are grounded in and explained by a particular technology of the intellect—the exclusive provenance of and reliance on oral speech. Similarly, modern forms of consciousness and thought are rooted in a different technology of the intellect: literacy and its democratic manifestation in the printing press. This is not only an alternation between two different instruments or means of expression but two different structures or templates of consciousness. If the human mind, or at least the brain, is everywhere the same, how does one account for the existence of different cultures? The answer is that the mind is mediated into culture through differing technologies which not only realize it but also constitute it. An analysis appropriate to this complexity must recover the structure in the technology for it provides a key to accounting for differences not only of degree but of kind. While Goody argues that he is not simply writing a great divide view of history, he does distinguish all societies that have acquired writing from those that have not. Different technologies of the intellect mediate between consciousness and other social institutions. How, by extension, do such technologies mediate between one stage of civilization and another?

The second tradition of technological analysis starts not from anthropology but from history, not from a structure of the primitive and literate mind, but a movement between two technologies of literacy. If for Jack Goody the transit from orality to literacy is critical to an understanding of the transit from primitive to modern, this second tradition attempts to assess the historical significance of the movement from script to print, from parchment and manuscript to paper and press, from a chirographic to a typographic tradition, specifically within the Western tradition. While both the anthropological and historical traditions are underscored by Havelock's pioneering scholarship on the growth of a literate and scholarly tradition in ancient Greece,[12] the historical tradition operates with a distinctive problematic: what were the consequences within the West of establishing ancient traditions of art and learning on a new technological base? The transformation in question is from one form of literacy—hieratic, ecclesiastical, working via manuscript in a monastic cell—to another—demotic, secular, exercised via paper in the typecaster's workshop. The effect of literacy here is not on a general human consciousness—at least not directly—but on the commonwealth of learning and through that commonwealth on other social institutions. It is here that Marshall McLuhan made his earliest and most striking contribution in *The Gutenberg Galaxy*, if not by the evidence he adduced then by the scholarship he inspired.[13] It was dissatisfaction with *The Gutenberg Galaxy* that led to Elizabeth Eisenstein's *The Printing Press as an Agent of Change*. That work, at its most modest level, sketches the role or the possible role that printing played in all those movements of life and thought which we identify with modernity in the West: the Renaissance, Reformation, the rise of Science and, as a promissory note, the emergence of republican politics. Her work, in part, is an extended essay in redressing the balance, changing the relation between figure and ground, an attempt to tell an untold or neglected part of the story: the role of the printing press in the characteristic movements of modernity.

But in telling that story she inserts into the narrative a critical shift of focus. There was nothing modern about modernism. Better: the features we identify

with modernism—*esprit de system*, rationalization, secularism, democracy—
are modern in the antique sense of the word: they were present from the
beginning. The features we identify with modernism have been recurring
aspirations, motives, tendencies throughout the Western tradition but they
lacked a basis in material practice by which they could be secured and made a
continuous part of the inheritance. The printing press was the material mediator
between recurrent but fragile ambition and permanent achievement and development. Printing did not cause the Renaissance or the Reformation or indeed
anything else. However, it allowed for a redistribution of psychic energies, a
redeployment of social forces, that allowed for the previously weak, unsecured,
and evanescent renaissances and reformations to coalesce into a permanent and
cumulative development. Printing, in short, changed the nexus of social forces
within which the very notion of historicality could operate.

A major part of Professor Eisenstein's achievement is to pick up and keep
in fruitful tension a distinction that one finds in McLuhan and many others. She
argues that print, and by my extension any other communications technology,
works its consequences along three relatively orthogonal dimensions: preservation, creation, and dissemination. Whereas attention has usually been fixed on
the dimension of transmission—it is in sending, imparting, distributing messages that the effects of technology occur—Professor Eisenstein emphasizes the
forms in which technology encodes and recodes a cultural tradition and thereby
fixes it or stores it. New forms in which a culture is created and preserved open
up, in turn, new possibilities of transmission and reception. Our attention has
been so fixed on a transmission model of communications—a model which is
bourgeois in the sense that it assumes transmission and exchange are the very
symbols of enlightenment and progress—that conditions of creation and preservation have been glossed over. But it is from these latter phenomena that many
of the sharpest consequences of technology flow.

This same argument was, as I mentioned, advanced by McLuhan, though
not surprisingly as an aphorism: one meaning of the "medium is the message"
is that it is culture "retrieved rather than received" that is of significance.
Professor Eisenstein, in emphasizing less the distribution or transmission of the
printed word, but the differences between storing culture in expensive, single, or
relatively few manuscript copies as opposed to storing culture in cheap, multiple
printed copies, opened a significant reorientation in the analysis of printing.
Similarly, the powers and limitations inherent in the capacity of the human
memory to store and retrieve culture *versus* print as an extrasomatic collective
memory allowed for catching both the distinctive qualities of the oral tradition
and the fresh possibilities opened by printing. Finally, the contrast between
creating through the memory arts, formulaically in performance, as opposed to
creation by composition, anterior to performance, suggests some of the distinctive qualities of thought, interaction, and power in the age of speech and the age
of print. While Professor Eisenstein did not at every point press this particular

gain, it remains one of the valuable legacies of *The Printing Press as an Agent of Change*.

The third subtradition I wish to treat briefly is even more distinctively North American and identified with the Canadian economist Harold Innis, though I would include within the analysis, if space permitted, a number of members of the Chicago School of Social Thought. This subtradition begins from neither history nor anthropology; in fact, it does not begin from any textual tradition at all but rather from lived experience, from a commonsensical encounter with a rapidly changing world.

Innis's interest in the oral tradition was not related to primitive life or wild modes of thought nor was that interest tied to the displacement of a manuscript tradition by way of the printing press. Innis, and he is representative of many, grew up in an oral tradition encysted within a larger industrial or industrializing civilization. It was the consequence for culture and politics of the extirpation of this tradition, of this transformed tradition, that arrested him. The attempt to preserve that tradition was realized not merely in politics, though there it provided strategy, but in intellectual life as well. The oral tradition was the continuing center—sometimes as ground, sometimes as figure—of scholarship.[14]

Innis's background is not atypical in North America. He was bred to the world of speech rather than print. The experience of growing up in communities, villages, neighborhoods was an experience in which speech was the mode by which daily life was conducted. Its idioms did not come from books, except scripture and catechism and perhaps Shakespeare. It was an oral medium whose expressions matched the rhythms of nature and the seasons, whose expressions were close to the hard surfaces of life. The contrast between such a world and that of the city and professional life—books, the printed world, sophisticated vocabulary and syntax, a genteel literacy, an urbanized style—provides a dialectic in experience, not in logic. It is not a dialectic which European intellectuals, bred in highly literate cultures and for whom the world of books was a first order experience, would find easy to share. This is above all a dialectic of politics for in the political traditions of North America, a democracy of the foot and tongue, the world of print is decidedly secondary and supportive. The transit, in short, from community to society, a transit not of theory but of lived experience, is simultaneously a transit from speech to print, and from an egalitarian democracy to a manipulated world of printed and later broadcast words.[15]

Innis's subject was the exercise of power through communications. That power—dual faced as is shown in the positive and negative valences of an empire—is not a product of ideology as normally understood; it is not a product of conditioning, attitude change, or influence. Rather, power is exercised through the production and reproduction of real social life. The structure of social life is given at a minimal level, in its fundamental coordinates: time and space.

Whether conceived of as history and geography, duration and extent, real life is lived in space and time. Media of communication provide the structure of time and space. They not only provide it; they imagine it. Technology reveals its intention, its aspiration, in the configuration of time and space. Moreover, these are cultural definitions; they are not found artlessly lying around in nature. All geography is social geography; all history is social history. Before time and space can coordinate and control human activity, they must be imagined as phenomena. Therefore, Innis's work was an attempt at cultural history and cultural geography: analysis of the various conceptions of time and space that have regulated social activity in relation to the media of communications that both expressed and realized, imagined and sedimented, these conceptions.

Innis's conceptions of communications assumes the interdependence of economics, politics, and culture. Power is the term which holds them together and is realized in control over time and space or the capacity to act in time and space. The very framework within which action is exercised or over which control is asserted are themselves imaginative or cultural constructs. Power to act and control time and space are exercised through technology. The shape of technology conditions the way in which power is exercised. But the technology makes assumptions about space and time. Technology engenders what it perversely pretends only to display. And, as technology changes, the conditions of power change along with the configuration of societies. This happens not because technology is a *prima causa* but because technology embodies the intentions and aspirations it realizes. Central to technology is communications because it is the instrument through which time and space are defined, duration and extent are sedimented. The medium is the message because it inscribes the materiality of culture. Innis was particularly concerned with the varying degrees to which technologies lent themselves to the generation of monopolies of knowledge or, conversely, the ways in which such monopolistic tendencies could be resisted. It is in the context of monopoly that he privileged the human voice and resisted the incursion of mechanical and more controllable forms of communication.

Innis's strategy of culture, a strategy of both theory and practice, a strategy won in experience, was to create a historical hermeneutic of technology that centers communication in relation to power. His strategy was to restore the voice, to restore it as a metaphor with which to unmask the pretensions of theory and to simultaneously restore a public counterforce to the "entrenched monopolies of communication" that present a "continuous systematic ruthless destruction of elements of permanence essential to cultural activity."[16] The metaphor of the voice provided the conditions, admittedly implicit and unarticulated, for what we would later call an "ideal speech situation." Print, as the symbol of the mechanization of communication, extirpated the voice by monopolizing all the effective temporal and spatial transactions and deprived the public from any political role. Here, with a vengeance, was a critique of a spectator theory of politics forged off a spectator theory of knowledge.

In setting out these three subtraditions revolving around the metaphors of talk and text, I disguise how hopelessly interlarded they have become. Virtually everyone who writes about these matters draws on each of the traditions, at least in erecting a framework of discourse and I divide, therefore, what has become increasingly indivisible. I wedge these traditions apart so that we can more easily recover the metaphoric basis of talk, text, and template and in order to attach these metaphors to the larger triangles woven through this text. There is a homology, in short, between talk, text, and template; conversation, textuality, and structure. This mapping of homologies becomes more critical the more one's concerns move from tracing the efficient causes of historical development to an hermeneutics of technology relative to politics and the "revolution" of computer technology. I shall return to these matters following a detour through the assessment of the primal metaphor of speech and print.

## III

Despite the attitude taken here, an attitude at once respectful and hopeful, towards the project of writing communications history off the metaphors of speech, print, and electronics, it is a field, a space, littered with landmines for the unwary and incautious. I will speak of two of the problems that this particular formulation of the subject matter presents.

The first is an essentially technical matter. The very progression of stages or phases of civilization with its implication of sharp ruptures and breaks seems to lock us into one of those dreaded historical determinisms with which we have been visited from Hegel forward. I will briefly treat the matter of determination in the next section. At this moment I need only observe that the division of history into neat phases suggests some iron law of determination of the phases as a whole that is easily and quickly absorbed into a law of development governing each and every detail of the historical process. This inevitably situates technology outside of history, as an archimedian point, from which the entire historical process is governed at large and in detail. Most disturbingly, the law of phases and the technological metaphors that drive the phases often become devices for disguising real complexity and, curiously enough, for closing down the detailed work of substantive historical investigation.

Let me provide a few examples. It sometimes appears that the transition from speech to print occurred uniformly and all at once; like the onset of the modern world in the old joke, it occurred precisely at 8:07 a.m. on September 17, 1906. The notion of rupture, of break, or, more alarmingly, revolution points to radical discontinuities in the historical process instead of the more glacial and unevenly phased movements that upon inspection are always the case. The oral tradition does not disappear at any level of social development. While much speech takes on the character of writing—it originates in writing—and is reinserted into the practice of speaking, the oral tradition persists among particular social groups, even functionally literate groups, and within the organization of certain social practices and institutions: music, humor, certain stretches of intellectual life. The oral tradition, then, is always a residual; even at the most advanced stages in the development of communications technology. But even the term residual is misleading. Raymond Williams injected this term, along with its companion, emergent, into the discussion of base and superstructure Marxist cultural theory.[17] He was arguing that the movements present in any given social formation cannot be directly explained by a given phase of development of productive powers; some social movements are the residual of an earlier phase of social development; some anticipate a future phase of social development. In this sense, the oral tradition is a permanent residual in literate societies. Groups conserve certain ways of speaking and living, a restricted code, for example, whose natural home is an older way of life. But such residuals are also emergents as the example of linguistically based national movements will attest. They are more than blind attempts at preservation. The defiant reassertion of an ancient tongue, both as a medium of group communication and as the framework of an entire way of life, is also an attempt to discover a new social order using as a resource practices of another era. Similarly, what appears at first blush as an emergent form often recapitulates through an alternative technological device quite traditional styles of thought, interaction, and power.

Literacy is an even more problematic phase and practice. Literacy soaks through a population unevenly and often refers to quite distinct habits and practices—reading rather than writing as opposed to reading and writing; functional literacy, craft literacy, sophisticated literacy—rather than a uniform skill or mentality. Many different literacies with quite differential consequences exist at the same moment of time and the practice of literacy is something that itself shifts historically to a degree that what is common to different literacies— the ability to shape letters in space or interpret a simple paragraph of prose—may be of less significance than what is different about them. For example, in Robert Darnton's studies of eighteenth-century France, he felt compelled to differentiate between two different types of literacy: intensive and extensive.[18] He wanted to account for different uses of the same underlying skill, though even the word "same" is suspect here. Some uses of literacy seem best understood on an information model and others on a ritual model. Some literate practices emphasize the superficial acquisition of a large number of printed texts—endless reading of the new, the novel, the original, and the different. Other literate practices work their effect by the continued repetition of the same texts, the habitual and ritualistic rereading of the same texts in the same order in harmony with the same outside events. Here literacy approximates the recitation of song, chant, and epic tale of the oral tradition—a private ritual for more individualistic times.

What is in its own way even more remarkable is the constancy and durability of cultural forms as they migrate between different media; that is, for a certain range of phenomena—speech, print, electronics and, if I can hazard a guess, the fully integrated computer-satellite television system—seems to make little difference. Culture, as a structure of intelligibility—politically effective, aesthetically pleasing, cognitively correct—resists much of the transforming power of technology. As an example, Robert Darnton's studies of the oral tradition of fairy tales seek to establish two compelling arguments. First, he demonstrates the transformations in structure and meaning of fairy tales as they migrate from one national culture to another, even though the medium, speech or print, remains the same. Second, he also demonstrates that the same fairy tales contain master tropes—devices of plot, narration, characterization, and theme— that persist virtually unchanged within the same national culture across time, media, social groups, and even across forms of writing. The same images recur in folk tales and philosophy, in myth and social science, in rhetoric and religion.[19] It is precisely this circumspective attention to detail, including the detail of media within the medium—book, newspaper, and periodical as differential modes of literate activity—that seems to be easily lost within the gross complexes of historical progression encased within the abstractions of speech, print, and electronics.

The problem of circumspection is a constant within any form of scholarly writing but there is an additional difficulty with the technological history of communications that is rather more particular and unique. There is nothing particularly radical about treating the history of communications as a series of episodes in the evolution of technology. Quite the opposite. Technology has long functioned as a master symbol in historical writing. The centering of technology is to act in concert with the deepest promptings of the culture. To center technology is an act of the profoundest piety. We are, after all, a technological civilization. Technology is the ground of our dreams and aspirations. It provides the fact through which we attempt to change history and the metaphor through which we inscribe history, society, and the self. It is, therefore, a particularly dangerous animal to grasp by the tail and attempt to control.

Because we are a technological civilization, we have a natural tendency to read our concerns and our achievements back into the historical records and to characterize all history as the steady advance of technology. If in our private imaginations, the machine is satanic and demonic, it becomes a story of the fall; if in our private imaginations, the machine is liberating and redemptive, it becomes a story of salvation and transcendence. The tendency is everywhere in our scholarship. Jack Goody has a line somewhere that "archaeology makes materialists of us all." He refers to the tendency to use the artifacts found in archaeological digs, primarily a record of tools, as the basic evidence of human advance, evolution, and achievement. In both these ways—reading backward and reading forward—history becomes technological history, the ages of metals, discoveries, inventions, and industries. All human nature and all human achievement are cast in the shadow of *homo faber*.

In *Technics and Human Development*, Lewis Mumford attempted to redress this unfortunate tendency, though it is one to which he had mightily contributed in his earlier *Technics and Civilization*.[20] The reduction of history to technological history not merely overlooks but undervalues and practically eradicates achievements in art, ritual, mythology, religion, language, moral regulation, and governance: the creation of a container of culture that permits and fosters technical and every other kind of creativity and keeps it, when we are lucky, from running amok. The danger in the structure of technical metaphors for capturing the history of communications is that they contribute to the very obsessions it is necessary to control; they contribute to the misshapen view of human possibility and purpose from which Vico and the expressivist tradition were trying to awaken us.

Technological development is the bet upon which we place our lives, our politics, our education, our society—indeed, our very selves. Therefore, the writing of history technologically appears to be only normal. We take our achievements to be technological so why not read the history of the human achievement as a technological one and organize the narrative around the innovations that define its advance? It is particularly appealing to Americans. At the 1876 Centennial, William Dean Howells celebrated the national genius that spoke through the Corliss engine and reassured his audience that "by and by" we would have "the inspired marbles, the breathing canvases, the great literature."[21] For the moment we had to be content with technology.

We are a long time waiting. But Howells testifies in his silence to the ways in which technology operates for us as fact and symbol. As Eric Leed has argued, the symbol of technology is used to grasp ineffable but highly important issues: to what extent are we gaining or losing control of ourselves, nature, our destiny, our environment?[22] Artifacts have a deep moral significance precisely because they can be made to speak to larger issues—our liberties and our enslavements. Technology enjoys privileged metaphorical status because it has played such a central role in our development, not only in the conquest of nature but as a means of social regulation. Again, Eric Leed: "Technology in the image of the machine has been and continues to be the most prevalent way in which we represent as objective and external those mechanisms of self-regulation which are the condition of our civility."[23]

In short, the act of adopting the metaphor of technology to describe communications leads us to the pious center of society and to adopt uncritically the very "structure of feeling" that technology engenders. It also leads to our participation in all the silences and evasions that are central to the culture, all the secrets cast into darkness by the metaphor of the machine.

The metaphor of technology, most damagingly, casts one back into an objectivist mode of thought so that technology is something that is merely lying about in the bosom of nature waiting to be discovered, locked into geological strata waiting to be released to work its good for humanity. Technological history becomes one long series of begats, a biblical litany of parents and progeny. Speech begat writing which begat printing which begat, and here we get more specific, the newspaper, magazine, cylinder press, telegraph, telephone, lithography, radio, and so on to infinity. Human history is one long voyage of discovery through a beneficent nature that slowly yields its treasures. It is the machines in nature that await us as we await our destiny in the machines. In this standard metaphorical rendering it is the machines that possess teleological *insight*. It is they that possess and prevision the future. Human action is therefore reduced to one dimension: the power to act in concert with whatever nature has planned for us. In this reduction, the fact of power as exercised, power as imagined in and through technology, is eclipsed and occluded. Power is made an incidental by-product of an essentially technological purpose. It above all removes power from politics and society and makes it incidental to the entire social process.

The metaphor of technology constantly verges into a romanticism and never more so than in the technological history of communications. This romanticism is clearest in celebrations of speech, talk, and the oral tradition. The very reaction which constituted the expressivist tradition is part of this romanticism—the attempt to root the idea of the nation in some notion other than the social contract suggested in the primordial ties of language, the encompassing linguistic milieu that defined the very identities operating within it.[24] It was Vico's discovery and Herder's achievement to develop a conception of language that could be used as a weapon against both the scientific enlightenment and the political enlightenment as well. The discovery of the oral tradition and the folk was the discovery of a world that was lost and losing, a counter-identity to assert in the face of a contract theory of the state. Much of the impetus for research into the oral tradition was to recover it for its political uses: to restore a past as prelude to a new future. The impulses that formed the idea of an oral tradition were set out by a literate intelligentsia, by intellectuals in the modern sense of the word. In the oral tradition they found a counter-mentality and a counter-identity—their own negation, as it were, and "rooted this negation in the medium of communication used by the people."[25] Traditional society, dissolving under the impact of political and industrial revolutions, was reborn in the construct of an oral culture. This construct was invested in the rural lower classes, in the old regime and in "primitive" non-European people.

We are usually quick to recognize this primitivism, this going native, this romanticism of the people and the medium of their existence—speech. But it is not the only form romanticism takes; it is merely its reactive form. The romanticism of modernity, of the printing press, is now so much a stock part of the culture, so normal and ordinary, that it is virtually unrecognizable except in philistine evangelism: the belief that you can remake social groups by making them literate. But far more problematic is the romanticism of the future. The future and past function as alternating constructs. The oral tradition was a "framing conceptualization of enormous usefulness to men conscious of the modernity of their age, for it pointed to the pathologies inherent in industrializing societies while at the same time allowing social theorists to describe those pathologies as characteristic of a transitory period between two forms of community, one unconscious and pre-industrial, the other consciously constructed and post-industrial."[26] Nostalgia for the future is equally useful and I think more potent than nostalgia for the past. This nostalgia is recognizable in its open forms—the linkage of Charlie Chaplin's tramp with a computerized space age. But the nostalgia for the future found in our most advanced social theories—structuralism and deconstruction—is less apparent to the clerisy charged with their cultivation and protection.

## IV

Despite such reservations, I think the metaphor of communications as talk, text, and template is worth pursuing, as long as the metaphors are continuously rotated so that, like a prism, they pick up the various dimensions of the phenomena. My only proviso is that technology be conceived not in the first instance as a reflection of nature but as a form of action: the means by which we carry on the conversation of our culture, inscribe nature, and structure being. It is in the image of the homunculus that these functions come together.

In contemporary usage a homunculus is a miniature human, a dwarf. In sixteenthth-century medicine, however, a homunculus was a theory stipulating that a fully formed human was already present in complete detail in the spermatazoa: the human writ small. In our context, to view technology as a homunculus suggests that certain technologies or certain artifacts compress into themselves the dominant features of the surrounding social world. The homunculus is the society writ small. It is also the human person writ small insofar as it serves not merely as a template for producing social relations but as a template for human nature as well.

> Men and women of the electronic age, with their desire to sweep along in the direction of the technical change, are more sanguine than ever about becoming one with their electronic homunculus. They are indeed remaking themselves in the image of their technology, and it is their very zeal, their headlong rush, and their refusal to admit any reservation that calls forth such a violent reaction from their detractors. Why the critics ask are technologists so eager to throw away their freedom, dignity, and humanity for the sake of innovation?[27]

The homunculus is found in the Jewish cabalistic tradition as the golem and it is that ambiguous presence that gives a both moral and intellectual force to Norbert Weiner's last reflections on *God and Golem*. As David Bolter has recently argued, it is the computer that raises the question of and serves the function of a homunculus.[28] It is also the computer that drives these metaphorical explorations in communications history. Voice and text take on significance when they serve as a source of understanding and as a source of critique of the electronic homunculus. It remains an open question whether the media—in the extended sense I have been using the term—function as a homunculus, whether they can serve as an adequate site for social analysis.

This is not, it should be quickly added, a question of determination or causality, at least in any normal sense. There is absolutely no suggestion that the computer or the printing press or even the voice cause or determine the essential features of society or human nature. But they do not, to use Raymond Williams's rewriting of the notion of determination, merely set limits or create pressures. When technology functions as a master symbol, it operates not as an external and causal force but as a blueprint: something that makes phenomena intelligible and through that intelligibility sets forth the conditions for its secondary reproduction. It sets limits, in Williams's sense, but the limits are boundaries to be explored and broken through. Further, technology exerts more than pressure. Once adopted as fact and symbol, as a model of an instrument for, it works its independent will not by virtue of its causality but by virtue of its intelligibility: its ability to realize an aesthetically pleasing, politically powerful, socially regnant order of things.

There seems to me no objection in principle to the practice of using an artifact as a homunculus. Even if the artifact is less than the essence of the whole social system, it does hold its valuable secrets and instructive possibilities. The totem functioned in this manner for Durkheim. For Marx it was the commodity that held the secret of the capitalist mode of production. Understanding how a commodity is produced, exchanged, and attains value illuminated the essence of capitalism. However, the commodity is simply not a defining technology in the sense set out by David Bolter:

> A defining technology develops links, metaphorical or otherwise, with a culture's science, philosophy, or literature; it is always available to serve as a metaphor, example, model or symbol. A defining technology resembles a magnifying glass, which collects and focuses seemingly disparate ideas in a culture into one bright, sometimes piercing ray. Technology does not call forth major cultural changes by itself, but it does bring ideas into a new focus by explaining or exemplifying them in new ways to large audiences.[29]

Henry Adams's image of the dynamo, a condensation symbol of a whole array of power technologies, better served as a homunculus than did the commodity. Power technology effected the very displacements—the removal of time, place, and vision—that laid the groundwork for the creation of the commodity form. But information technology, by the time of the *Grundrisse* and *Capital*, had already begun its displacement of power technology as the homunculus of industrial and capitalist civilization. This was the argument I applied to the telegraph in an earlier paper: the separation of communication from transportation and their reintegration through a switched circuit provided the model of social organization for the 1840s onward. Like the telegraph, the computer does not work itself; it directs work, a technology of command and control. For example, the essence of the American space shuttle program is the computer that controls every aspect of its operation. The computer leaves intact older power technologies but it integrates them into a new complex of social engineering and that is why the central engineering problem is the economy of a signal and why electrical engineering is the central discipline. Power machines are no longer agents on their own, subject only to direct human intervention; now they must submit to the hegemony of the computer that coordinates their effects. And that is why

> As a calculating machine, a machine that controls machines, the computer does occupy a special place in our cultural landscape. It is the technology that more than any other defines our age.   .   .   . For us today, the computer constantly threatens to break out of the tiny corner of human affairs (scientific measurement and business accounting) that it was built to occupy, to contribute instead to a general redefinition of science to technology, of knowledge to technical power, and, in the broadest sense, of mankind to the world of nature.[30]

David Bolter here catches one other phenomenon in the prism. The computer had a long history as a producer good before it broke free, worked its way into the consumer market where it could function not only as a template of and for industrial and business activity but as a model of an entire way of life. This movement from foreground to background, from producer good to consumer good, reenacts, it seems to me, the general history of communications technology and provides a clue for searching through its consequences. But by the time it arrives as a consumer good, the game is pretty well over. At that point the symbol becomes a cause, the part a whole, the expression an essence. Having reorganized the way we work and produce, conduct politics and education, it can then redefine the very notion of a skill and the practices of daily living.

Our interest in the history of communications makes sense diagnostically as an attempt to understand the present, the present as a product of the past and as a seedbed of the new. But that understanding must be more than a ratification in theory of the achievement of practice. The evolution of communications from speech to print to electronics provides a series of benchmarks on which to inscribe a theory, elucidate a set of practices, and perform a critique. It is hard to be agnostic about all this. The choice of a metaphor from which to unhinge communications history privileges a moment in the entire process and casts loyalties in the political struggles of our time. Similarly, there is nothing innocent in the contest in philosophy and social theory over the metaphors of conversation, text, and structure. Here philosophy, technology, and history are merged and together they point toward the future we desire. If we privilege the computer or structure, we shall arrive at a destination that is predetermined. My own loyalties are clear, at least at the level of history and theory. Speech and conversation are privileged and therefore a marker against which to measure historical movement, technological innovation, and political practice.

## NOTES

1 James W. Carey, *Communication As Culture* (Boston: Unwin and Hyman, 1989), Chap. 3.

2 John Lukacs, *Historical Consciousness* (New York: Harper and Row, 1968).

3 Lukacs, 13.

4 Lukacs, 13.

5 Elizabeth Eisenstein, *The Printing Press as an Agent of Change* (Cambridge: Cambridge University Press, 1979), v. i.

6 Isaiah Berlin, *Vico and Herder* (New York: The Viking Press, 1976), xvii.

7 Berlin, xviii.

8 Pierre Bourdieu, "Symbolic Power," *Identity and Structure: Issues in the Sociology of Education*, ed., Denis Gleeson (London: Uffington Books, 1977), 112-119. The same argument is developed more elegantly, though with a different emphasis, by Paul Ricoeur, *Lectures on Ideology and Utopia* (New York: Columbia University Press, 1986).

9 Jack Goody, *The Domestication of the Savage Mind* (Cambridge: Cambridge University Press, 1977).

10 Goody, 10.

11 Goody, 11.

12 Eric Havelock, *Preface to Plato* (New York: Grosset and Dunlap, 1967).

13 Marshall McLuhan, *The Gutenberg Galaxy* (Toronto: University of Toronto Press, 1962).

14 Harold Innis, *The Bias of Communication* (Toronto: University of Toronto Press, 1951).

15 I borrow this argument, and part of what follows, from Eric Havelock, "Harold Innis: A Man of His Times," *ETC* 38 (1981): 242-267.

16 Harold Innis, *Changing Concepts of Time* (Toronto: University of Toronto Press, 1952), 15.

17 Raymond Williams, *Marxism and Literature* (Oxford: Oxford University Press, 1977), Chap. 8.

18 Robert Darnton, *The Great Cat Massacre and Other Episodes in French Cultural History* (New York: Basic Books, 1984), 249.

19 Darnton, Chap. 1.

20 Lewis Mumford, *Technics and Human Development* (New York: Harcourt, Brace, Jovanovich, 1966); Lewis Mumford, *Technics and Civilization* (New York: Harcourt, Brace and World, 1934).

21 Dee Brown, *The Year of the Century 1876* (New York: Scribner, 1966), 130.

22 Eric J. Leed, "Voice and Print: Master Symbols in the History of Communication," *The Myths of Information*, ed. Kathleen Woodward (Madison: Coda Press, 1980), 41.

23 Leed, 42.

24 Leed, 47.

25 The argument here is a paraphrase of Leed's incisive formulation.

26 Leed, 48.

27 J. David Bolter, *Turing's Man* (Chapel Hill: University of North Carolina Press, 1984), 14.

28 Bolter, 209-210.

29 Bolter, 11.

30 Bolter, 8-9.

# Part Two

# Culture and Ideology

# Representation of Interests and the New Communication Technologies: Issues in Democracy and Policy

*Stanley Deetz*

The rapid development of new communication technologies is having a significant impact on our society and world. Discussions of these changes range from hype for a grand futuristic world, to gloomy forecasts of dehumanization, to carefully detailed accounts of the effect on productivity. The potential significance of the changes projected in even the most balanced of reports has initiated debate on policy issues. An important issue in these debates has been the concern that present and anticipated technological developments enable certain groups to have far more influence than others on the social construction of meaning and on public decision making.[1]

The development and implementation of specific technologies have been uneven. Developments have been more rapid where there is a ready market, where the technology supports basic values, and where technical change is most cost effective. Natural fears have arisen that these developments will lead less to change than support existing power configurations and sustain current differences between groups. Such developments have consequences for democracy in the public arena as well as in the private, corporate, and what collectively might be called the cultural arena. Writers such as Habermas have detailed this "colonization of the life-world."[2]

As a response to this situation many groups have argued for greater emphasis on equal access, new national and international policies, and various forms of resistance.[3] Such positions are often misguided and ill-formed because of superficial understandings of the relations among communication, power, and technology. Many new forms of meaning domination as well as extension of existing forms of domination are possible with these new technologies. Systems must be analyzed in light of a theory that can account for new forms of power and that can provide a new normative foundation for policy development.

In this essay, I will use aspects of Habermas's theory of communicative action to pull together the divergent theoretical literatures on these issues more for the purpose of demonstrating a possible common agenda rather than to propose a new direction. To accomplish this, the essay will discuss the current forms of technological development and the relation of communication technologies to collective social development. Further, these relationships will be considered in light of the moral issues they raise as well as the effectiveness and technological issues. Finally, a normative foundation based on a discussion of moral issues will be applied to three areas: decisions regarding communication policy, the development of specific communication technologies, and the design of communication systems.

## THE CURRENT SHAPE OF COMMUNICATION TECHNOLOGY DEVELOPMENT

The current development of communication technologies is both rapid and uneven. The rapidity of development is clear in a number of areas. Computer assisted communication systems have greatly increased the speed and quantity of data transmission, the storage of large quantities of potential information, and the efficient retrieval of stored data by relevant users. In addition, the development of optical fiber and satellite transmission capacities has greatly expanded the size, integration, and possible interactivity of these systems. Further still, technological developments have greatly reduced the cost of such systems so that both corporations and individuals can own or have access to equipment necessary for participation in these larger networks. Put together these technological advances have had great impact on the data available through mass media, the nature of home entertainment, and the increase in size and control by corporations. These developments are clearly transforming aspects of modern life, even though debate continues on the nature and extent of the changes.[4] Few aspects of the everyday life world and culture are left untouched. New technologies replace older ones before most consequences of the prior implementation can be understood.

Equally clear is the uneven nature of this development. As certain types of technologies move ahead rapidly, others lag behind. The choices are not random, but follow identifiable patterns. Some of these are so obvious and appear so insignificant as to escape notice. For example, despite great strides in mediating and technologically extending our hearing and vision, smell and motion sensation have few accompanying technologies. High fidelity transmission of visual images and sound enter into the home daily, but we lack even the most elementary extension of other senses. While certain movie productions such as "Earthquake" try to technologically extend a sense of bodily motion, I know of no instance of a smell track. Few studying communication would suggest that these other senses are trivial even in the U.S. culture. While it can be argued that there are technological reasons why such development is limited, no one two hundred years ago thought that the sight and sound technologies we have today could be developed. There are cultural reasons why development has focused on certain senses, a focus that has cultural consequences. Clearly a culture can be conceived where other developments could have occurred. This sense that it could have happened differently helps us focus on the central issue: the choices made.

If posing the issue at such a fundamental level is problematic, attention can turn to less grand but still significant choices. Currently, technologies are being developed where there are pre-existing markets, where the technology is in line with existing concepts and values, and where technological developments are cost effective. Rarely are human needs considered when financial resources are limited. Rarely are social, noneconomic factors assessed in choosing where to place resources for development.[5] For example, the large and sophisticated data bases for investment firms can readily be contrasted with the slower development of computer assisted data retrieval systems for nonpromotional consumer information. This is not surprising in the U.S. since the government has played a reactive rather than a proactive role in making policies. Most development choices have been made by profit-making companies, with differences settled in the marketplace. Understanding this relation can make clearer the direction technological development takes and the motives that drive it.

The directional development of technologies has political consequences. Power differences already exist among different groups in society. Groups differ in their ability to influence public policy and to construct messages and meanings. These differences can be increased or decreased by the ways in which particular technologies are developed. New technologies also have the potential to lead to decentralization and greater democratization of meaning production and transmission. Communication technologies can aid the expansion of democracy or further positions of control and domination. Certainly the motives and structures for determining which technologies get developed have an effect on this political balance. Currently little is known about these effects, despite their critical social importance. As long as communication research focuses primarily on the efficiency and effectiveness of new technologies without a clear concern with "whose" and "what" ends are being advanced, issues of the political and cultural effects of technological development cannot be addressed.[6]

## THE POLITICS EMBEDDED IN COMMUNICATION TECHNOLOGIES

The social distribution of power and influence, particularly in regard to expression opportunities and the effect on the social construction of meaning and personal identity, should be the central communication policy question.[7] The concept of the politics of meaning and identity construction is obscured in modern society through linguistic and social forces. A distinction between two central conceptions of politics may aid in revealing the processes and in introducing a broader political understanding of communication technologies.

For the sake of easy identification I will designate such distinctions as capital "P" *Politics* and small "p" *politics*. Small "p" *politics* is the ordinary notion of politics that we discuss in everyday life. It is the politics of presidential elections, debates over policies, decision making, the giving of information, and influence in the explicit, direct sense of persuasion. Capital "P" *Politics* is less easily understood and discussed. This is the *Politics* that exists in social formations, institutional forms, the distinctions made by language, and the particular technologies available. This is the *Politics* of consensus and consent rather than direct influence.[8] For example,the greatest censorship comes in what is never thought and the forces that make some things unthinkable rather than in restrictions on what can be printed.

Consent is always the hidden side of power. Every teacher intuitively knows this. I teach a class of over 300 people. Everyday that I step into it I know that I do not teach this class primarily by authority or power, but rather by consent. At any moment the students can end the class. Fortunately (for me), students hold a number of beliefs about how they should act or what will happen to them if they don't that control their behavior. In addition, a number of social habits and unconscious everyday practices leave alternative behaviors unthinkable. Similarly in corporate organizations, despite all the discussions of forms of power, managers only manage by the willful consent of the managed.[9] No organization is powerful enough to regulate the behaviors of its employees any more than a state could enforce its laws, without organized consent. The organization of consent rather than the organization of power is central to understanding *Politics*. This is the *Politics* of the background practices of everyday life.

In looking at the organization of consent, a number of important processes are revealed. Habermas identifies a number of examples of what can be called the progressive colonization of the life-world.[10] Most basic to his analysis is the historical emergence of "rationalization." In the progressive "liberation" from traditional values and institutions, efficiency and effectiveness become the primary criteria for the evaluation of all of life. The "non-rational" aspects of life become inconveniences to be ordered and brought under control. For example, in the "rationalized" world view even children can be seen as potentially disruptive and intrusive. The building of systems and institutions to order child bearing and rearing suppress the potential conflict with the productive world.[11] The attempt to accomplish practical ends through what Habermas called technical-instrumental reasoning has become a guiding mode of rationality superceding all other possible forms of reason.[12] Weber accounts for this through the rise of the protestant ethic, the working out of the enlightenment project, and the development of the methodic form of life.[13]

The presence of a "modern" mind and set of practices leading to a postmodern crisis hardly seems controversial. This is not to claim a distinct new age or to hold that past forms of reasoning are gone. Competing forms of reasoning and discourses do coexist, but not as equals. Technical/scientific/ instrumental reasoning processes have become privileged; others have been marginalized. Such domination creates a two-sided problem: one regarding the formation of self-identity and the conception of self-interest and another regarding the social opportunity for the expression of self-identity and self-interest. Such opportunities involve both the capacity and availability of the appropriate medium. Habermas calls this problem systematically distorted communication.[14]

The problem of systematically distorted communication and the *Politics* of discourse incorporates more than just the domination of one form of reasoning over others. Systems of thought, expression, and communication medium may contain embedded values that are at odds with the person's own values, if such a person could openly assess them. Yet, the embedded values can be reproduced in expression without awareness. For example, when the woman on the plane next to me referred to the flight attendant as a stewardess, she expressed values that probably were not her own. It is not as if she said or felt anything negative nor did she seem to offend the flight attendant. The term "stewardess" is not degrading alone, but enacts gender as a valuational distinction and relevant dimension of occupational classification. Similar practices are common. If people could work back though the concepts they utilize, they would often find a gap between what they reflectively think and feel and what they unwittingly express. A similar form of possible distortion exists in the gap between personal identities and the images people live. One can look in a mirror and evaluate the body based on a host of external images while never carefully considering how the person feels or what is personally desired. If this is clear we can see why, ironically, the inability of people to distinguish between *Politics* and *politics* can lead them to be most powerless at the moment they think they are powerful. For example, an image may be constructed to enable one to succeed. Yet at the moment of success the image rather than the person has the success, a success which means that the individual didn't get expressed at all.

Numerous detailed descriptions of the distortions arising from values embedded in language and socially constructed images have been provided. Marcuse provided perhaps the most vivid conception in his discussion of the "closing of the universe of discourse."[15] Post-structuralists and feminists, however, have added great sophistication to the analysis. As Weedon has suggested, the most important aspects of this are to open up the politics of personal experience through a rejection of the conception that "language is transparent and expresses already fixed meanings" and the disjunction of private and public discourse.[16] The meaning of personal experience is the central and most difficult issue. As Weedon expressed it: "The meaning of experience is perhaps the most crucial struggle for meaning since it involves personal, psychic, and emotional investment on the part of the individual."[17]

Beyond the individual, *Political* implications are embedded in media products. Gans and Tuchmann have shown how standard media "frames" structure the content and presentation of news, thereby providing a concept of human responsibility, "objective" accounts of events, and particular conceptions of the causal relations among events.[18] Gitlin and Hall, through use of Gramsci's concept of "hegemony," penetrated even deeper into the processes by which consent is engineered through media.[19] Media messages elaborate ideology into common sense and everyday practices by reproducing social conflict in terms derived from the dominate ideology. Lafort has added a different angle by showing the emergence and function of central "imaginary significations" in society.[20] Such imaginary representations of the unity of society block perception into historically produced social divisions by producing a "closed discourse which, masking the conditions of its own engendering, claims to reveal that of the empirical social reality."[21]

The *Politics* of discourse, whether seen in self-expression, interpersonal interaction, or mediated forms, harkens back to the dual issues of identity formation and social construction of the world. Since a technology is at root a materialized ideology, contemporary analysis of power and ideology interpellation provide an initial look at the *Politics* of communication technologies. Pêcheux, following Althusser, argued that ideology "interpellates individuals into subjects" through complex, "forgotten" interdiscourses whereby each subject has a signified, self-evident reality which is "perceived-accepted-submitted to."[22] As Thompson presented Pêcheux's analysis: the hidden-forgotten discursive formation "creates the illusion that the subject precedes discourse and lies at the origin of meaning. Far from this being the case, it is the subject which is 'produced' or 'called forth' by the discursive sequence; or more precisely, the subject is 'always already produced' by that which is 'preconstructed' in the sequence."[23]

In a similar fashion each communication technology can be said to create a particular "subject" with an identity and reality as part of a discursive sequence. For example, the telephone as a technology situates a "self" in a definable role as speaker and posits a distinct perspective on communication as having a fixed starting and stopping point, a source and destination; it is not just a medium, it provides a perspective on both the interactants and the process itself. It is an embodied subjectivity which is reclaimed as the identity of the user. In using a telephone I implicitly enact the value preference of certain sensory experiences over others and the desire for convenience over competing needs. The implied concept of communication and standard telephone scripts, as Hopper discusses in this volume, defines rights and responsibilities, structures the identity of self and other, and gives preference to certain topics and types of messages. The bodily presence of the other and the use of nonverbal elements might all change the nature and direction of the conversation.[24]

Moreover, the subject formed is not accidental since the "world" and "identity" arise in power-laden discursive structures. I might not choose to use the telephone *if* I reflected on the way I understand communication, the production of meaning, and the values I have regarding human beings. But the significance of this choice is misunderstood if thought of only in terms of convenience, effectiveness, or effect. The inability to reflect and make such choices is the issue. The fundamental issues of the choice are often hidden and not considered and the choices that are considered are themselves positioned out of the background of the forgotten. Reflection cannot reclaim a genuine choosing agent. Only the creation of new discursive structures can. While Pêcheux's work might be accused of underestimating human imagination and creative potential, the relative expertise and physical requirements of many technologies leave the technological aspect of the discursive sequence more resistant to examination and change than linguistic expressions.

In sum, each technology structures sensory advantages, provides ways of knowing the world, privileges certain notions of what is real, and posits personal identities. A technology posits a subject, has an epistemology, and structures value choices. In this, I am not suggesting that technology is somehow bad. Opening the *Politics* of technologies is an important step in critically assessing the relation of technological development to the representation of personal and group interests. A technology is never *P*olitically neutral. Technological developments give preference to forms of expression and thought some of which reflect a group's actual interests and thus enable individuals to represent them in public forums. But others extend domination and the interests of others are advanced at the the expense of those of one's own group in self expression. It is important to examine the ways a technology aids or hampers the expression of interests and collective decision making toward maximally meeting the needs of the various segments of society.

## THE MORALITY OF INTEREST REPRESENTATION

When examined closely, *P*olitical issues are at root moral issues. They are not simply choices of who will have what or how much, they are issues of the collective formation of culture and human identity. Carey's essay reviews the literature related to the fragmentation of identity in postmodern society and finds that a primary issue confronting the society is the loss of community and personal identity through community. Or as Habermas argued, language-based, life-world institutions that provide consensual understanding are overloaded by the need for ever more system coordination. As they fail they are supplemented by the development of system-steering media that coordinate by instrumental values like money and power.[25] Once set in motion the process creates further domination by social systems but is simultaneously weak and leads to dysfunctional motivation and legitimacy problems.[26] The cycle is both endless and a crisis. The struggle is to reclaim identity in new ways either through a reformation of community itself or to create new forums for meaning and identity formation.

The old foundations for stable meaning and identity have passed away. Nietzsche in proclaiming the death of God was one of the first to identify this feature and its significance in Western society. The gradual erosion of certainty derived either from a universal concept of human nature or a motor of history metaphor is both a liberation and curse.[27] In the modern context, concepts of human nature, technological determinacy, and segmented or evolutionary history are rhetorical ploys to privilege certain meanings and gain political advantage. Each "moral fiction" [28] is an attempt to regain a collective identity and a particular identity. Once we understand that every attempt to proclaim a foundation is a disguised *P*olitical maneuver to advance a particular group's view of the world, we place in question both the utopian futurism of technological advocates and the dire forecasts of the technological determinists. What we see is that communication technologies enter into the collective debate and decision of what we willfully or unwittingly become.

The lack of any possible non-arbitrary, non-privileging foundation in nature can easily lead to giving up on moral questions altogether or relinquishing such questions to the private, subjective, or nonrational realms. No *a priori* standard for evaluating the effects of certain technologies or the goodness of our collective movement into the future seems possible. Power seems to be the only possible arbitrator among positions. The loss of a foundation seems to justify those with advantage extending it into further domination. The various nihilist positions emerging from this perspective have provided human disasters in world politics (Hitler) and in the ecology. Clearly such positions are problematic on both practical and theoretical grounds.

An alternative way to see our situation offers interesting opportunities, however. If a foundation is lacking, power need not be the only arbitrator. Instead, we can argue for a different type of moral conception. The moral issue changes from what is good or right to the process by which we will determine the good and the right.[29] Morality is a communication issue of ethical knowledge formation rather than a religious or private one.[30] How can we engage in communication that enables equal participation in our collective formation? Morality need not privilege one stance over another. Instead, morality can denote discourse that is freed from domination. This is not totally new to us. Our conceptions of democracy in the *p*olitical realm demonstrate a partial public commitment to such ideals. Unfortunately, such conceptions have not adequately accounted for the relation of language and technology to experience and have not been extended to the *P*olitical realm. Before building these corrections, a brief examination of the moral foundations of *p*olitical democracy may be helpful.

Western concepts of *p*olitical democracy rest on a "natural" right to have one's interests fairly represented in matters that affect one's well-being and pursuit of happiness. Such ideals are operationalized in election processes and state institutions. With the development of mass and computer-mediated communication networks these ideas are extended through the principles of free speech and the free marketplace of ideas. Even if we retain the human nature foundation and supplement it with conceptions of equal access to media, such an approach still leaves out essential concerns with the representation of interests.

If there is a gap between what we think and what we really think or a systematic distortion in our expression, free speech does not help. In fact, the concept of free speech helps guide us away from careful examination of the misrepresentation of interests. Having a right to expression cannot assure that one's interests have been taken into account. For example, if equal access to television programming simply means that some groups who get air time lack the resources and skills to use the media well, the free marketplace of ideas fails. Even more significantly, if people have access to a medium and do use it well, yet the medium itself carries with it certain vested interests, the marketplace fails and its failure is even more deeply hidden. The action-at-a-distance of television and its reliance on the visual image, for example, may undermine the best televised presentation of feminist conceptions of human interconnectedness. I do not wish to enter the debate as to whether television or any other communication technology could be made value neutral and used to express all messages. The point is that it does not and the possibility of such seems weak given the comparative ease with which it conveys some messages and the existence of real comparative economics and time. Advantage rather than necessity is the issue. While an adequate reconception is still to be worked out regarding the mass media, there is some general awareness of the problems and social responsibilities. This is largely because the mass media are conceptualized, at least partially, as part of the public arena. Far less attention exists in areas conceptually cast into the private. The public/private distinction is a significant *P*olitical act in regard to the extensions of democracy.

Compared to studies of media, relatively few analyses have focused on the effects of business corporations and their implementation of communication technologies on cultural production apart from the most abstract capitalist reproduction thesis. The corporate context and work experience have unique effects on personal identity and contemporary mass culture, making them potentially far more direct and interesting than media analyses. The turn in Marxist thought from work relations to ideology has led at times to the avoidance of the ideology of the workplace.[31] This is a point Habermas seems to have missed in his own maintenance of a private/public conception.[32] Modern writings on the eclipse of the state by corporate organizations show both the inadequacies of the private/public split and the pervasiveness of the modern corporation.[33] The state is no longer the primary institution in society, even in the United States. The U.S. is one of a small handful of nations that has a gross national product larger than the gross product of several international corporations. But structure, as well as size and power, is at issue.

While the state's power is exercised primarily through restriction, corporate organizations make most decisions as to technological development, utilization of resources, and working relations among people. Corporate values and practices extend into non-work life through time structuring, educational content, economic distributions, product development, and the creation of needs through advertising. Such a situation is likely to grow since the state lacks the ability and resources to collect data, to engage in large scale monitoring, and to generate large scale value consensus *even if* it desired to create values alternative to corporate organizations. In short, the primary decisions regarding individual and collective development are being made in the corporate world. Virtually no socially shared concept of free speech or rights to represent one's own interests in decisions that affect those interests appears within the corporation or in relation to the larger public.

Corporations thus *de facto* make most policy decisions regarding the development and implementation of new communication technologies. Moreover, the implementation of communication technologies within corporations has important effects on the size and structure of corporations and their ability to represent the interests of different groups both internal and external to corporations. Both of these point to a need for a normative foundation that can extend to an assessment of corporate choices and systems in regard to participation in our collective formation. Corporations do not make simple commercial decisions but important *P*olitical ones hidden and protected from public debate and control. The need remains for the means to analyze these and develop public policy directives that represent the various publics' interests in these decisions.

## THE POLITICS OF EVERYDAY LIFE

When technology and culture are carefully examined, a very important *P*olitics of experience is evident. Every technology extends our sensual self in some direction.[34] In this sense, technologies are extensions of subjectivity. McLuhan popularized such a position years ago.[35] And, despite the many debates over the dramatic claims following from his work, few would deny the initial premise. A technology is not a simple tool to be picked up and utilized as something apart from the individual user. Its use provides a point-of-view or way for the self to be engaged in the world. In watching a television program the self is extended to another time and place with a particular situated perspective. The camera angle becomes the self's vantage point highlighting and bringing certain aspects into view and hiding and passing over others.[36] While as a critic I may step back and curse the camera angle, as an engaged subject I strain to see as I would strain to hear the whispers in the back of the room while giving a lecture.

Technology as an extension of senses positions us in the world through the same processes that our senses do in the absence of technology. But as different animals are positioned in the world differently owing to different bodies and different sensory acumen, different technologies position differently, as, for example, when the telephone extends the voice and hearing but not sight.[37] The *Politics* of experience is first a *Politics* of sensuality. In what ways will our personal and collective subjectivities exist? In what way will our insides be extended out and in what ways will the external world be allowed in? When a technology is utilized, its particular subjectivity is encountered at the meeting of the inside and outside as the shape of perception and expression, both of which are systematically skewed.

I am not using the concept of subjectivity in a metaphorical sense only. Understanding the fundamental connection of subjectivity and its extensions is useful in keeping the moral character of the issue clear. If someone were to argue that only certain classes of people should be allowed to have glasses or hearing aids or, more radically, that everyone's eyes should be poked out, we would all readily recognize the moral and political issues involved. Yet, technologies are developed that privilege one sense over another, that enable only some people to have access to some data, and that make only certain types of data available. These developments are treated as different kinds of issues lacking moral implications. Communication technologies each have their own way of positing/ extending human senses. Modern computer-assisted communication technologies have changed the way human beings are in the world and thus have moral/ *Political* consequences.

First, the electronic connection competes directly with the transportation connection as the primary way of being with others. The data is fundamentally different in each case: to walk through the earthquake rubble is different from seeing it on television; to talk with another is different from the most complete dating service data bank; to face the person to be layed-off and to know his/her personal situation is different from his/her file. On the surface this looks like a simple gain since we now have an additional form of contact and have not lost an old one. This gives people advantages in getting in touch with certain types of data. They can access data that was not available before. But with the institutionalization of preferences and the assessment of economic gains without consideration of human costs, the deck is stacked.[38] And as long as short-term effectiveness is the measure, the electronic connection can show a marginal advantage without an adequate calculation of social costs, particularly when there is no one in the decision process who represents certain publics and the general public's interests. At times such costs affect economic revenues. For example, when Johnson and Johnson substituted electronic ordering for its hospital sales force, numerous other sources of social contact had to be implemented to maintain sales. While this appears to be a natural check on the degree of electronic substitution, it leaves out all effects not translatable to the product being sold and all social interests that are distorted in the economic equation.

The relative cost savings of substituting electronic data for physical association easily leads to the substitution of one form of sensually being-with-others for another. In an analogy to Gresham's law (bad money drives good money out of circulation), cheap data substitutes for expensive data, particularly when cheap forms of data provide additional benefits to power elites. This can be seen in the substitution of multiple choice for essay examinations in universities. The technology makes possible an institutional structure which in turn makes it unthinkable to do anything but use the technology. As long as the discussion sticks to efficiency criteria coupled with a procedural concept of fairness, neither the history of the change nor the manner by which certain forms and expressions of knowledge become preferred can be discovered.

Further, the substitution of the electronic for the transportation connection changes the speed of decision making. Electronic connections make possible and reward spontaneous, action-oriented decisions. In contrast, transportation connections tend to encourage reflective critical thought. For example, if you are angry at another and a phone is available, the anger can be expressed in a virtually instantaneous manner. If you must travel to express the anger, the time of the journey allows reflection, rehearsal, and reconsideration. In addition, numerous events and competing systems can intrude during travel, thereby contextualizing and recontextualizing the anger. Through these processes the anger may be increased, diminished, or changed in character and the decision as to when and how to express it can be considered and reconsidered. Electronic connections encourage focus and reaction. Transportation connections encourage holism, connectedness, and proaction. Electronic connectedness makes possible high-speed management as well as exaggerated stock market swings.[39]

The intuitive public understanding of this difference accounts for the fear in a nuclear age. The electronic connection rather than the sheer force of the bomb is the issue. Human beings have always been able to annihilate the species. What has changed is the speed at which an irreversible decision can be communicated and the action carried out without the possibility of rethought and other systems intervening. Annihilation with clubs, arrows, or guns is slow and both its sensuality and the possibility of environmental and decisional intervention makes completion unlikely. First, the development of a fast transportation system and then the addition of electronic connection make the modern systems different.

Neither a transportation nor an electronic connection is necessarily better nor is either neutral. They differ in the way they put us sensually in touch with others and in the relation between spontaneous and reflective thought. The trade-off between narrow sensuality, speed and efficiency, *and* disparate sensuality, reflective thought, and intrusion are experienced throughout modern life. This is manifest in the relatively simple choice of using an online search rather than going to the library. In each case the trade-off is similar: Getting what you think you want fast or having what you want reformed constantly in light of intrusions and unexpected findings. The *Political* questions do not arise simply with the trade-off but, especially, when efficiency becomes the sole criterion and one type of connectedness replaces the other without free and informed engagement in the choice. How are different interests represented? To what extent are different human interests and different group interests privileged with advantage given to one type of connectedness? And, where is the forum for such decisions of public consequence?

Similar issues arise in a careful exploration of the development of data bases in connection with communication technologies. A communication technology, including a data base, is an extension of thinking and memory. Each embodies a way of thinking and a kind of memory. Each gives preference to certain forms of knowledge and communication. Turkle has, of course, shown this in regard to male and female students' use of computers but the issue can be framed in more general ways.[40] Book smartness is easy to store, street smartness is difficult. Codified, classified "knowledge" is easily stored and retrieved in such systems; intuitive, conceptual insights are not. Mitroff, following Jung, demonstrated that in organizations, different individuals have identifiable preferences for different types of data in their decision making.[41] For example, simple distinctions can be made between individuals based on their placement along thinking/feeling and sensing/intuiting axes. In doing so, four thinking styles can be identified. Individuals might be classified by data preferences as: thinking-sensing, thinking-intuiting, sensing-feeling, and intuitive-feeling. Existing communication technologies give preference to thinking-sensing. In fact, the concept of data in modern society is often reduced to this single type. What Weber and Habermas have called rationalized knowledge holds a certain primacy over potentially competing forms of reason.[42]

Ironically, studies of corporate decisions demonstrate a high reliance on intuitive data forms.[43] The good story that carries intuitive thinking and feeling data has high impact on decisions. In other words, insight into the process or organization of events is often more critical than codified information. Emerging value structures and implementation of new communication technologies tend to depreciate this preferred form of thought. Managers try to get around this by making intuitive decisions and justifying them after the fact by codified information. It is hard to see how the insightful story could be stored or retrieved.

Structural changes in the use of electronic connection and efficiency evaluations, added to the implementation of specific technologies, privilege a type of data and, hence, people who prefer that form of data. Individuals with other preferences find their data less valued and available. Some of these data preference differences may be divided roughly along gender lines if we apply analyses like that of Gilligan.[44] If the possibility exists that some thinking styles are more generally used by women and others by men, communication technologies and attendant storage and retrieval systems have implications for sexual politics. Continued development in current directions aids some groups over others by reproducing and accentuating existing power differences. Even if preferences change and different groups become equally proficient in the use of the preferred rationalized data form, we are still left to ask what has been lost and in what ways have decisions (and ultimately, our collective development) been altered. Would we have chosen this direction had we been given the opportunity to participate in the decision?

Finally, the current preference given to certain technologies along the lines of the dominant form of rationalization leads to a continuation of the process Braverman called "deskilling."[45] By this Braverman means that the conception of problems and possible responses becomes severed from the execution of the chosen action. Work in this special sense becomes "thoughtless." This does not mean that modern jobs are not frequently quite skilled in an ordinary sense. High-tech jobs may require high degrees of training, but what is learned is rules of execution based on decision chains devised by absent decision makers. Resultant alienation and procedural ideology without responsibility have personal and collective costs.

Various technologies have different effects on the expression of human interests, the prevailing form of knowledge, and the structure of decision making. Such technologies are active components of the human dialogue. The question then becomes what guidance we should have to determine the policy decisions that will shape this dialogue.

## TOWARD A NORMATIVE FOUNDATION FOR COMMUNICATION TECHNOLOGY POLICY

Human choices about the development and implementation of different communication technologies influence representation of interests and collective human development. Given modern discussions of ideology and power, the concept of choice is a relative one. Both the choices and the basis for choice are highly structured. While freedom through self-reflection and agency through will can only serve as covers for domination, I believe Habermas was right when he argued that through communication and collective education an arena of autonomy and open social formation can occur.[46] The only other even partially positive alternative would appear to be the discovery and fostering of sites of resistance. Unfortunately, I suspect that we will generally find what Burowoy concluded in his studies of organizations—most acts of resistance ultimately play back into systems of domination.[47] Perhaps all social hope is unwarranted, but the act of writing presupposes it. Hope cannot come from new utopian ideals but through collective choices grounded in neither images nor persuasion.

On what basis should such decisions be made? Are there criteria beyond free speech, the free marketplace of ideas, and effectiveness that might both be communally agreed upon and overcome the shortcomings of such criteria when applied to the modern context? And, more to the point, can criteria be established that do not provide a basis for merely a new and perhaps better disguised form of group advantage? Finally, if such principles are not to be found, can we define the conditions of public debate by which such principles could be created or decisions regarding their creation and implementation could be discussed? I think we can work toward answering such questions and since they inevitably involve moral questions regarding our collective development, I think we should.

Fortunately, these are not distinct questions. In fact if these questions are examined carefully, it becomes clear that if the last question can be answered we have made considerable progress toward answering all of them. For in determining the essential conditions of public decision making, we have said much about the basic value presumptions on which decisions about the design of communication systems and communication technological development should be based. Allow me to summarize briefly here what issues such a standard must address. I will then conclude with a brief review of these value presumptions and the communication practices they suggest.

1) Since no technology is neutral, the development and implementation of communication technologies are collective concerns. Communication technologies influence the way we are as sensual creatures, the forms of knowledge and data available and valued, and the participation by different groups in decision making.

2) The current distinction between the private and the public tends to be misleading when put within the context of the new communication technologies. The privacy of individuals and a realm of private meaning not collapsed into the collective requires new forms of protection. And, institutions such as the entertainment media and business corporations need to be assessed in light of their public role. Further, new forms of public participation in media and corporate decisions need to be considered.

3) Communication research and analysis has often been limited by the acceptance of effectiveness and efficiency as sole criteria for systems assessment. If such criteria are to be used, questions such as "whose and what goals are being advanced" need to accompany them. More significantly, alternative values and non-instrumental assessments are essential for understanding the social impact of the new communication technologies.

4) The public has little awareness of the means by which personal identity and experience are formed in regard to communication systems. Means of understanding and overcoming self-deception and systematic distortion are as important today as freedom of speech and understanding propaganda were in their time. Forums for open debate over developmental values and policies need to be implemented within real decisional chains.

The task that is left is the attempt to describe a model of public discussion that can be utilized in public debate and that can serve as an initial normative standard for assessing communication systems and technological development. Numerous scholars have addressed aspects of this need in regard to media and organizations.[48] In each case guidance is given toward developing models of domination and of communication in which open discussion might take place. Perhaps no one has offered as extensive a scheme as Habermas.[49]

In his development of universal pragmatics, a development that is extended and offered as a moral framework by Apel, Habermas provided a useful place in which to situate such discussions.[50] While his approach has been justifiably criticized on several philosophical and practical grounds, it can be instructive as an initial basis for discussing the moral foundations for communication technology policies. I advance it here not as a necessary foundation but as a preferred or arguable position toward further debate. In this sense, it is a position that can be derived out of the community itself and thus serve as a moral foundation. Since Habermas's position is well known, I will be brief.

Basically, Habermas argued that every speech act can function in communication by virtue of common presumptions made by speaker and listener. Even when these presumptions are not fulfilled in an actual situation they serve as a base of appeal as conversation turns to argumentation regarding the disputed validity claims. The basic presumptions and validity claims arise out of four shared domains of reality: the external world, human relations, the individual's internal world, and language. The claims raised in each are respectively: truth, correctness, sincerity, and intelligibility. Thus we can claim that each competent, communicative act represents facts, establishes legitimate social relations, discloses the speaker's point of view, and is understandable. Any claim that cannot be brought to dispute serves as the basis for systematically distorted communication. The ideal speech situation must be recovered to avoid or overcome such distortions. It should be clear that this conception applies not only to the everyday and ordinary acts of communication but also models our collective decisions as to what our society will be and what kind of people we will become—our moral responsibility.

The motivational framework of the ideal situation is founded on a concept of communicative action. In modern society communication has been reduced in conception to one of several means by which we engage in rationalized strategic action; that is, acts oriented toward the accomplishment of some end. In conceptualizing communicative action Habermas reveals the more fundamental nature of communication as structured toward reaching understanding. It is within this conception that new communication technologies and the decisions to develop and implement them must be considered. In the end, a technology needs to be evaluated on the grounds of its contribution to the equitable representation of human interests and contribution to reaching communicative understanding. Unfortunately, most communication researchers have only studied communication as influence within the strategic action model. Comparatively little is known about communication as participation.[51] Even studies of participation in decision making have tended to be structured around concepts of effectiveness and influence. Examining the structure of interaction aimed at reaching understanding reveals much about our moral commitment. Habermas has demonstrated the characteristics of the situation that is freed from barriers that would obstruct the process of communication.

First, the attempt to reach understanding presupposes a symmetrical distribution of the chances to choose and apply speech acts. In interpersonal communication this would specify the minimal conditions of skills and opportunities for expression. When we extend these through a consideration of new technologies, initial focus needs to be on equal access, distribution of training opportunities, or development of technologies with ease of access and low skill requirements. It also draws up issues of information availability and suggests consideration of information as a public utility, including public management.

In addition, ownership and managerial structures, both as in control of and as products of technologies, need to be investigated within the concept of symmetrical distribution of chances.

Second, the understanding and representation of the external world needs to be freed from privileged preconceptions in the social development of "truth." Ideally, participants have the opportunity to express interpretations and explanations with conflicts resolved in reciprocal claims and counter-claims without privileging particular epistemologies or forms of data. The freedom from preconception implies an examination of ideologies that would privilege one form of discourse, disqualify certain possible participants, and universalize any particular sectional interest. Communication technologies need to be examined with regard to how they function ideologically to privilege certain perceptions and forms of data and obscure historical processes.

Third, participants need to have the opportunity to establish legitimate social relations and norms for conduct and interaction. The rights and responsibilities of people are not given in advance by nature or by a privileged, universal value structure, but are negotiated through interaction. Acceptance of views because of an individual's privilege or authority or because of the nature of the medium represents a possible illegitimate relation. Authority itself is legitimate only if redeemable by appeal to an open interactional formation of relations freed from the appeal to authority. Values and norms legitimately exist in society by the achievement of rational consensus, subject to appeals to warrants supporting the assumed social relations. To the extent that particular technologies embody values, hide authority relations, or reify social relations, they participate in domination.

Finally, interactants need to be able to express their own authentic interests, needs, and feelings. This would require freedom from various coercive and hegemonic processes by which the individual is unable to form experience openly, to understand the self, and to form expressions presenting them. Technology can aid in the formation of self or other as images establish distance that denies the formation of "otherness" and the interrogation of self. The examination of technology in its structuring of the interior would be important to understanding its effect on the accomplishment of such an ideal.

This is just a sketch of issues needing to be addressed in the construction of a moral foundation for communication technology policy and collective self-formation. I have not considered any of the issues of putting such a conception into practice or the actual implementation of different decisional systems. Such an agenda requires that the discussion be framed in a useful way so that we first understand the problems. It is doubtful that clarity or a rational foundation will mobilize public action, but they can serve as points of departure and landmarks for discussion. The ideal communication situation is a fiction, but perhaps Habermas is correct that "on this unavoidable fiction rests the humanity of relations among" people.[52]

# NOTES

1 David Burnham, *The Rise of the Computer State* (New York: Random House, 1983); Irving Horowitz, "New Technology, Scientific Information and Democratic Choices," *Information Age* 5 (1983): 67-73; and E. Barber, "The Second American Revolution," *Channels* 62 (1983): 21-25.

2 See especially, Jürgen Habermas, *The Theory of Communicative Action, volume 2: Lifeworld and System,* trans. Thomas McCarthy (Boston: Beacon Press, 1987), pp. 332ff. For review, see: Oscar Gandy and Charles Simmons, "Technology, Privacy and the Democratic Process," *Critical Studies in Mass Communication* 3 (1986): 155-168.

3 For review, see, Jennifer Slack, *Communication Technologies and Society: Conceptions of Causalities and the Politics of Technological Intervention* (Norwood, NJ: Ablex, 1984); and Marjorie Ferguson, ed., *New Communication Technologies and Public Interests* (Newbury Park: Sage, 1986).

4 See, for example, Michael Dertouzos and Joel Moses, eds., *The Computer Age* (Boston: M.I.T. Press) and Michael Traber, ed., *The Myth of the Information Revolution: Social and Cultural Implication of Communication Technology* (Newbury, CA: Sage, 1986).

5 C. Fisher, "Studying Technology and Social Life," in *High Technology, Space, and Society,* ed. M. Castells (Beverly Hills: Sage, 1985), 284-300; and R. Walton, "Social Choice in the Development of Advanced Information Technologies," *Human Relations* 35 (1982): 1073-1084.

6 Kim Cameron and David Whetten, eds., *Organizational Effectiveness: A Comparison of Multiple Models* (New York: Academic, 1983); and Stanley Deetz, "Critical-Cultural Research: New Sensibilities and Old Realities," *Journal of Management* 11 (1985): 121-136.

7 See, for example, Jean Lyotard, *The Postmodern Condition* (Minneapolis: University of Minnesota Press, 1984); Pierre Bourdieu, *Language and Symbolic Power* (Cambridge: Polity Press, 1989); Jean Baudrillard, *The Mirror of Production,* trans. Michal Poster (St. Louis: Telos Press, 1975); T. van Dijk, ed., *Discourse and Communication: New Approaches to the Analyses of Mass Media Discourse and Communication* (Berlin: de Gruyter, 1985); and Habermas, 1987, 105 & 311ff.

8 See, Stewart Ranson, Bob Hinings, and Royston Greenwood, "The Structuring of Organizational Structures," *Administrative Science Quarterly* 25 (1980): 1:17; Michael Burawoy, *Manufacturing Consent: Changes in the Labor Process Under Monopoly Capitalism* (Chicago: University of Chicago Press, 1979); Stephen Lukes, *Power: A Radical View* (London: Macmillan, 1974); and Fredric Jameson, *The Political Unconscious: Narrative as a Social Symbolic Act* (Ithaca, NY: Cornell University Press, 1981).

9 For development, see Stanley Deetz and Dennis Mumby, "Power, Discourse, and the Workplace: Reclaiming the Critical Tradition in Communication Studies in Organizations," in *Communication Yearbook 13,* ed. Jim Anderson (Newbury Park: Sage Publications, 1990):18-47

10 Habermas, 1987; *Communication and the Evolution of Society,* trans. Thomas McCarthy (Boston: Beacon Press, 1979); and *Legitimation Crises,* trans. Thomas McCarthy (Boston: Beacon Press, 1975).

11 Joanne Martin, "Deconstructing Organizational Taboos: The Suppression of Gender Conflict in Organizations." Paper presented at the annual meeting of the Academy of Management Meeting, Anaheim, California, 1988.

12 Habermas, *Knowledge and Human Interests,* trans. Jerome Shapiro (Boston: Beacon Press, 1972); and *The Theory of Communicative Action, volume 1: Reason and the Rationalization of Society,* trans. T. McCarthy (Boston: Beacon Press, 1984).

13 Max Weber, *Economy and Society,* trans. and ed., G. Roth and C. Wittich (Berkeley: University of California Press, 1978).

14 Habermas, "On Systematically Distorted Communication," *Inquiry* 13 (1970): 205-218.

15 Herbert Marcuse, *One Dimensional Man* (Boston: Beacon Press, 1964); see also, Michel Pêcheux, *Language, Semantics and Ideology: Stating the Obvious,* trans. Harbans Nagpal

(London: Macmillan, 1982); H. Davis and P. Walton, *Language, Image, Media* (Oxford: Blackwell, 1983); and James Chesebro, "The Media Reality: Epistemological Functions of Media in Cultural Systems," *Critical Studies in Mass Communication* 1 (1984): 111-130.

16 Chris Weedon, *Feminist Practice and Poststructuralist Theory* (Oxford: Basil Blackwell, 1987), 85.

17 Weedon, 79.

18 See for example, Herbert Gans, *Deciding What's News* (New York: Pantheon, 1979) and Gaye Tuchman, *Making News* (New York: Free Press, 1978). Such "routines" are not simply given with the media, but are supported by ownership and organizational characteristics. See, Nina Eliasoph, "Routines and the Making of Oppositional News," *Critical Studies in Mass Communication* 5 (1988): 313-334.

19 Significant here are: Todd Gitlin, *The Whole World is Watching* (Berkeley: University of California Press, 1980); Stuart Hall, "Signification, Representation, Ideology: Althusser and the Post-Structuralist Debates," *Critical Studies in Mass Communication* 2 (1985): 91-114; and Farrel Corcoran, "Television as Ideological Apparatus," *Critical Studies in Mass Communication* 1 (1984): 131-145.

20 See, for example, Claude Lafort, *The Political Forms of Modern Society* (Boston: Polity Press, in press); and Stuart Ewen, *Captains of Consciousness: Advertising and the Roots of the Consumer Culture* (New York: McGraw-Hill, 1976).

21 Cf. John Thompson, *Studies in the Theory of Ideology* (Berkeley: University of California Press, 1984), 25.

22 Louis Althusser, "Ideology and Ideological State Apparatuses" in *Lenin and Philosophy and Other Essays*, trans. Ben Brewster (London: New Left Books, 1971); and Pêcheux.

23 Thompson, 236.

24 The most complete discussion of the role of the "face-to-face" in recovering the "other" from objectified thought is: Emmanel Levinas, *Totality and Infinity*, trans. A. Lingis (Pittsburgh: Duquesne University Press, 1969).

25 Habermas, 1984, 342; see also, Habermas, 1987, 395ff.

26 Habermas, *Legitimation Crisis*, trans. Thomas McCarthy (Boston: Beacon Press, 1975); Herman Lubbe, "The Loss of Tradition and the Crisis of Progress: Social Change as a Problem of Orientation," *The Human Context* 7 (1975): 49-60; and Stanley Deetz, "Social Well-Being and the Development of an Appropriate Organizational Response to De-Institutionalization and Legitimation Crisis," *Journal of Applied Communication Research* 7 (1979): 45-54.

27 Peter Berger, Brigitte Berger, and Hansfried Kellner, *The Homeless Mind: Modernization and Consciousness* (New York: Random House, 1973); and Habermas, 1975.

28 Alasdair MacIntyre, *After Virtue: A Study in Moral Theory*, 2nd ed. (Notre Dame, IN: University of Notre Dame Press, 1984).

29 Karl-Otto Apel, "The *a priori* of the Communication Community and the Foundation of Ethics: The Problem of a Rational Foundation of Ethics in the Scientific Age," in *Towards a Transformation of Philosophy*, trans. Glyn Adey and David Frisby (London: Routledge & Kegan Paul, 1979); and Deetz, "Ethical Considerations in Cultural Research in Organizations," in *Organizational Culture*, ed. P. Frost, L. Moore, L. Louis, C. Lundberg, & J. Martin (Newburg Park, California: Sage, 1985) 251-269.

30 Apel, 226-240; Habermas, 1987, 91ff.; and Deetz, "Keeping the Conversation Going: The Principle of Dialectic Ethics," *Communication* 7 (1983): 263-288.

31 See Deetz and Mumby.

32 Habermas, 1984, 507ff.

33 Herbert Schiller, "The Erosion of the National Sovereignty by the World Business System," in *The Myth of the Information Revolution: Social and Cultural Implication of Communication Technology*, ed. Michael Traber (Newburg, California: Sage, 1986) 21-34; and Immanuel Wallerstein, "The Withering Away of the State," *International Journal of the Sociology of Law* 8 (1980): 369-378. For review see: Sandra Braman, "Vulnerabilities: Information and the Changing State," unpublished paper available from the author.

34 The concept of technology as an extension of the body is based in the work of Merleau-Ponty. See for development, *Phenomenology of Perception*, trans. Colin Smith (London: Routledge & Kegan Paul, 1962), 98-147 & 207-298.

35 Marshall McLuhan, *Understanding Media: The Extensions of Man* (New York: McGraw Hill, 1964).

36 Vivian Sobchack, "Towards Inhabited Space: The Semiotic Structure of Camera Movement in the Cinema," *Semiotica* 42(1982): 317-335; and Beryle Bellman and Bennetta Jules-Rosetle, *A Paradigm for Looking* (Norwood, New Jersey: Ablex, 1977).

37 Erwin Straus, *Phenomenological Psychology* (New York: Basic Books, 1966).

38 See Fisher.

39 Donald Cushman, *High Speed Management* (Albany: SUNY Press, 1989).

40 Sherry Turkle, *The Second Self: Computers and the Human Spirit* (New York: Simon and Schuster, 1984).

41 Ian Mitroff, *Stakeholders of the Organizational Mind* (San Francisco: Jossey Bass, 1983).

42 Weber; Habermas, 1984.

43 Ian Mitroff and R. Kilmann, "The Stories Managers Tell," *Management Review* 64 (1975): 18-28.

44 Carol Gilligan, *In a Different Voice: Psychological Theory and Women's Development* (Cambridge, MA: Harvard University Press, 1982). For general reviews of literatures relating women and technologies, see: Kramarae, ed., *Technology and Women's Voices: Keeping in Touch* (London: Routledge & Kegan Paul, 1989); and Lana Rakow, "Gendered Technology, Gendered Practice," *Critical Studies in Mass Communication* 5 (1988): 57-70.

45 Harry Braverman, *Labor and Monopoly Capital: The Degradation of Work in the Twentieth Century* (New York: Monthly Review Press, 1974); and W. Heydebrand, "Organizational Contradictions in Public Bureaucracies," in *Organizational Analysis: Critique and Innovations*, ed. Thomas Benson (Beverly Hills, California: Sage, 1977), 85-109.

46 Habermas, 1984.

47 Burowoy, 199. For further discussion see, Habermas, 1987, 393ff.

48 For example, Lyotard; Baudrillard; Hall; Gitlin; and Burowoy.

49 Habermas, 1984 & 1987.

50 Habermas, 1979, 1984 & 1987; Apel, 1979; Deetz, 1983.

51 Deetz, "Conceptualizing Human Understanding: Gadamer's Hermeneutics and American Communication Research," *Communication Quarterly* 26 (1978): 12-23.

52 Thompson, 267.

# Breaking into Silence: Technology Transfer and Mythical Knowledge Among the Acomas of *Nuevo Mexico*

*Alberto Gonzalez and Charmaine Bradley*

Perhaps influenced by the growing suspicion of ethnocentrism within the academy,[1] communication scholars have begun to chart an ethic for symbol-sharing in intercultural contexts.[2] The diffusion of Western technological values and thinking is now a subset within the general concern for intercultural sensitivity and awareness. Pilotta and Widman identify how the practices of technology transfer are problematic to intercultural communicative competence. They state, "The central difficulty is that technology's current mode of rational-scientific legitimation encourages a style of intercultural delivery and implementation that is insensitive to cultural exigencies and, thus, subverts existing social meanings and practices."[3] As remedy, they advocate a dialogic process through which the technology donor comes to understand the cultural relevancies of the recipient for the purpose of "aligning"[4] the technology with the recipient's system of cultural relevancies.

The need remains to examine the ways in which the holders of specific cultural relevancies symbolically interact with technology initiatives. In this essay, we argue that the implicit values of a recently implemented program—*The Messenger*, a federally funded newspaper at the Acoma Pueblo—are at odds with the historical, mythic symbols of the Acoma people. Analysis suggests that in the aftermath of the transfer of this new technology, a technology designed to disseminate information and cultural values, the discourse of sacred "real knowledge" is diminished while the social pragmatism of secular knowledge is amplified. Additionally, *The Messenger* appears to undermine the culturally prescribed patterns for symbolic expression, ironically, even as it is used rhetorically to strengthen cultural awareness on the reservation.

Section one describes technological transfer as symbolic valuing and discusses its implications for intercultural dialogue. Section two provides information about the Acoma pueblo and, drawing from Jean Gebser's work on the mythical consciousness,[5] describes the mythic values of the Acoma. We establish the specific cultural relevancies upon which the transferred communication technology operates. Then we are in a position to show how the differing technological values and the cultural values become embodied in the language of *The Messenger*. Section three assesses *The Messenger* as it represents what were traditionally two complementary Acoma discourses, secular and sacred, that now appear as contradictory.

## TECHNOLOGY TRANSFER AS SYMBOLIC VALUING

As an ancilliary activity of developed societies, technology transfer "means direct foreign investment and turnover of local employees who have mastered the technology to implement it in local organizations."[6] Gruber and Marquis acknowledge the communicative process upon which the transfer of technology relies. They state, "Technical information, whether embodied in words, pictures, or material form, must be transferred from person to person, from place to place, and eventually into use if an adequate return is to be obtained from the investment in research."[7] At the same time, this perspective clearly reveals the underlying objective of transfer: to maintain and enhance the economic welfare of the social structures initiating and supporting research. The profit objective of the technology donor may not erect an inherent block to symmetrical dialogue. However, in the context of such donor-recipient relationships, Sau sardonically observes that historically "unequal exchange is a perennial feature of capitalism."[8]

In this economic context, we assume that unequal exchange is a perennial feature of technology transfer. Pilotta and Widman outline the implications of continued asymmetrical exchange: "The chief effect of this on developing countries is a contraction of their experiential horizons that seriously limits the range of available social, cultural, economic, and political possibilities, thus thwarting future development. A particular society, in this sense, is subordinated to the norms operating to organize the global [telecommunication] order."[9] As currently practiced, transfer is implemented without resources to investigate and/or mitigate possible disruption of the cultural life of the recipient community. The unconditional adaptation by the recipient culture to the new technology is assumed by the technology donor to be inherently good and necessary. Concerns have arisen that transfer then becomes a method of deculturization (or assimilation) through which the ideologies and power structures of the donor are reproduced in the recipient. The potential for the fractionalization of the recipient culture is increased as the donor mobilizes powerful resources to protect its investment from those who resist unilateral adaptation.

In the case of Acoma, technology transfer is intrasocietal and intercultural. It is not the Acoma leaders, but the U.S. government that has promoted *The Messenger* at the reservation and supplied the Acomas with monetary incentives to sustain its operation. The first attempt to publish a reservation newspaper failed from a lack of interest in the early 1980s. Recent federal grants have resurrected *The Messenger* and it continues to be published, in English, on a (more or less) monthly basis. At Acoma, the state has unilaterally prescribed and exported to the reservation a particular communication technology without reference to the specific cultural relevancies of the recipient. The state has established what it sanctions as an "official" source of information for, and about, the Acomas. This can be viewed as a political intervention by which the power hierarchy exports a rationalistic process of structuring knowledge, which it alone has legitimized, into another cultural setting. Implicit in this transfer is intolerance toward and deemphasis of the traditional mythic knowledge of the Acoma people and its oral mode of transmission.

The situation which confronts us, as Samli explains, is this: "While the technology itself, the transfer-related issues, and the actual transfer process have been the focal point of many studies, the aftermath of technology transfer has been neglected. However, the direct and indirect impact of the transferred technology must be singled out and evaluated so that future attempts will be more successful. Furthermore, if the aftermath cannot be identified, the overall assessment of the technology cannot take place."[10] The consideration of aftermath directs attention beyond the economics of technology transfer. It extends to, and places equal importance upon, the possibility of rupture in the arenas of cultural signification and adaptation in the host society. We contend, and hope to illustrate, that this "aftermath" of technology transfer, at least in part, can be identified as a (recipient's) symbolic response to a (donor's) strategy of technological diffusion. If the Western "culture of technology" is to eliminate ethnocentric and repressive practices in transfer, then donors must allow host societies to co-determine the pace and method for accommodating technological initiatives.

## MYTHICAL KNOWLEDGE AT ACOMA

We turn now to a description of the mythic knowledge of the Acomas. Our method of presenting these mythic features follows Jean Gebser's "method of pointing out the structures of consciousness . . . on the basis of their peculiar modes of expression in images as well as in languages, as revealed in valid records. This method seeks to illustrate, portray and make visible, sensible and audible, the different structures of consciousness from within."[11]

Acoma, known historically as the oldest continually inhabited city in the United States, is one of the 19 Pueblos in New Mexico. It is located about 65 miles west of Albuquerque and 10 miles from the Rio San Jose. The name Acoma is from a *Keres* word, *aku*, meaning "people of the white rock." The population of Acoma is approximately 3,900 and approximately 2,900 Acomas reside on the Acoma Reservation. The present spelling of Acoma was first used by a Spaniard named Espejo in 1583. The traditional language spoken at Acoma is a dialect, *Keresan*, which is shared by six other Pueblos in New Mexico. Acoma has 19 clans. The Antelope clan is the most influential. For a Native American to claim Acoma heritage, the individual must be one quarter or more Acoma blood and be enrolled on the Acoma Tribal Census Roll.

The cultural discourse of the Acoma displays two important categories of social knowledge: (1) secular, which is based upon social and political structures and governs pueblo affairs with the "outside world,"[12] and (2) sacred, which is based upon the kinship and clan social structure. The primary function of sacred knowledge is to affirm and transmit the cultural meaning in myths, ceremonies, rites, and traditions. Historically, the two kinds of social knowledge are distinct but interrelated and are highly dependent upon each other for interpretation.

### Secular Knowledge

Most of the people at Acoma live in four villages on an interstate highway fourteen miles from Old Acoma, or Sky City. These villages were established when a nearby railroad and interstate were built. Most of the homes in these villages have running water and electricity. At Old Acoma, however, tribal officials have attempted to preserve the traditional lifestyle. Old Acoma has no running water or electricity. This is where most of the sacred ceremonies are held and where the sacred tribal administration is housed. About twelve families permanently reside at Old Acoma throughout the year, principally the War Chiefs and their families and some *caciques* (any male of the Antelope clan) and their families. The War Chiefs and the *caciques* are perceived as sacred symbols that represent and protect the traditional culture of the Acoma people.

Election of tribal officers is held once a year—as it was held even before the invasion of the Spaniards in the 1500s. The Tribal Governor and other officials are appointed by the *caciques* who oversee the cultural and governmental structure within Acoma society. The *caciques* are more priest than chief; they counsel rather than command and are considered the highest religious officers as well as political heads, although they have little communicative interaction with the non-Acoma world. The appointed secular organization, consisting of the Tribal Governor, two Lieutenant Governors, a Secretary, Treasurer, and Interpreter are appointed by members of the Antelope clan, the War Chiefs, and Tribal Council Men. As tribal representatives, the officials must "understand our society through our usage of skills, knowledge, and self-confidence," to accomplish "justice, fairness, and social stability."[13]

The Tribal Governor and his staff serve as the medium through which social, environmental, and economic concerns of the people are heard, and if necessary, addressed to the respective U.S. federal and state agencies. A coordinator's duties and responsibilities are to oversee the actions of the Tribal Governor and his staff and also to watch over the general operation of the pueblo and act as liaison between the sacred and secular administrations. To that end the Acoma Governor, early in 1988, encouraged the coordinators to "have good relations, high expectations and for the programs to live up to certain standards."[14] The secular organization coordinates its business with the sacred organization frequently throughout the year. This interrelationship is intended to reinforce the traditional values and beliefs that prevail within the social and cultural structure of Acoma today.

## Sacred Knowledge

The sacred knowledge at Acoma is dependent upon the Acoma's historical awareness of the clan and kinship structure, myths, legends, and other ritualistic ceremonial awarenesses regarding religion. The sacred organization is maintained through a clan and kinship structure membership into which is derived through maternal lines. The *caciques* transmit their wishes and concerns to the people through the War Chiefs. The War Chiefs direct the religious and sacred operations of the Acoma world. Their duties are to preserve and teach the old customs and rites. Much of the important, culturally relevant information is orally directed to the medicine societies and the headmen of each *kiva* (shrine) who then orally disseminate the sacred teachings to the Acomas.

Sando states of general pueblo values: "The tradition of religious belief permeates every aspect of the people's life; it determines mankind's relation with the natural world in which his fellow man lives. To maintain such a relationship between the people and the spiritual world, various societies exist, with particular responsibilities for weather, fertility, curing, hunting, and pleasure or entertainment of the people."[15] Gebser describes this cultural orientation as a "mythical structure" that explains a phase of human civilization. For Gebser, "The mythical structure is typified by imagination [whose] symbol is the circle."[16] The mythical mind uses imagination to direct introspectively its interpretation of nature. The cyclical "motion" of the moon, seasons, and such, are the basis for a birth-to-death cyclical metaphor. The concept of time emerges as a nature-bound element. The word assumes special significance within the mythical structure because it is produced from the human "interior" and therefore is directly animated by the spirit force.

Mickunas and Pilotta elaborate Gebser's conception of the mythical consciousness: "The world of the mythical consciousness-structure closes into a circle empowered in its motion by polarities. Mythical man reveals this movement in psychic imagery in terms of posited polarities of life-death, light-dark, water-air; at the same time these images reveal a lack of three-dimensional space. The mythical temporicity, as a rhythmic circularity, is a movement common to phase-like events and to psychic images. While one is depicted in moons, seasons and repeated rituals, the other is depicted in psychic images."[17]

For the Acoma, the mythic consciousness is seen in attempts to engage the spiritual world, or what Gebser calls "psychic reality." Several rituals—fasting, pilgrimages, smoking—unify the individual with the spirit world and provide the psychic images which become the enactments of mythic awareness. The mythic mind of the Acoma is illustrated by recitations in the *kiva*, where reverent narrations (rhythmic breathing) transmit spiritual knowledge through the arts of ritual oratory, prayers, and songs. As Sando notes, "These observances are not spontaneous outpourings, or outbursts of the troubled heart, but are carefully memorized prayerful requests for an orderly life, rain, good crops, plentiful game, pleasant days, and protection from the violence and the vicissitudes of nature."[18] The prayers, songs, and ceremonies of the Acomas are motivated by a desire to achieve and maintain harmony with nature. Symbolizing the cyclic observance of the seasons and of the four cardinal directions, ceremonial song-prayers are sung in four choruses and dances are performed to the four directions. Natural rhythms are also represented in the drum beat (symbolic of the heartbeat of Mother Earth) or the sound of rattles and deer hoofs beating on turtle shells (symbolic of the sound of rain) and reinforce the interrelatedness of human existence with nature-bound elements. The origin and focus of the mythic knowledge is in the perception of cyclical temporicity which is symbolically expressed in the veneration of natural rhythms.

The mythic relevance of the periodic visits of the "spirit rain makers," or *ka'tsinas,* is that the rain-drought polarity fundamental to Acoma culture is reinforced. This polarity is an intrinsic element in Acoma cultural identity. The group of Kat'sina dancers produces the rhythm and the vocal and movement pattern that prays for rain for the entire community and affirms the traditional value of respecting natural elements. Consistent with the mythic orientation are sacred narratives about *Shi'pap* (place of origin), *We'nimatsi* (a sacred shrine of the Acomas), the twin warrior gods, *Masewi* and *Oyoyewi* that are orally transmitted to the younger generations through stories, song, dance, and prayer.

## Cultural Relevancies

The integrated knowledge of the secular and sacred organizations reveals cultural values that constitute "real knowledge" among the Acomas. We describe the "real knowledge" in terms of four cultural values that reflect the mythic consciousness: (1) orientation toward the present;[19] (2) harmony with nature;[20] (3) giving;[21] and (4) cooperation.[22] While these values hold generally across Native American cultures our description concerns how they are manifest among the Acomas.

1. *Present Orientation.* Traditionally, the Acoma's perception of time is different than the dominant Anglo perception.[23] Anglo perception of time is linear, with scheduled punctual activities being the norm. In the traditional Acoma perception events occur "whenever people are ready; when every one arrives the activities/events will commence."[24] In this view, the present is not experienced in isolation, but as movement that has its culmination in a completed cycle. Past and future are embodied in the present.

   This variable attitude toward time has been referred to as "Indian time" and has been an important contributing factor in the development and maintenance of an Acoma identity.[25] In contrast to rationalistic notions of time, the cyclical perspective has provided Acomas unstressed, unpressured time to teach younger generations the "proper way" to affirm family and clan relationships, and to exchange stories and information of cultural relevance.

2. *Harmony with Nature.* Mythical knowledge relevant to the interpretation and representation of natural events is most evident in Acoma religious ceremonies. Harmony with nature is shared by many other Native American tribes in rituals that emphasize the continuity of a harmonious relationship with Mother Earth.[26] For example, offerings of white corn meal represent the fertility and continuity of life and are accompanied by prayers and songs to the four (forming a circle) directions. Dances paying homage to animals such as the deer, the buffalo, the eagle, and the butterfly are performed. Respect and thankfulness for all of nature's offerings are recited in the songs and prayers. Children are encouraged to participate and learn the processes and procedures governing the religious rituals.

3. *Giving.* The value of giving is represented as part of many religious rituals and is stressed within the family and clan structure of the Acoma people. Emphasis is placed on providing for the well-being of all: family, community, and tribe. Through working together and the giving of time and labor, Acomas strive to maintain the continual observance of the cycles of Mother Earth. Ritual giving includes inviting friends and relatives to partake in the celebration of giving a newborn infant his/her "Indian name," the rewards of a successful deer hunt, the beginning of planting season, and even the death of loved ones.

4. *Cooperation.* Consistent with previous sociological research on Native Americans, the Acoma stress cooperation with others.[27] Competition is regarded with disdain by traditional Acomas. In fact, according to Attneave, "Whenever the needs or goals of a group conflict with individual decisions and preferences, the group will take precedence whether the group reference is the tribe, the band, the family, or any other coherent cluster of people."[28]

   Individual achievements by Acomas are viewed as contributions to the group, not to personal glory.[29] Excelling for personal fame is disapproved by traditional Acoma society because it sets the individual above and apart from others.[30] The focus on the individual is perceived to disrupt group cohesiveness and the harmonious balance of tenacious cultural ties. The solidarity of Acoma and the group cohesiveness that Acomas value is supported and reinforced by transmission of the "real knowledge."

Transmission of both forms of knowledge is perceived by the Acomas as important for the continuance of their cultural identity. Historically, this knowledge has been orally transferred from generation to generation through family, clan, *kiva,* and general public meetings. But since the 1590s, and the first interactions with Spaniards and subsequent other alien cultures, the Acomas have become subservient to the written word. They have been indoctrinated to recognize and respect the power of the written word. They have reluctantly learned to use it and incorporate it in the teaching of sacred knowledge, though with discretion and selectivity. Vivid recollections and memories of the damage and injustices perpetrated by the Anglo-European manipulation of the power of the written word and the Acomas' ignorance of this power have caused a disturbance between the two forms of organization that many Acomas feel endangers the future survival of their culture. The secular organization quickly learned that in order to save the Acomas from further destruction, literacy in the dominant language (English), must be acquired to communicate the Acomas' needs and demands to the governing agencies.[31] Traditionally, important information about tribal affairs and the actions adopted by the tribal governing system were disseminated at regularly held General Council meetings to which all Acomas were invited. However, attendance at these meetings has decreased and it has become more difficult for the secular organization to inform the Acomas of its decisions. This has resulted in "utilization of the community newsletter to a higher degree."[32] *The Messenger* has become a medium through which the Tribal Administration and other tribal departments can make announcements and provide news of recent occurrences and information about upcoming events.

## THE MESSENGER AND COMMUNICATIVE ACCOUNTABILITY

Nine 1988 issues were used in our analysis of the discourses presented in *The Messenger*. Each issue contained approximately six pages. The text that accompanied advertisements and general announcements was excluded from analysis. We focused specifically upon thirty-eight articles that delivered "factual" information or commentary on a variety of topics. Articles reported five general topics: education, alcohol abuse and health-related issues, tribal administrative decisions, employment opportunities, and Acoma law. The analysis draws freely from the expressions employed to explain the topics reported as being important to the Acomas. Using the articles as a text, our goal was to identify the "terms" or vocabulary by which secular and sacred meanings were expressed and how the terms were interrelated.

### Terms of Secular Discourse

The secular discourse in *The Messenger* reproduces the values of rationalistic, linear culture. Generally, these values are characterized by dualistic thinking. The central terms are "self" and "achievement." In contrast to the clan-self polarity given in traditional Acoma teachings, the self is viewed as primary in *The Messenger*. In place of the clan is found a self-other duality which is made manifest in the quest for economic advancement. For example, an article on alcohol abuse closes with the encouragement to "value yourself, your life, your health. These are your greatest wealth!" The "bottom line" of a job training program is "helping people help themselves." These terms signify a movement away from the value of giving. Traditionally, tribal membership was a condition for unqualified assistance and "sharing" with others. Requiring assistance did not carry social stigma. Indeed, the reverse was true. Not to seek assistance from family and friends ran counter to Acoma expectations. Such behavior would be perceived as a rejection of the clan.

The "self"-oriented terms also diminish the value of harmony with nature. In this discourse, nature is coveted, not conjoined. Acoma leaders have begun lobbying politicians and have threatened litigation to retain water and management rights to lands being leased from the government. However, the leaders have "reached a point where they are so determined to retain the land that they are revealing some of their religious secrets."[33] The competitive "achievement" of retaining the lands replaces the motive of preservation of the sacred. The fact that this new political approach is an institutionally sanctioned response to a history of Anglo territorial incursion does not disguise, indeed it emphasizes, the axiological shift apparent in the rhetoric of the secular organization.

The secular discourse depicted the Acoma in a competitive relationship with the Anglo world. The skills needed to be successful were attitudinal, such as being "optimistic," and political, with an emphasis on "production" and "networking." A summary of tribal concerns in one article listed, "land acquisition, livestock, political networking, and political identification." Reflecting the focus on the self, education was a related means for personal achievement because "depth . . . in skills can determine the individual's capacity to succeed in life." Education was described as a tool "to achieve both individual opportunity and social progress."

Articles in *The Messenger* engage secular discourse in reporting the activities of Acoma officials. As used in the articles "self" and "achievement," along with the attending "education" and "skills" are secular terms that stand in profound contrast to the mythical "knowledge" emphasized in Acoma tradition.

### Terms of Sacred Discourse

In the available articles, we discovered only two invocations of sacred knowledge. An article reported several men building a basketball court. According to the report, they worked with "great strength" and "mighty power" to complete the job. These descriptors are references to Pieyatiamo, the "rainmaker" god in Acoma myth. The rainmaker god commands water, the regulator of all life, especially in the desert. Associating the men with the rainmaker god is a powerful sign of thankfulness from the community. The efforts of these men revealed the values of giving and cooperation.

A second instance involved Simon Ortiz, former poet laureate at the pueblo. At a meeting with students, he stated, "As young people, as members of our Acoma Community, we are told to be strong people, to have courage and to always seek knowledge." "Knowledge" as used by Ortiz did not refer to secular "education." It referred to the spiritual knowing gained from and reflected in harmony with nature. In a short essay, Ortiz describes the mythical process of knowing: "[It] is an opening from inside yourself to outside and from outside of yourself to inside but not in the sense that there are separate states of yourself. Instead, it is a joining and opening together."[34]

What we have called the "real knowledge," the cultural values the Acomas see as central to their identity and seek to preserve, is occasionally used to justify tribal decisions and practices. However (perhaps to underscore that this knowledge is tacitly understood by the writers), this is done simply by allusion. For example, tribal officials asked that all livestock be removed from the Acoma valley grazing area. The justification stated, "Traditionally, no livestock is allowed in the area during the planting season and not until the harvest season is complete." In another instance, the outcome of 1988 tribal appointments was reported by stating, "Through a traditional process of filling tribal lead positions numerous male individuals were selected."

In other articles, the justification is less clear. Prior to certain religious ceremonies at Old Acoma, residents are informed that no automobiles will be allowed on the mesa. Further, "no hats, recorders, radios or cameras are allowed . . . to ensure the safety of all and to avoid unnecessary problems." None of these justifications reveals the symbolism of the action specified. The vagueness of these references would be expected if we concluded the symbolism to constitute tacitly understood knowledge. In *The Messenger*, Acoma culture is a hidden entity, supreme and unquestioned, yet without form and identity.

Faced with this scenario one might legitimately suppose that the unquestioned status of certain beliefs is a condition enjoyed by all cultural formations at their height. However, though less so than other Native Americans, the Acomas have relaxed traditional rules and become more open to "the outside world." Recently, because of alcoholism and poverty on the reservation, the Acomas have been under severe pressure to reconsider and return to traditional choices. According to editor Kim Victorino, a primary goal of the newspaper is to reinforce traditional teachings among the younger Acoma, who are perceived as increasingly less "knowing" and less committed to Acoma ways.[35] If, however, the writers are reluctant to specify the symbolism of "real knowledge" because it is secret and told only in ceremonial settings, then *The Messenger* becomes irrelevant as a medium for reinforcing cultural values. The newpaper's coverage defaults almost exclusively to stories that convey secular values in isolation from their basis in sacred knowing. In an ironic twist, the secular discourse promotes assimilation to escape poverty and sickness even as the Acomas hold assimilation responsible for spiritual loss. Absent a sacred discourse in *The Messenger*, the secular values gain a powerful legitimacy.

## CONCLUSION

There is clear evidence of resistance to the secular emphasis in *The Messenger*. The tribal Governor's persistent admonitions to residents that they rely upon the newspaper for information created controversy within the community.[36] This suggests that the very convention—objectivity— prescribed by the elite and intended to foster social coherence was least capable of facilitating it.

In fact, *The Messenger* insures its own irrelevance as writers continue to print outdated and partial information. The Acomas will not publish obituaries, birth announcements, or details of events occurring at the pueblo. Such information is perceived to violate the values of giving and cooperation that traditionally have been achieved through a Native American rhetoric of nonrecognition or "silence."[37] Yet simultaneously constrained by the convention of objectivity, the Acomas cannot accept sacred knowledge as a legitimate topic. At Acoma, the residual elements of the mythical consciousness do not blend easily with a technological consciousness that privileges the immediate and discrete.

We disclaim that the central tensions of telecommunication transfer at Acoma stem exclusively from an orality/literacy opposition. It is not the case that the print medium precludes expressions of the mythical consciousness. Nor is it that Native Americans are technologically incompetent to manipulate print effectively. For example, the tribal newspaper of the Sisseton-Wahpeton Sioux of South Dakota, the *Sota Iya Ye Yapi*, appears to present traditional knowledge in journalistically unconventional though culturally understood ways.[38] The central tensions surrounding *The Messenger* stem from the absence of donor and recipient accountability for the communicative consequences of this project. Simply put, the problems are more those of intercultural mediation than technological competence.

Accountability in technology transfer can be facilitated only when the notion of "aftermath" becomes an integral dimension of the transfer process. Consideration of aftermath occurs as donor and recipient explore, prior to implementation, the range of social and cultural alterations implied by a telecommunication project. Thus, socially responsible technology transfer is initially discursive. From a perspective that assumes symmetrical dialogue, three phases of transfer aftermath are required: significance, translation, and integration.

*Significance* addresses several questions relating to the project's goals and attempts to identify areas of material change in the social life of the community. What are the expressed perceptions toward both the target need(s) within the community and the corresponding project? For whom is the need significant? For whom is it not significant? What is the degree of community commitment to the project? What is the assessment of the community about the project as a permanent fixture in the social landscape?

Questions that identify patterns of accommodation point to issues of *translation*. Who is the visible sponsor for the project and what is the sponsor's relationship to the community? By extension, what existing symbolic structures will the project influence and in what manner are they to be influenced? Is the manner by which the community comes to experience (directly and indirectly) and understand the project encompassed by the symbolic structures? What are the traditional roles of instrumental and relational discourses and how will these discourses interface with the project? Is competence with the technology intended to be realized evenly throughout the community?

*Integration* refers to issues that contemplate the final disposition of the project vis-á-vis the communicative patterns of the community. What measures are best suited to detect how the community has internalized or incorporated the initiative within its discursive frameworks? What are the available socio-cultural structures compensatory to an undesired rupture in tradition? What resources can be applied in the event of rupture?

Technology transfer is repressive to the extent that it exacts a unilateral aligning of economic interest and social practice. The transfer of telecommunication initiatives is ethnocentric and repressive when technically advanced cultures assume that the rationalistic method by which information is conceived, valued, and disseminated can be accommodated unproblematically within the existing communicative patterns of other cultures. At Acoma, a technocratic elite has introduced and legitimized a medium for expression that subverts the perceived cultural relevancies of the community. In a sense, the imposed convention of "objective" news presentation mirrors the belief in an objective technology. Finally, the recipient was not extended, and failed to exact, equal participation in the development of the project.

The notion of a culture of technology allows communication scholars to approach technology transfer as an intercultural phenomenon. This approach affords at least three opportunities for further research. These opportunities focus on the reculturizing potential of the communication patterns valued by and implicit in telecommunication projects, the discursive images and representations from other than Eurocentric perspectives that arise to advocate the legitimation of technology, and the mediatory rhetorics employed by recipient societies to resist technological hegemony. Case studies centered upon these areas may lead to further descriptions of how discourse is structured in transfer; specifically, how discourse in the receiving culture accomplishes the validity, rejection, or integration of particular projects.

## NOTES

1  James W. Carey, "Communication and Culture," *Communication Research* 2 (1975): 173.

2 See for example,  Molefi K. Asante, "Intercultural Communication: An Inquiry and Overview into Research Directions," in *Communication Yearbook 4* , ed. Dan Nimmo (New Brunswick, NJ: Transaction, 1980), 401-410; Joseph J. Pilotta, "The Phenomenological Approach," in *Intercultural Communication Theory: Current Perspectives,* ed. William B. Gudykunst (Beverly Hills: SAGE Publications, 1983), 271-282 and; Stella Ting-Toomey, "Rhetorical Sensitivity in Three Cultures: France, Japan, and the United States," *Central States Speech Journal* 39 (1988): 28-36.

3  Joseph J. Pilotta and Tim Widman, "Overcoming Communicative Incompetence in the Global Communication Order: The Case of Technology Transfer," in *The Underside of High-Tech: Technology and the Deformation of Human Sensibilities,* ed. John W. Murphy, Algis Mickunas, and Joseph J. Pilotta (New York: Greenwood Press, 1986), 161.

4 Pilotta and Widman, "Overcoming Communicative Incompetence," 167.

5 Jean Gebser, "The Foundations of the Aperspective World," trans. K. F. Leidecker, *Main Currents in Modern Thought* 29 (1972), 80-88.

6 Charles T. Stewart, Jr. and Yasumitsu Nihei, *Technology Transfer and Human Factors* (Lexington MA: Lexington Books, 1987), 3.

7 William H. Gruber and Donald G. Marquis, eds., *Factors in the Transfer of Technology* (Cambridge: M.I.T. Press, 1969), 3.

8 Ranjit Sau, *Unequal Exchange, Imperialism and Underdevelopment: An Essay on the Political Economy of World Capitalism* (Calcutta: Oxford University Press, 1978), 63.

9 Pilotta and Widman, "Overcoming Communicative Incompetence," 160.

10 A. Coskun Samli, ed., *Technology Transfer* (Westport, CT: Quorum Books, 1985), 13.

11 Gebser, "Foundations of the Aperspective World," 81.

12 "Education Committee to be Established," *The Messenger*, December 1988, 1.

13 "Minorities and Higher Education," *The Messenger*, November 1988, 3.

14 "Coordinators Meeting," *The Messenger*, January 1988, 2.

15 Joe Sando, *The Pueblo Indians* (San Fransisco: The Indian Historian Press, 1976), 22.

16 Gebser, "Foundations of the Aperspective World," 82.

17 Algis Mickunas and Joseph J. Pilotta, "A Phenomenology of Culture: An Introduction to Jean Gebser," *Reflections: Essays in Phenomenology*, (Spring 1981): 92.

18 Sando, *The Pueblo Indians*, 23.

19 Further explanations for these values can be found in Ira Eyster, "Culture through Concepts: A Teacher's Guide," (ERIC Document Reproduction Service No. ED 176 928), 1980 and; Larry Faas, "Cultural and Educational Variables Involved in Identifying and Educating Gifted and Talented American Indian Children," (ERIC Document Reproduction Service No. ED 255 010) 1982.

20 William D. Hanson and Margaret D. Eisenbise, "Human Behavior and American Indians," (ERIC Document Reproduction Service No. ED 231 589) 1983.

21 Barbetta Lockhart, "Cultural Conflict: The Indian Child in the Non-Indian Classroom," (ERIC Document Reproduction Service No. ED 195 397) 1978 and; Dorothy L. Miller and Anthony Garcia, "Mental Issues among Urban Indians: The Myth of the Savage-Child," (ERIC Document Reproduction Service No. ED 129 485) 1974.

22 Danielle Sanders, "Cultural Conflicts: An Important Factor in the Academic Failures of American Indian Students," *The Journal of Multicultural Counseling and Development* 15 (1987): 89-90 and; Betty M. Skupaka, ed., "The 'Holding Power' Workshop," (ERIC Document Reproduction Service No. ED 194 284) 1972.

23 See Carolyn Attneave, "American Indians and Alaska Native Families: Immigrants in their Own Homeland," in *Ethnicity and Family Therapy*, ed. M. McGodrick, J. K. Pearce, and J. Giordano (New York: Guilford Press, 1982), 55-83; Faas, "Cultural and Educational Variables"; Hanson and Eisenbise, "Human Behavior and American Indians"; Sanders, "Cultural Conflicts."

24 Floy C. Pepper, "Teaching the American Indian Child in Mainstreaming Settings," in *Mainstreaming and the Minority Child*, ed. R.L. Jones (Minneapolis: Council for Exceptional Children, 1976), 133-158.

25 Brooks Anderson, Larry Burd, John Dodd, and Katharin Kelker, "A Comparative Study in Estimating Time," *Journal of American Indian Education* 19 (1980): 1-4, and; Lockhart, "Cultural Conflict."

26 Sando, *The Pueblo Indians*, 22-25.

27 Margot M. LeBrasseur and Ellen S. Freark, "Touch A Child—They Are My People: Ways to Teach American Indian Children," *Journal of American Indian Education* 21 (1982): 6-13.

28 Attneave, "American Indians and Alaska Native Families," 66.

29 Pepper, "Teaching the American Indian Child," 133-158.

30 Karlene George, "Native American Indian: Perception of Gifted Characteristics," in *Face to Face with Giftedness* (World Council for Gifted and Talented Children, 1983), 220-249, and; Lockhart, "Cultural Conflict."

31 "Indian Education," *The Messenger*, November 1988, 3.

32 "Acoma Governor Histia Conveys Information to His People," *The Messenger*, September 1988, 1.

33 "El Malpais Update," *The Messenger*, October 1987, 3.

34 Simon Ortiz, *Song, Poetry and Language—Expression and Perception* (Navajo Community College Press, 1977), 8.

35 Kim Victorino, Personal Interview with Charmaine Bradley, 17 March 1989.

36 "From the Governor's Office," *The Messenger*, November 1988, 1.

37 K.H. Basso, "'To Give Up on Words': Silence in Western Apache Culture," *Southwestern Journal of Anthropology* 26 (1970): 213-230.

38 "History as Rhetoric: Political Activism among the Sioux of Sisseton, South Dakota." Beth Bader and Alberto Gonzalez, paper presented at the Central States Speech Association Convention, Cincinnati, OH, 1986.

# Structuring Closure through Technological Discourse: The Mormon Priesthood Correlation Program

*Tarla Rai Peterson*

Heisenberg's Uncertainty Principle, announced in 1925, established modern quantum mechanics and allowed the development of sustainable nuclear fission reactions. The Uncertainty Principle postulates a fundamental randomness and chaos in nature. Although order may exist, certainty is never 100 percent, and some phenomena happen spontaneously. Perhaps the most disconcerting aspect of this principle is that "not everything that happens has a cause."[1] Scientists had previously assumed that the ultimate impediment to accurate prediction and explanation of the universe was the ability to precisely measure and predict phenomena, and that this difficulty could eventually be overcome with appropriate mathematical models. However, the Uncertainty Principle defines nature as intrinsically unpredictable. While accepting a reality independent of human observation, it postulates that observation influences "reality" in an unknowable way, and that when humans make conclusions about their environment "it is the theory which decides what can be observed."[2]

When applied to the study of organizational behavior, this perspective emphasizes the precarious nature of organizational reality and the related interplay between members' assumptions and their symbolic enactments. Communication enables organizational members to structure their institutions through strategies of social reproduction that simultaneously create, and are created by, institutional culture. Such reproduction allows institutions to adapt to their environments while maintaining relatively stable hierarchies.[3] However, the precarious authority relationships within any hierarchy remain vulnerable to those adaptations. When organization leaders perceive a potentially dislocating change in established patterns of domination (whether from the external environment or from within the organization) they often attempt to preserve current relations by controlling communication, a primary means of social reproduction.

This frenetic attempt to produce stability in an uncertain world can be viewed as a modern "psychosis." Kenneth Burke describes *occupational psychosis* as a response to basic production routines shaping, while emphasizing certain aspects of, other experience "analogously to the patterns of work."[4] The modern emphasis on causality and order led Heisenberg to warn that the threat of chaos would tempt western society to invite technology to "take on [the] role of bringing order into our thought."[5] The "technological" has indeed achieved the status of "master psychosis," or has become the "point of reference by which to consider questions of valuation."[6]

Working from Burke's assumption that "the state of the communicative medium [is] affected by the 'master psychosis,'" I propose in this essay that *technological discourse* provides institutions with a primary means for both articulating and rationalizing technological psychosis.[7] This discourse enables organizational leaders to draw upon cultural apprehension about chaos to induce cooperation with domination. Further, while organizational attempts to control all human activity constitute the potential for alienating any member, they have especially alienated marginalized groups such as women. After explaining my use of the term *technological discourse*, I will examine the Church of Jesus Christ of Latter-day Saints' (Mormon) *correlation* program as an example of such discourse, focusing on its articulation of the relationship between women and men. Finally, I will show how *correlation* reifies current hierarchical patterns (based on patriarchal authority), and makes the creation of alternative structures increasingly difficult.

## DEFINING TECHNOLOGICAL DISCOURSE

*Technological discourse, or language used to structure human action according to rules of closed systems*, can damage both the organization that relies on such discourse and individuals who are part of that organization. Through its idealization and precise definition of organizational hierarchy, the immediate connection with reality demonstrated by creative communication is lost. My definition of *technological discourse* is primarily derived from concepts articulated by Jacques Ellul and Henry W. Johnstone, Jr. Ellul argues that technology mediates between humanity and the environment (as well as between individuals), results from a conscious design of means to achieve some explicit end, and reduces life to "manageable" fragments. Experience that cannot be appropriately "managed, manipulated, utilized, is rejected and discarded as worthless."[8] Prompt denigration of "inappropriate" experience stabilizes organizational structure and prevents individuals from pursuing creative alternatives.

Johnstone shares Ellul's negative perspective toward technology, arguing that "technologies that are substituted for creativity are felt as a threat to human communication."[9] However, while identifying technology as a threat he goes beyond Ellul's determinism (which is paradoxically consistent with technological psychosis), suggesting that the problem of communication and technology is essentially a matter of substituting mindless, automatic systems for creative thought and communication. Johnstone differentiates between "creative" and "technological" communication as follows:

> A process is creative if it consists of a series of steps none of which is strictly determined either by the project that the steps contribute to or by the preceding steps in the series, but each of which, once taken, is seen to have been a fitting sequel to its predecessors. . . . A process is technological . . . when it is a series of steps in which either a given step or the project as a whole determines the sequel to the given step or else the question whether the successor is fitting to its predecessors does not arise.[10]

Thus, creative dialogue consists of communication that makes sense after the fact but is not entirely predictable before the fact. Conversely, in technological discourse the means are determined by the end and each step predetermines another, or acceptance of prevailing interaction patterns is so complete that questions regarding appropriateness do not occur.

Technological discourse flourishes within organizations as techniques become more sophisticated and technologies take on purposes of their own. As Burke argues, "social relations were first ascribed to nature, and then 'derived' from it."[11] Culturally validated myths justify an original tool use and then extend the tool until it obtains a supporting context of habits, norms, language, and the like. Carolyn Miller adds that technological consciousness promotes the value of efficiency and "the tendency to conceive of the world as a closed system."[12] Closed systems entice people with the promise:

> All problems are solvable, or conceivably solvable. . . . Closed-system thinking substitutes 'effective procedures' for invention and self-contained knowledge of the system (isolated expertise) for dialectical discovery of agreements. In a closed system, there is a correct solution discoverable by one who knows the system.[13]

Technological discourse, then, masks its own temporality, while provoking participants to forget "the role of imagination in supplying the background against which all thinking, however abstract, must find its place and meaning."[14] As awareness of imagination is suppressed, technological discourse is progressively judged by rules reflecting technical values and assumptions, and generated by a concern for technique.

Such discourse inhibits women more than men, because women often demonstrate a "profound reticence about formal systems."[15] Turkle proposes that women find formal systems incompatible with formative experiences in closely bonded relationships, and with the "web of connectedness" Gilligan found to characterize women's adult lives.[16] Women have also acted primarily in subordinate roles, playing by rules structured and enforced by men. Thus, while they have long participated in technological discourse, their specific opportunities have differed from men's. Women's status as simultaneously marginalized and integral players in technologized society makes them both vulnerable to alienation from bureaucracy and potentially dangerous to that bureaucracy.

Technologies must operate through people who both "succumb and resist, whose identities are both created by the dominant discourse of power and knowledge and simultaneously create themselves in opposition to that discourse."[17] Oppositional self-creation provides women (and other marginalized persons) an opportunity to use their experience as marginalized participants to politicize organizational discourse, thus challenging the dominant discursive framework by denying its neutrality. The more completely technological discourse permeates an organization, however, the more difficult it is for members to invent or articulate alternative visions. As bureaucracies technologize language, communication becomes more a means for codifying and articulating rules and less a means for oppositional self-creation. Technological discourse inhibits oppositional self-creation to the degree that it is judged by rules of its own making. Although the resulting stability protects the organization from environmental turbulence, it also erodes the potential for innovation.

## DEVELOPING A TECHNOLOGICAL DISCOURSE

The Mormon Church's *correlation* program is an omnipresent technological discourse that is judged by rules of its own making and continually regenerated by a concern for technique. *Correlation* refers to an institutional management plan established in 1960 for defining hierarchical relationships and coordinating interaction among groups within the church. Originally implemented as a means of achieving organizational stabilization and efficiency, it has since become the standard against which all action is measured. It codifies hierarchical relationships within the organization and precludes oppositional self-creation by its ubiquitous nature. Women have been especially vulnerable to *correlation* because their subordinate status, as in many religious organizations, expedites systematic domination.

### Precursors to Correlation

Mormon rhetoric has not always echoed the conservative Christian voice regarding gender. On February 11, 1846, church leaders and the first four hundred families commenced the Mormon trek across the plains and mountains to the valley of the Great Salt Lake.[18] In their isolated home the Mormons built a kingdom that included ecclesiastical control of politics, a communal economic system, and polygamous marriage.[19] Organizational goals ranged from basic survival to kingdom-building, both of which required a fully committed and participating membership. Maximum utilization of every member's potential was required in order for the organization to survive within an environment that was both culturally and physically hostile. For example, (church president) Brigham Young suggested that women (who tended to be smaller than men) take over the medical and teaching professions; as well as jobs in printing, telegraphy, and retail so that men could be put to use in the fields and in heavy construction.[20] Both men and women were expected to dedicate their lives to building God's kingdom.

The Victorian image of womanhood was inappropriate to this communitarian frontier society. Female kingdom builders were encouraged to develop occupational and professional skills, while continuing to bear children. They had their own newspaper that advocated both woman suffrage and polygamy, dominated the medical profession, managed public institutions and their own households, and irreverently blurred traditional distinctions between public and private spheres.[21] These women formed a powerful female hierarchy "whose form and jurisdiction [from church-wide to local level] paralleled the male priesthood hierarchy." Eliza R. Snow, whose power was partially due to marriage connections (she was a plural wife of both Joseph Smith and Brigham Young—the church's first and second presidents), built "a union of the women of the Church," that controlled the Relief Society (women's organization); Young Ladies Mutual Improvement Association (female youth organization); and Primary (children's organization).[22]

Following the Civil War opposition to practices such as polygamy intensified to catastrophic proportions. Faced with disincorporation, confiscation of real property, and imprisonment of church leaders, the Mormon church began shifting toward an accommodation with Victorian society. Attempts to build a new kingdom were replaced by attempts to preserve the existing hierarchy. Alterations in both life style and world view accompanying this transformation exacted a heavy toll on Mormon women's power relations.[23]

The Mormon church is administered by an ordained lay priesthood. Exclusion from priesthood authority means exclusion from organizational authority, for "all offices in the Church derive their power, their virtue, their authority, from the priesthood."[24] Mormon women, who originally participated equally in priesthood "ordinances" such as performing healings or prophecies, have been progressively excluded from this source of organizational influence. Women to whom church founder Joseph Smith had "turned the key [of priesthood authority]," proclaiming that "part of the priesthood belonged to them," were told in 1880 that women "hold the Priesthood, only in connection with their husbands."[25] By 1907, President Joseph F. Smith wrote that "a wife does not hold the priesthood in connection with her husband, but she enjoys the benefits thereof with him."[26] In 1954, Apostle John A. Widtsoe explained that God had provided motherhood to replace priesthood authority. Although she could no longer participate directly in the organizational hierarchy, "woman has her gift of equal magnitude—motherhood."[27] The Mormon women's relationship to the priesthood hierarchy never had been defined formally, and thus these women became vulnerable to progressive marginalization.

Between 1908 and 1922, church president Joseph F. Smith established a committee to select, write, and publish standard theology manuals for priesthood quorums. Smith's "priesthood reform movement" began defining organizational relationships, in part by terming the women's organizations "auxiliaries" and clarifying their roles as "supporting" the priesthood.[28] The distinction between priesthood quorums (groups of males who had been ordained to the priesthood) and auxiliaries (support groups dedicated to specific populations within the church such as women, children, and youth), resulted in an increasingly superior-subordinate relationship between the two, "with the priesthood hierarchy supervising the women's auxiliaries."[29]

## The "Correlation Executive Committee"

The "Correlation Executive Committee" distinguished itself from preceding reforms by its carefully implemented church-wide restructuring. Although the 1960 committee began by attempting to coordinate the church's teaching curriculum, its chair, Harold B. Lee of the Quorum of the Twelve Apostles (the church's presiding body), explained that "consolidation and simplification of church curricula, church publications, church buildings, church meetings" would include "many other important aspects of the Lord's work."[30] Lee was intimately acquainted with the unwieldy church bureaucracy. He preferred centralized control, stressed obedience to current authority, and expressed suspicion of intellectuals.[31] By the late 1950s, he became concerned that the church's rapid growth was leading to loss of control and possible doctrinal dilution. He was equally worried that social changes would threaten the sanctity of the family.[32]

In 1960, he persuaded church president David O. McKay to establish a new committee, the Correlation Executive Committee, to evaluate the curriculum of all church programs. As committee chair, Lee immediately began "a major and prolonged effort to reorganize the Church."[33] Like previous reform movements, correlation emphasized the duties of those in priesthood authority positions. Lee attempted to establish the priesthood as the power center of the Church. He told coworkers that he wanted to "bring the priesthood back where it should be, according to revelations, and then determine the relationship of the auxiliaries to the priesthood."[34] He further proclaimed that "we must wake the priesthood up to assume their responsibility and we must place greater emphasis on leadership at all levels."[35] "Correlation" quickly became "priesthood correlation."

Auxiliaries no longer raised and managed their own funds. After writing their own lessons for decades, they were required to send suggestions to a central correlation committee where their lesson material was rewritten. Auxiliary magazines were discontinued and consolidated into three "official" church publications. Relief Society Social Services and the Primary Children's Hospital, previously administered by women, were now to be administered by male-dominated Church departments.[36] At the close of her twenty-three year tenure as church Primary president, LaVern Watts Parmley said:

We don't have as much responsibility. We don't have the *Children's Friend*. We don't have the Primary Children's Hospital. We don't write our own lessons. We don't sell— we used to sell our own supplies. We used to do everything. We used to do all our editing and do all our printing. We did everything. . . . I have at times just jokingly said, "I don't know why they need a president now. We're just told what to do and when to do it and how to do it!"[37]

The women's organization, the Relief Society, also lost the autonomy it had cherished since its formal organization in 1842. Because only men receive priesthood ordination all authority reverted to men "who not only made policy, managed, and planned, but developed centralized programs as well."[38]

When Lee became church president in 1972, the Department of External Communication (later renamed Public Communications) was created, the Correlation Committee became the Division of Internal Communications, and professional managers were placed in charge of most administrative functions.[39] In 1972, the Corporation of the President moved its growing staff of experts into a new twenty-six story office building. New administrative layers were added churchwide. By 1982, the church had grown to a population of over 5.5 million and the majority of its members no longer resided in Utah. It was the largest religious organization in the states of Utah and Idaho, and the second largest in Arizona, California, Hawaii, Nevada, Oregon, Washington, and Wyoming. The Mormon Church was the fourth largest in Tonga and Samoa, and is expanding rapidly in South America.[40] "Although plagued by the problems of centralization common to all multinational corporations, the Church effectively . . . maintained highly centralized control through the hierarchy of line authority, both at headquarters and within the body of the Church."[41]

By 1973 (when Lee died), the correlation program had produced "a standardized and sanitized instructional curriculum."[42] The reform process continued under Lee's successors, Spencer W. Kimball and Ezra Taft Benson. Wiley claims that by the 1980s, "many young members of the Church had internalized doctrine and organizational procedure to such an extent that there was a new orthodoxy in the ranks of the Church."[43] "Flowcharts [now] reveal the framework of an institution whose lifeblood circulates through meetings, committees, quorums, classes, and councils."[44] Thus, the "essence of Mormonism" can be found in the language of teaching manuals, training guides, and bulletins produced by correlation committees.

## DISCOURSE THROUGH CORRELATION

As within any large administrative structure, some power must be delegated. With delegation comes "some discretion over the performance of the task. . . . Information that is passed through bureaucratic channels is [also] selectively screened along the way by various persons."[45] Although performance discretion and information screening preserve uncertainty in technocracies, their potential is severely limited when communication becomes technologized. As members communicate according to rules structured by and for correlation, they forget what ends the technology was originally meant to serve. Instead, the emphasis is on sustaining the correlation program for its own sake.

Training guides and bulletins promote correlation as an end in itself, instilling the "principles of correlation" in the minds of church leaders. To successfully serve as a bishop (ecclesiastical leader for local church units), a man should "systematically study the *General Handbook of Instructions* to familiarize himself with the organization and policies of the Church. [He should also] review . . . the other handbooks published by the Church."[46] These include publications (prepared by correlation committees) such as *Bishop's Guide, Bishop's Training Course and Self-Help Guide, Bishopric Training In-Service Booklet*, and the *Melchizedek Priesthood Handbook*.[47] These publications explain that priesthood correlation provides a means for helping "individuals in the balanced programs of the Church." A worksheet in the *Bishop's Training Course and Self-Help Guide* directs Bishops to "circle ideas above which help you understand the principles of priesthood correlation." The available options extol correlation's virtues. For example, "to avoid duplication, all Church programs are unified through correlation."[48] Bishops read that their "best sources of ideas are ward leaders and other members who have become thoroughly familiar with handbooks and manuals provided by the Church."[49] Publications are sprinkled with frequent references to the value of correlation. Correlation makes bishops more organized, efficient, and effective. Further, it provides an "opportunity for the Bishop to train the leaders of each priesthood group."[50] Bishops are encouraged to use the teaching opportunity created by correlation to teach correlation because "correlation is an essential principle to teach all auxiliary officers and teachers."[51]

## Defining Gender

Members are taught from correlated instruction manuals each Sunday. Lessons from the manuals used during 1987 and 1988 in the Relief Society and in the adult priesthood quorums, as well as those used for teenagers, present a unified message regarding appropriate gender roles.

First, men and women are fundamentally different. Women learn that "being born as women brings to you many endowments that are not common to men and therefore make you unique." Further, "these are eternal differences— with women being given many tremendous responsibilities of motherhood and sisterhood and men being given the tremendous responsibilities of fatherhood and the priesthood."[52] Lessons for teens attempt to justify differences between males and females:

> For a very wise purpose, man and woman were created by God to play different roles. Boys and girls and men and women are different in many ways. . . . Difference between the sexes was designed to provide individuality to all God's children. . . . Differences occur in special interests. Young men like physical activities and masculine things as they

prepare to assume their roles as husband and father and provider. They become protectors. They become leaders. Young ladies, by comparison, are feminine in nature. They do the things they like to do to prepare for the role of mother in the family.[53]

Training guides direct bishops to help male and female teenagers develop appropriate differences. They must provide boys with "opportunities to lead, to plan, and to be engaged in activities which challenge the physical, emotional, and spiritual powers of their budding manhood."[54] Girls "should be helped to understand how to live more satisfyingly and fully in their social lives. Efforts should be made to help them build self-confidence and respect."[55] The *Bishopric Training In-Service Booklet* teaches that "the home and home life form the first problem of a woman, whether it be her childhood's home, or that which she builds with her mate. . . . Therefore young womanhood should be taught how to establish happy homes."[56]

It is essential that church leaders teach members to cooperate with these divinely ordained distinctions, for although each woman has "the right and the responsibility to direct [her] own life [she should not be] deceived; [she] must also be responsible for [her] choices. . . . God is unchanging, and . . . He has entrusted to his daughters the great responsibility of bearing and nurturing children."[57]

Another lesson explains that women will naturally choose to be mothers because the "roles of wife and mother are in the soul and cry out to be satisfied. It is in the soul to want to love and be loved by a good man and to be able to respond to the God-given, deepest feelings of womanhood—those of being a mother and nurturer. Fortunately, women do not have to track a career like a man does."[58] Instead they can "follow the noble, intuitive feelings planted deep within [their] soul by Deity."[59] A lesson titled "Privileges and Responsibilities of Sisters" states that "young women should plan and prepare for marriage and the bearing and rearing of children. It is [their] divine right and the avenue to the greatest and most supreme happiness."[60]

Church members learn that women "have the greater responsibility not only of bearing children but of caring for them through childhood."[61] Another lesson reminds men that "the noblest calling of women is to be honorable mothers." "Quorum training suggestions" at the end of the lesson include a suggestion to "have some brethren report on ways in which they can supplement family income without having their wives work outside their homes."[62]

Correlation also trains bishops to counsel members who experience financial needs. Bishops should advise against accepting government assistance if any type of work is available. They "should use the opportunity to teach the importance of work. However, every effort should be made to see that a mother stays in the home with her children."[63] Even if a woman is "forced" to earn a living "the roles of wife and mother are still the most important."[64] At-home child care, where women "can teach and care for their own children at the same time they are caring for other children," is suggested.[65]

Although few lessons in the series for male teens are devoted wholly to the role of women, many provide indications of role expectations. Women are portrayed exclusively as mothers, for "women do not possess the priesthood any more than men attain motherhood, the feminine equivalent of the priesthood." Young men are encouraged to limit their positive regard for women to those who choose to act as mothers because "motherhood is the great talent and calling given to women, and upon their magnification and use of this calling depends their exaltation."[66] Young women's lessons teach how to become wives and mothers. Lessons such as "Power of Creation," "Motherhood: A Divine Calling," and "Preparing for Motherhood" seek to help young women "develop greater respect for the power of creation," and to show "that motherhood is the noblest of all [their] callings."[67]

### Self Subordination by Women

Priesthood is now "the correlating factor in the church" and offers all male members the opportunity to play God on earth.[68] Women should support the priesthood hierarchy because "this patriarchal order has its divine spirit and purpose, and those who disregard it under one pretext or another are out of harmony with the spirit of God's laws. . . . It is a question largely of law and order."[69] Training guides teach members their appropriate hierarchical position by charting the "priesthood line of authority" from God to the individual. Authority moves from God to the prophet and his counselors, to the twelve apostles, to other general authorities, to regional representatives, to stake presidents, to bishops, to priesthood leaders in local units, to home teachers, to parents ("usually headed by a father"), to individuals. Because God's power flows through priesthood, women are part of this hierarchy only when "the absence of the father" requires them to "assume the role of leader in the home."[70] Women learn that God determines priesthood ordination, or the lack thereof, by a person's gender. As one lesson states:

> There are a few women in the Church who complain because they do not hold the priesthood. I think the Lord would say to you, 'murmur not because of the things which are not given thee.' . . . [Men have not] set the rule concerning those who should receive the priesthood. That was established by him whose work this is, and he alone could change it.[71]

Because God, rather than men, has determined woman's lack of authority within the church hierarchy, men are not responsible for their preferred status. Their authority needs no justification, for they simply have the right to command others.

One lesson warns that insubordination is self-destructive:

> God will not ennoble a person, man or woman, who refuses to uphold by faith, prayer, and works those whom God has called and ordained to preside over them. . . . You will want to sustain the priesthood authority. . . . [The belief that] it is subversive to [one's] free agency to be directed by the power of the priesthood . . . comes from misunderstand-

ing. . . . Following the priesthood of the Church is an expression of faith in the Lord's continuing guidance of his Church.[72]

By "supporting the priesthood" women treat the hierarchy as a "real" order of relationships, reconstituting its foundation as an expression of domination.

The truly strong woman is able to withstand "worldly" influences that would prevent her from following the priesthood. One lesson tells of a woman who "had excelled in education, earning advanced university degrees." Upon receiving a priesthood blessing that indicated "her life's work led in another direction" she chose to "leave those pursuits and marry." Despite her professors' attempts to dissuade her, the woman chose to pursue a life more consonant with her God-given "feminine" qualities. "Passage of time confirmed that the decision was right."[73] This woman's "willingness to acknowledge the role of the priesthood in her life" exemplifies the appropriate "expression of a woman's faith."[74]

The *Relief Society Handbook* implicates female Relief Society leaders in the effort to normalize female subordination. The brief Relief Society history included in the handbook neglects to mention early Mormon women's church-sanctioned performance of functions now reserved for (male) priesthood bearers. Rather, women are urged to accept the "blessings" and teach their sisters "the purposes and functions, of the priesthood." Women trained by the Relief Society should become "a better resource to the priesthood."[75] Relief Society leaders are urged to accept "training and counsel from their priesthood leaders," and promptly extend this training to any new leaders:

> A Relief Society leader should be given her first training a week after she is called. This training should teach her the mission of the Church and the purpose and goals of Relief Society, the priesthood line of authority, and basic Relief Society responsibilities and resources.[76]

Those responsibilities center on supporting the priesthood. As an "auxiliary" the Relief Society is one of the "'aids' which are subordinate to the priesthood to sustain and support the priesthood."[77]

Lessons also explain how teens should enact superior/subordinate roles upon reaching adulthood. One lesson reaffirms the male's superiority by explaining, "as a young man matures, he may find some empathetic girl companion who will listen to his hopes and aims and desires. He will develop complete confidence in his sympathetic listener. He may even be willing to heed her counsel and advice."[78] Consideration of a woman's counsel is an optional courtesy because "the man is the head of the family unit by virtue of his priesthood." Although he should "seek the interests of his wife and confer with her on matters of mutual concern. . . . All counsel completed, however, the decision is with the father."[79]

Young women reify their subordinate relation to men by learning how to "support the priesthood." In a lesson on "Women and Priesthood Bearers" the teacher is to ask the girls how they can "support a young man who holds the priesthood." Suggested responses are:

1. Concentrate on his good points.
2. Supply ideas.
3. Treat him as you want him to become.
4. Have a listening ear.
5. Be honest in your praise.
6. Support him in projects and callings.
7. Be a counselor, when asked.
8. Do what is delegated to you.
9. Sustain him with your prayers.

A "true" story about a young female leader who was frustrated because her male counterpart in the church youth organization failed to carry out his responsibilities tells both how to implement these suggestions and how one should feel towards the priesthood (and by extension, all those who are eligible for ordination). "Lynne [began to treat the young men as] responsible priesthood holders. Once they knew the young women were depending on them and would support and encourage them, they led out. . . . Lynne began to be led by the priesthood. It was a wonderful feeling!"[80] Thus, a woman's happiness is achieved by learning the art of followership.

Lessons detailing appropriate behaviors for both women and men extend the importance of accepting male dominance to intimate relationships. A lesson on marriage states that "as no man can be perfect without the woman, so no woman can be perfect without a man to lead her."[81] A lesson titled "Sustaining the Priesthood" states that despite woman's required sacrifice of self "the father and husband has always had the responsibility of caring for and directing the affairs of the family." Women are told that "your husband, as the priesthood bearer, is the head of the home. You, the helpmeet, are not the head, but just as important—the heart of the home."[82] As head of the family, the father is "the legislator, the judge, the governor."[83]

Men and women within the family exist in a parent/child relationship, for "the righteous priesthood bearer seeks for his family the same goals that God seeks for his children."[84] Husbands are urged to "be kind and affectionate, teaching [their wives and children] with mercy and kindness and justice and in righteousness all the day long."[85] Each bishop learns that "only as [he] prepares [his wife and children] to reign with [him] in all eternity will [he] be doing [his] best."[86] Because of her husband's leadership role, the bishop's wife must provide an exemplary model of womanhood. She "is to be friendly with all the ward members, but she should remain in the background. . . . She should be active in all activities that she can possibly attend to show she supports [her husband]."[87] Every "family should be thought of as an eternal entity presided over by a righteous bearer of the Melchizedek Priesthood with a wife who has been sealed to him for time and eternity."[88] Ultimately, there is "nothing sweeter than a home where a man is living his religion, magnifying his priesthood, with his wife supporting him in every way."[89]

In sum, correlated lessons teach that women and men are fundamentally different; this difference fits women to be mothers and men to be (priesthood) leaders; and women should follow and support their priesthood leaders (who happen to be men). Accordingly, a woman's gospel study will "help her implement . . . [the] . . . principles [of] spiritual living, compassionate service/ social relations, and instruction for strengthening the home and family" in her life.[90] In contrast, men study "to become perfected in [Christ], . . . that [their] use of the holy priesthood may be more effective."[91] Women's appropriate activity is limited to supporting or "sustaining" others, especially those who wield priesthood authority. Her passive, receptive function contrasts sharply with the men's active, achievement-oriented role in the mission statements of corresponding Relief Society and priesthood manuals. Women must "have faith, . . . strengthen the families, . . . give compassionate service, and sustain the priesthood."[92] Men, on the other hand, should "be worthy of receiving the full power of the priesthood and stand forth as noble fathers, husbands, and priesthood bearers."[93]

## LIMITATIONS OF TECHNOLOGICAL DISCOURSE

Correlation in the Mormon Church narrowly defines members' organizational roles (based on gender), and risks alienating over half the organization's population. If women's lack of priesthood is divinely ordained, they remain forever guilty of an inferior status. Burke argues that guilt "comes not from the breaking of the law but from the mere *formulating* of the law."[94] By formally removing women from the direct line of institutional authority, correlation potentially undermines the ability it claims to promote, for all those who are destined to "support" and "sustain" the institution have been rendered perpetually weak.

Additionally, while correlation's specific directions for all task performances (including cognition) may promote organizational stability, its precision presents two major dangers. First, members are discouraged from differentiating between organizational goals (basic doctrines) and their enactments (traditional behaviors). In this sense correlation works as a "deskilling" process. Burke argues that a master psychosis can function as "trained incapacity," creating a situation where abilities "function as blindness."[95] Persons who are unable to differentiate between basic doctrines and task performances may leave the organization despite their commitment to organizational goals because they oppose specific behaviors imposed by correlation.

Correlation has also fundamentally altered organizational goals. While it was originally implemented as a means to "help the Lord bring to pass the eternal life of each member," correlation has now become an end in itself.[96] Because correlation is invoked as the means for determining the appropriateness of any action, behaviors that sustain correlation are valued and encouraged, while others are devalued and hidden. Thus, attention is drawn away from the organization's original goal of building an organizational structure that facilitates the "eternal life of each member." Correlation focuses energy, instead, on fortifying existing power relationships within the organization.

Systems of technological discourse, often precipitated by concern for institutional stability, exert similar influence within many organizations. Competency tests recently implemented by several states' education departments exemplify the same potential difficulties as correlation. Public attention to students' reading (and other) deficiencies has motivated accountability movements in education. Nancy Wood claims that when public officials become concerned about accountability "they invariably mandate more tests." Although "reading comprehension is increasingly regarded as an idiosyncratic process that is influenced not only by the particular reader's current skills but also by his or her purpose, values, knowledge, and beliefs," existing technology does not provide efficient means for evaluating these aspects of reading.[97] "Testing," however, becomes more crucial than learning because testing stabilizes institutional configurations by making each participant accountable to the institution as well as to all other members.

To prepare for testing, educators have attempted to codify literacy by creating lists of skills that students need. However, when a group of Texas educators was convened by National Evaluation Systems (the testing company that received the Texas test contract) members learned that the testing company had their own list. Educators could delete from, but not add to, the provided list. Testing companies involved in competency programs like those in Texas, New Jersey, or Minnesota now make the lists that guide curriculum and materials development. Ironically, these programs were implemented in reaction to fears that schools were not teaching students critical or creative thinking skills. As Wood points out, "critical thinking, creative thinking, and even memorization are not well represented on testing lists."[98] Thus, competency movements have diverted attention from the original concern with students' learning and refocused it on the technology of testing.

Burke argues that technological language is popular largely "by reason of its low anthropomorphic content. It is designed for machines."[99] Organizations designed for people will be limited more by what technological discourse leaves out than by what it focuses on. Because technological discourse includes merely those aspects of experience that can be clearly phrased with its terms, it is devoid of the complexities involved in organizational life. Although refusal to admit uncertainty may produce temporary stability, uncertainty does not disappear from a technologized organization. Until the deliberate selection of alternatives required to establish and perpetuate the technology is specifically stated, it merely lies hidden beneath the ramifications of the system. Although "the hierarchic principle is indigenous to all well-rounded human thinking," a rigid hierarchy can interfere with the very processes that it was intended to facilitate.[100]

Technologized discourse, which provides an efficient means of codifying correct behavior within a closed system, discourages integrative approaches to organizational predicaments. Conversely, resigning ourselves to the absence of certainty that simultaneously induces and constitutes creative dialogue requires abandoning the aspect of perfection implicated in prediction and control, and frees us to pursue previously unforeseen options.

## NOTES

The author wishes to thank David Williams for providing crucial texts for this analysis.

1 Victor J. Stenger, *Not by Design: The Origin of the Universe* (Buffalo, NY: Prometheus, 1988), 108-109.

2 Werner Heisenberg, "Remarks on the Origin of the Relations of Uncertainty," in *The Uncertainty Principle and Foundations of Quantum Mechanics*, ed. William C. Price and Seymour S. Chissick (London: John Wiley & Sons, 1977), 5; 3-6.

3 Tarla Rai Peterson, "The Rhetorical Construction of Institutional Authority in a Senate Subcommittee Hearing on Wilderness Legislation," *Western Journal of Speech Communication* 52 (1988): 259-276.

4 Kenneth Burke, *Permanence and Change* 3rd ed. (Berkeley: University of California Press, 1984), 39.

5 Werner Heisenberg, *Across the Frontiers*, trans. Peter Heath (New York: Harper & Row, 1974), 65.

6 Burke, *Permanence and Change*, 45.

7 Burke, *Permanence and Change*, 49.

8 Jacques Ellul, *The Technological System*, trans. Joachim Neugroschel (New York: Continuum, 1980), 35-46; 46.

9 Henry W. Johnstone, Jr., "Communication: Technology and Ethics," in *Communication Philosophy and the Technological Age*, ed. Michael J. Hyde (University: University of Alabama Press, 1982), 38.

10 Johnstone, "Communication," 40.

11 Burke, *Permanence and Change*, 274.

12 Carolyn R. Miller, "Technology as a Form of Consciousness: A Study of Contemporary Ethos," *Central States Speech Journal* 29 (1978): 232.

13 Miller, "Technology," 234-235.

14 William Barrett, *The Illusion of Technique: A Search for Meaning in a Technological Civilization* (Garden City, NY: Anchor Press/Doubleday, 1978), 336.

15 Sherry Turkle, "Computational Reticence: Why Women Fear the Intimate Machine," in *Technology and Women's Voices: Keeping in Touch*, ed. Cheris Kramarae (New York: Routledge & Kegan Paul, 1988), 56.

16 Carol Gilligan, *In a Different Voice: Psychological Theory and Women's Development* (Cambridge: Harvard University Press, 1982).

17 Kathy E. Ferguson, *The Feminist Case Against Bureaucracy* (Philadelphia: Temple University Press, 1984), 22.

18 Brigham H. Roberts, *A Comprehensive History of the Church of Jesus Christ of Latter-day Saints: Century I*, Vol. 2 (Provo, Utah: Brigham Young University Press, 1930/1965), 541.

19 Leonard J. Arrington, *Great Basin Kingdom, Economic History of the Latter-day Saints, 1830-1900* (Lincoln: University of Nebraska Press, 1958).
20 Tarla Rai Peterson, "Justifying Ideological Conflict: Mormon Abandonment of Polygamy." Paper delivered at the Annual Meeting of the Speech Communication Association, Boston, November 1987, 11.
21 Peterson, "Justifying Ideological Conflict."
22 Jill Mulvay Derr and C. Brooklyn Derr, "Outside the Mormon Hierarchy: Alternative Aspects of Institutional Power," *Dialogue: A Journal of Mormon Thought* 15 (Winter 1982): 31.
23 Peterson, "Justifying Ideological Conflict."
24 "The Oath and Covenant of the Priesthood," in *Come Unto Christ* (Salt Lake City, Utah: Corporation of the President of the Church of Jesus Christ of Latter-day Saints, hereinafter referred to as Corporation of the Church, 1983), 32.
25 Relief Society Minutes of Nauvoo, 28 April 1842, p. 40; Sarah M. Kimball, Relief Society Minutes of Salt Lake, 22 June 1878, Historical Department, The Church of Jesus Christ of Latter-day Saints; John Taylor, 8 August 1880, *Journal of Discourses* 26 Volumes (Liverpool: Franklin D. Richards et al., 1855-86) 21:367.
26 Linda King Newell, "The Historical Relationship of Mormon Women and Priesthood," *Dialogue* 18 (Fall 1985): 26.
27 John A. Widtsoe, *Priesthood and Church Government*, 2nd ed. (Salt Lake City: Deseret Book Company, 1954), 90.
28 Derr and Derr, "Outside the Hierarchy," 32.
29 Derr and Derr, "Outside the Hierarchy," 25.
30 Derr and Derr, "Outside the Hierarchy," 27.
31 Armaund Mauss, "Assimilation and Ambivalence: The Mormon Reaction to Americanization," *Dialogue* 22 (Spring 1989): 44.
32 Peter Wiley, "The Lee Revolution and the Rise of Correlation," *Sunstone* 10 (Jan. 1985): 20.
33 Wiley, "The Lee Revolution," 20.
34 Wiley, "The Lee Revolution," 20.
35 Derr and Derr, "Outside the Hierarchy," 28.
36 Derr and Derr, "Outside the Hierarchy," 28-29.
37 Derr and Derr, "Outside the Hierarchy," 29.
38 Derr and Derr, "Outside the Hierarchy," 29.
39 Wiley, "The Lee Revolution," 21.
40 D. Michael Quinn, "From Sacred Grove to Sacral Power Structure," *Dialogue* 17 (Summer 1984): 9.
41 Derr and Derr, "Outside the Hierarchy," 30.
42 Mauss, "Assimilation," 44.
43 Wiley, "The Lee Revolution," 22.
44 David J. Whittaker, "An Introduction to Mormon Administrative History," *Dialogue* 15 (Winter 1982): 17.
45 Ferguson, *The Feminist Case*, 18.
46 *Bishop's Guide* (Salt Lake City, Utah: Corporation of the Church, 1984), 4.
47 The following training guides were used in this analysis: *Bishop's Guide*, (Salt Lake City, Utah: Corporation of the Church, 1984); *Bishop's Training Course and Self-Help Guide* (Salt Lake City, Utah: Corporation of the Church, n.d.); *Bishopric Training In-Service Booklet* (Salt Lake City, Utah: Corporation of the Church, n. d.); *General Handbook of Instructions* (Salt Lake City, Utah: Corporation of the Church, 1985); *Melchizedek Priesthood Handbook* (Salt Lake City, Utah: Corporation of the Church, 1984); *Relief Society Handbook* (Salt Lake City, Utah: Corporation of the Church, 1988).
48 *Bishop's Training Course*, I-7.
49 *Bishop's Training Course*, I-25.
50 *Bishop's Training Course*, I-11.
51 *Bishop's Training Course*, II-71.

52 James E. Faust, "Message to our Granddaughters," in *Learn of Me* (Salt Lake City, Utah: Corporation of Church, 1986), 74.

53 Shane B. Inglesby, "Priesthood Prescription for Women: The Role of Women as Prescribed in Aaronic Quorum Lesson Manuals," *Sunstone* 10 (March 1985): 32.

54 *Bishop's Training Course*, III-29.

55 *Bishop's Training Course*, III-29.

56 *Bishopric Training*, 18-19.

57 Spencer W. Kimball, "Privileges and Responsibilities of Sisters," in *Learn of Me*, 126; 128.

58 Faust, "Message," 76.

59 Faust, "Message," 80.

60 Kimball, "Privileges," 123.

61 Gordon B. Hinckley, "Cornerstones of a Happy Home," in *Come Unto Me* (Salt Lake City, Utah: Corporation of the Church, 1987), 180.

62 "The Role of Women," in *Come Follow Me* (Salt Lake City, Utah: Corporation of the Church, 1983), 150; 153.

63 *Bishop's Guide*, 26.

64 "Being Personally Prepared," in *Come Unto Me*, 376.

65 "Being Personally Prepared," 378.

66 Inglesby, "Priesthood Prescription for Women," 30.

67 Karla S. Gunnell and Nicole T. Hoffman, "Train up a Child in the Way He Should Go," *Sunstone* 10 (March 1985): 35.

68 *Bishop's Training Guide*, III-21.

69 "Turn Your Heart to your Children and Fathers," in *Come Unto Christ*, 15.

70 *Bishop's Training Course*, I-8.

71 Hinckley,"If Thou Art Faithful," in *Learn of Me*, 88-89.

72 Faust, "Message," 78-79.

73 "Being True to Our Foreordained Missions in the Last Days," in *Come Unto Me*, 113.

74 "Sustaining the Priesthood," in *Learn of Me*, 160.

75 *Relief Society Handbook*, 1.

76 *Relief Society Handbook*, 28.

77 *Bishop's Training Guide*, II-71.

78 Inglesby, "Priesthood Prescription for Women," 32.

79 Inglesby, "Priesthood Prescription for Women," 31.

80 Gunnell and Hoffman, "Train up a Child," 35.

81 "They . . . Shall Be One Flesh," in *Come Follow Me*, 187-188.

82 "Sustaining the Priesthood," in *Learn of Me*, 159.

83 Harold B. Lee, "A Father's Most Important Work," in *Come Unto Christ*, 114.

84 "Elijah's Mission: The Family," in *Come Unto Christ*, 10.

85 "They . . . Shall Be One Flesh," 188.

86 *Bishop's Training Course*, I-28.

87 *Bishop's Training Course*, I-27.

88 "Elijah's Mission," 9-10.

89 "Sustaining the Priesthood," 161.

90 "Preface," *Come Unto Me*, v.

91 "Message from the First Presidency," *Come Unto Christ*, v.

92 Barbara W. Winder, "The Mission of Relief Society," *Learn of Me*, 5.

93 "Message from the First Presidency," in *Come Follow Me*, iv.

94 Kenneth Burke, *A Rhetoric of Motives* (Berkeley: University of California Press, 1969), 228.

95 Burke, *Permanence and Change*, 7; 7-11; 49.

96 *Bishop's Guide*, 2.

97 Nancy Wood, "Codifying Literacy: Identifying and Measuring Reading Competencies in Statewide Basic Skills Assessment Programs," *Journal of College Reading and Learning* 22 (Fall 1989): in press; and "Standardized Reading Tests and the Postsecondary Reading Curriculum," *Journal of Reading* 32 (December 1988): 224.

98 Wood, "Codifying Literacy," 7-13; 13.
99 Burke, *Permanence and Change*, 58.
100 Burke, *Rhetoric*, 141.

# Modern Discourse on American Home Technologies

*Karen E. Altman*

At one time in the past, the American home was widely considered a place of moral uplift, a sanctuary for spiritual life, a refuge from the new industrial world of greed, grime, and machine. In the "Victorian era" of separate spheres, the home, domesticity, and family comprised the private domain while industry, the state, and politics comprised the public domain. With historical hindsight, Christopher Lasch called the home and family of this period a "haven in a heartless world."[1] A characterization of today's home from Jean Baudrillard contrasts sharply with Lasch's. In the current "era of hyperreality," the habitat or domestic universe has been converted into a "satellitization of the two-room-kitchen-and-bath put into orbit in the last lunar module." Baudrillard claims that "[we] are here at the controls of a micro-satellite, in orbit, living no longer as an actor or dramaturge, but as a terminal of multiple networks" in a space void of anything once imagined as a separation of public and private spheres.[2] Sometime between the haven and the micro-satellite, there occurred an "industrial revolution in the home" during which, Ruth Schwartz Cowan argues, technological systems, economic institutions, and social practices changed the character of the home, its work, and its relations with industry and commerce.[3]

The American home is a technological site. Some of its earliest manual and mechanical forms, such as the egg beater, maintain their place alongside of microwave ovens and interactive electronic games. Similar to other sites such as the corporate organization, the military, and the human body, the home will continue to be increasingly technologized and human interpretations of technologization will continue to be debated: technology as liberation versus technology as colonization.

Gaining a historical understanding of the home as a technological site is a difficult task due to certain traditions in the study of technology. One major tradition has been to examine particular types of technologies based on their form

or function:  communication technology, transportation technology, architec-
tural technology, mechanical technology, and housework technology.  Given
these divisions, interconnections among technologies and practices in the home
remain opaque.  Literature on communication technology, for example, builds
extensive connections among print media, radio, television, and computers, but
does not develop relations between media and other technology in the home such
as indoor plumbing, bottle-feeding, or sewing.[4] Literature on household technology, on the other hand, traces the mechanization of ironing, cooking, cleaning,
and transporting, but without examining concurrent developments in communication media.[5] In contrast to researchers, manufacturers, advertisers, industry
trade journalists, and retailers create connections among diverse technologies in
the home.  Indeed, radios and refrigerators often have been displayed side-by-
side in a single advertisement or in the same store window, and, moreover, were
widely found one-on-top-of-the-other in millions of American kitchens.  By
switching the focus of study from separate forms or functions of technology to
a principal site or location of technology, we gain another view of technologies
in the American home.

A second tradition also has obscured, if not distorted, understanding of
technologies: a malestream bias in scholarship.[6] Among the many critiques of
this bias, three are prominent:  malestream bias emphasizes technologies of
interest to men, especially weaponry, to the exclusion of those of importance to
women, particularly reproductive and housework technologies;  malestream
bias assumes that technology is neutral to social differences between the sexes
in terms of access and uses, or worse, denies that women are even involved with
technology;[7] and malestream bias fails to acknowledge that contemporary
technological values and practices were developed by men within extensive
organizations of white male profit and social control.  A growing feminist
literature has begun to reinterpret established accounts of technologies and to
offer alternative perspectives by featuring women or gender relations in analyses
of technology.[8] For example, some feminists have started to rethink technology
in terms of communication systems that encourage some human interactions
while discouraging others. Contributors to the volume *Technology and Women's
Voices:  Keeping in Touch* discuss the political, economic, and technological
structures that organize social interactions, on the one hand, and women's
struggles to make a living, care for families, and communicate with one another
through, or perhaps despite, technology, on the other.[9]

There is little doubt that many technologies currently produced for and
used in the home such as video recorders and computers are different in
technological composition than strictly mechanical apparati.  But these technologies do not exist outside of the social relations in which they are produced
and the discursive or representational practices for talking, envisioning, or
thinking them. The significant difference between microwave and conventional
gas ovens for understanding American culture is not in the technological
specificities of their apparati, but in how they are imagined, talked about, and put

into use within the material constraints of cost, availability, and access among diverse social groups of people.

My analysis concentrates on discourses and images that circulate widely in American culture and construct ways of seeing, imagining, or valuing technology in the home. Repeated discourse themes and images about technology circulate widely through advertising, public relations, journalism, advice books, and other mass communication media. Such widely available discourses offer specific ways of seeing and valuing technology and its everyday uses. Public discourse and images tend "to make some forms of experience readily available" and to make certain practices and values seem natural or taken-for-granted.[10] Moreover, public discourses and images articulate patterns among technologies, social relations, and cultural practices, and thereby construct certain realities and knowledges about technology in the home. It is in discourse that technologies are made to be meaningful within the social relations of a particular historical time.

Discourse and images on home technologies have changed across American history, but I will not be focusing on the changes. I will focus here on one distinct theme or discourse formation—that is, modern discourse on home technologies. Modern discourse and images inscribed very particular social relations onto home technologies and embedded home technologies in particular social relations. My analysis of social relations of the home is drawn from feminist theory which, as I will discuss, includes the relations of gender, work and leisure, and public and private.[11] My larger purpose is to contribute to a cultural history of technology and the American home.

## MODERN DISCOURSE

A recent Black and Decker campaign describes its toaster, can opener, and clock-radio as "space makers" and its coffee maker as "space maker plus."[12] Toasters, can openers, and coffee makers are not new technologies; they have been in American homes for many decades. There is an important difference, however, in the discourse. For most of this century, these household appliances have been called "labor saving devices," with time saving and other efficiency features. Current discourses and images of space saving and style coordination constitute a wholly other way of constructing household technology.

One superordinate theme, a type of discourse formation, organized discourse and images of home technologies throughout much of the twentieth century. I call this superordinate theme "modern" or a "modern" discourse formation. I do not use the term "modern" to mean the philosophical project of "modernity" associated with the Enlightenment or Age of Reason, nor do I mean the literary theory and practice of "modernism."[13] Instead, I use the term "modern" as it appeared in the discourse of advertising, journalism, home economics, and public relations. Hence, my analysis is not about theories of modernization or modernism, but on the use of the term "modern" in a particular historical practice: discourses on the home and its technologies.

Raymond Williams has emphasized the importance of examining key-words in the study of culture. His work on five keywords in British literature and politics in *Culture and Society, 1780-1950* and his glossary in *Keywords: A Vocabulary of Culture and Society* opens a perspective for thinking about those words crucial to discourse used about the home and its technologies.[14]    In hundreds of ads, advice columns, design books, and popular press articles, several keywords appear repeatedly.    These include variations of the term "modern" (modernize) and the terms "science," "technique," "standardization," and "efficient." Keywords are similar to ideographs in political discourse which, according to Michael McGee, are those terms or phrases that provide guides for thinking and warrants for action without the burden of propositional argument or exhaustive chains of reasoning.[15]

There is no exact beginning or ending date of the modern theme or discourse formation, but it can be distinguished from both a previous moral discourse that justified increased technologization of the home in spiritual terms and from an emerging contemporary discourse that is characterized by some loosening of, if not collapse between, constructed oppositions such as mascu-line-feminine, work-leisure, and public-private. For example, sewing machines and religious symbolism were dissociated from one another around the turn of the twentieth century, and today sons and fathers "reach out and touch" each other after seven decades of associating woman with the home telephone.[16]

By the 1920s, much of American life was being shaped in the name and imagery of modernity.[17] Automobiles and skyscrapers were called modern and watches, baby powder, and linens were stylized as modern.[18] Variations on the term "modern" and modernist visual aesthetics began to construct everyday realities, knowledges, and values of the American home. Modern discourses and images eventually constructed meaningful places for technologies as diverse as vacuum cleaners, telephones, bathroom fixtures, and television sets in the home.

Advertisers marketed modern, labor-saving devices over old-fashioned ways. "Why in the world do you continue to empty the filthy contents of a cleaner bag," asked the headline of a 1929 ad, "When Modern Sanitary Methods Are Available in the Air-Way Sanitary System?"[19] Modern methods and techniques, it was repeatedly claimed, saved time, saved labor, and created healthier and happier living. The "modern, automatic" 1950 Hotpoint electric dishwasher was "the greatest time- and labor-saving appliance ever invented for the home."[20] Electric ranges, irons, washing machines, and refrigerators operated by scientific methods, offered scientific cleanliness or cooking, simplified work techniques, and saved time.

Much advertising and product promotion of modern household technolo-gies was informed by scientific home management, the major theory of home economics into the 1940s, through such business publications as Christine Frederick's *Selling Mrs. Consumer.*[21]    Based on Taylorization and time and motion studies from industry, scientific home management, it was claimed, could replace housework by the housewife's application of scientific principles and by the use of technologies.[22] Home economists, journalists, and advertisers emphasized the application of scientific principles to household tasks through applied techniques and labor-saving technologies to ensure greater efficiency, standardization, cleanliness, and health.

Communication technologies, especially radio, also came to be called modern. In contrast to *fin-de-siècle* amateur broadcasting which involved both transmission and reception through such mechanical components as crystal sets with headphones, radio receivers were standardized and stylized for reception only. The broadcasting industry presented receivers as technically advanced and stylish. RCA Radiola made radio "not only greater, but simpler" by "engineers" in "laboratories."[23] But trade articles warned radio dealers "don't talk circuits. Don't talk in technical terms," and home economist and popular journalist Christine Frederick, advised radio dealers to take a lesson from washing machine, vacuum cleaner, and other household appliance sellers to define the modern home based on ease and style rather than machinery.[24] Communication and housework technologies alike were standardized and sold in the name of efficiency and simplicity.

One form of advertising enabled manufacturers or retailers to display many of their appliances, equipment, and other devices within one advertisement. Such all-in-one advertisements combined coffee makers, washing machines, and telephones in the American home. For instance, Western Electric displayed its sewing machine, toaster stove, telephone, and other electrical devices under the claim "electricity does the work."[25] In 1951, Firestone used the all-in-one advertising form to combine refrigerators, stoves, washing machines, radios, phonographs, and television as "home appliances."[26] Crosley also used the all-in-one form and filled it with a patriotic theme: "The American Way." Crosley's ad featured the title of its essay contest, "What the American Way of Life Means to Me," flanked by images of stoves, TVs, radios, and other major appliances.[27] Advertising integrated technologies as commodities in the home rather than separating them according to technological function.

Modern visual images often complemented modern discourse. Modernist aesthetics shaped and stylized many commodities of the 1920s, including technologies.[28] Vacuum cleaners, for example, were not only efficient, scientifically engineered, and easy to use, they often were stylish. The elongated, sleek shapes and radiant beams of modernist design characterized many Hoover company ads.[29] Among the most opulent images, however, were those of a new 1920's technology: bathroom fixtures. Tubs, shower stalls, sinks, and toilets were standardized in mechanical function, but highly stylized in color and design.[30] Hence, modern discourses and imagery combined scientific principles and technique with a sense of style.

Modern keywords of science, technique, efficiency, and standardization and modernist images of technology were configured, moreover, with very specific social practices and relations among people. In other words, discursive and visual constructions inscribed particular social relations onto technologies and embedded technologies in particular social relations.[31]  Technologies became saturated with very specific social relations of gender, work and leisure, and public and private.

## SOCIAL RELATIONS OF THE HOME

The home has been equated with domesticity and leisure as opposed to politics and work throughout American history.  It has been called the private sphere and woman's place by ministers, philosophers, domestic scientists, and historians, among others.[32]  This view of the home mystifies the complexity of social organization.  When the home is thought of as a refuge or haven from the work world, it is difficult to think that the cooking, cleaning, hauling, and caretaking done there is hard human work, especially when it is done by women without the economic pay that distinguishes work in business practices.  Feminist theory, by contrast, offers an alternative perspective on social realities.  Joan Kelly argues that the "doubled vision" of feminism sees

> not two spheres of social reality [i.e., public v. private, man's place v. woman's place],
> but two (or three) sets of social relations.  For now, I would call them relations of work
> and sex (or class and race, and sex/gender).  In a Marxist analysis, they would be termed
> relations of production, reproduction, and consumption.  In either case, they are seen as
> socially formed relations . . . making "men" and "women" social categories.[33]

From the perspective of feminist theory, the home is a site of multiple social relations, not a separate sphere.

The concept of articulation provides a way to understand how keywords are combined or linked with particular social practices or relations in discourse.  Articulation involves the building of linkages, patterns, or formations that are not necessary, natural, nor economically determined.[34]  Patterns or formations articulated at one moment are not necessarily articulated at another.  Through discourse and images, certain relations in a historical moment become linked or united.  Stuart Hall argues:

> the term [articulation] has a nice double meaning because "articulate" means to utter, to
> speak forth, to be articulate.  It carries the sense of language-ing, of expressing, etc.  But
> we also speak of an "articulated" lorry (truck): a lorry where the front (cab) and back
> (trailer) can, but need not necessarily, be connected to one another. . . . It is a linkage which
> is not necessary, determined, absolute and essential for all time.[35]

'Language-ing' or 'image-ing,' then, construct links or patterns among technology and social relations in the home.

Social relations of gender, work and leisure, and public and private were commonly articulated with home technologies in modern discourse and images.  The modern theme or discourse formation specifically linked technology with scientific technique, standardization, and efficiency as well as with clearly differentiated gender, work and leisure, and public-private relations.  Everything had a demarcated social place in modern discourse based on sets of oppositions.  The particular articulations of technologies and these social relations were neither natural nor mirror reflections of some underlying or objective reality; they were cultural and rhetorical constructions.

## "PLEASE . . . LET YOUR WIFE COME INTO THE LIVING ROOM!"

The term gender calls attention to the culturally constructed character of human practices and values which seem to be natural: masculinity and femininity, sexuality, and family. Gender means the social practices of and relations between man and woman. Furthermore, I use the terms 'man' and 'woman' to refer to social categories in discourse rather than the experiences of historical or living men and women.[36] Gender was the major social relation inscribed onto home technologies in the modern discourse theme.

The central figure in modern discourses and images was the white woman as housewife and mother. The housewife was not simply a role, but a position in the social relations of marriage and family. Images of woman sweeping carpets, scrubbing floors, cooking dinner, and washing clothes articulated very specific practices and values of technology's place in maintaining a happy, healthy, nuclear family. Vacuum cleaners, stoves, and washing machines were said to "do the work" while woman gave her time to operate them in the love of her children or husband. It was through the relations of marriage that woman and technology were meaningfully linked in discourse.

Woman as housewife and mother served as foundation or mediator of technological systems in the home. Housework and communication technologies intersected through woman; they were a system.[37] In 1915, a woman united all of Western Electric's devices as she ironed, made breakfast, heated baby's bottle, sewed clothes, and talked on the telephone to her friend, Ethel. As she told Ethel, "My housework! Oh that's all done. . . . I just let electricity do my work nowadays." Her husband, Jack, approved: "Jack says they are the best that are to be had [*sic*].[38] The specific technologies changed, but the housewife grounded the system, as pictured, for example, in a 1942 public relations photograph where a woman used an electric blender and stove to make a cake while talking on the phone and caretaking.[39] The housewife at the center of multiple technologies was the most pervasive figure of modern images and discourse.

Other images of woman also were linked with home technologies. In contrast to the housewife, images of the flapper articulated newly attained political and sexual freedoms with technology in the 1920s.[40] The flapper—the modern symbol of choices and style—did not unite or mediate technological systems. Instead, she was the human embodiment of the shape, form, style, or color of specific technologies such as vacuum cleaners, radios, telephones, and cameras. In "meeting the modern demand for COLOR in bathroom fixtures," for instance, the thin body and robe of the long-legged, short-haired woman in the Montgomery Ward ad complemented the line, pattern, and color of the bathroom design, and in Hoover's promotion of "more d.p.m.," where d.p.m.

stood for dirt per minute, the shape and posture of the woman pushing the vacuum modelled the shape and form of the Hoover.[41] The flapper embodied a modernist aesthetic and linked modernity with technology.

The young female child also appeared in public discourses and images on home technology. In a Maytag washer ad headlined, "She will buy a new Maytag before her mother needs another," one-half of the full-page copy is a photo of the back of a young girl looking over her shoulder at the reader. The text placed the child in a genealogy of female use of washing machines by claiming, "We have hundreds of three-generation Maytag families on record where the original machine (grandmother's) is 30 to 40 years old and still working."[42] The uses and social memories of laundering were written through the lineage of the mother.

Discourses and images of a lone man or group of men rarely appeared in conjunction with home technologies with one notable exception: television. Man relaxing before the television set was a major figure of the 1950s. This does not mean that man was absent from the social relations of home technology. Rather, man was constructed in the social positions of suitor in courtship scenes or as family breadwinner. As breadwinner, man's pleasures were claimed to derive from his wife's use of technological devices. A Hotpoint dishwasher ad pleaded, "Please . . . let your wife into the living room!" The family wage-earner could save his wife from "kitchen exile" by purchasing the automatic dish-washer: "It also protects your family's health by doing dishes the sanitary way. . . . And it saves your wife at least an extra hour every day . . . that she can devote to happier homemaking," such as watching tv with husband and the kids.[43] Although man did not use housework technologies, he typically purchased, surveyed, and approved of their use by his wife. A spotless house, clean and pressed clothes, good cooking, healthy children, and a beautiful wife were among the many pleasures man would gain from woman's use of technological devices. Romantic and sexual relations between man and woman were woven through home technologies. Labor-saving devices, it was claimed, saved woman's physical energy and time so that she could beautify herself for her husband or engage in activities to keep her mind interesting for him. A Simplex ironer ad asked:

> Are you still the attractive, alert, up-to-date woman he married? Are you keeping up with the interesting things in life as he is, or are you devoting all your time, strength, and thought to housework?"[44]

Simplex eliminated work and gained hours for woman so that she could devote herself to companionship for her husband.

Couples pictured in romantic home settings gave a sexual allure to technologies. Images of elegant romance, for example, surrounded radio receivers and television sets as sources of entertainment. Freed-Eisemann, the "radio chosen by America's Aristocracy," portrayed spectacle, fashion, and romance in a ballroom of one of "America's Finest Homes."[45] In other displays, young couples sat entwined before the broadcast speaker or screen.[46]

The nuclear family presented the most fundamental relations for situating technology's place in the home. According to Roland Marchand, advertising used the "family circle" to introduce many commodities into the home during the 1920s. Images of father, mother, and children formed a circle into which commodities were inserted as additional family members.[47] Variations of the family circle extended through the 1950s, particularly with respect to television sets. In dozens of ads, the screen of the set displayed a human face or a pet as joining the family via television.[48] This humanizing of the set was one way to quell anxieties about television's potential threat to the family given the larger public debates about broadcasting's ability to bring the outside world into the home.

In modern discourse and images, no single or unmarried persons owned or operated technology. No single parents or families other than white ones appeared. Moreover, the white woman did not work outside of the home, especially for a wage. Discourses and images, therefore, inscribed technologies with particular gender positions in white family relations.

As Ann Gray argues, "[w]hen a new piece of technology is purchased or rented, it is often already inscribed with gender expectations."[49] Specific gendered practices and values place technologies in larger social organizations between man and woman. Modern discourse offers clear evidence of Lana F. Rakow's argument that "technologies in a gendered society are not gender-neutral."[50]

## "A MAN'S CASTLE IS A WOMAN'S FACTORY"

Human work and leisure in the home were entangled with gender relations. Work in the home meant woman's work whether it was called management or homemaking, and leisure was constructed with very different meanings for man and woman.

The most fundamental term used for work in the home was "housework." As Cowan notes, however, woman's work only came to be called "housework" in the twentieth century. "Housework" became associated with household tasks in an industrialized economy.[51] Housework could be replaced, home economists claimed, if the housewife applied scientific principles and used efficient, labor-saving technologies. Advocates believed that "the housewife should separate her managerial tasks from her physical labor"[52] and that, with her supervision, technology should do the labor. Indeed, the most widely used term for ranges, irons, vacuum cleaners, and dishwashers was "labor-saving device." Time and labor-saving devices, however, were misnomers. While they eliminated drudgery, these technologies actually changed time and labor in the home rather than saved time and labor.[53] Furthermore, the keyword of efficiency was linked to a particular organization of domestic work—individualized housework. When housework was associated with women performing the same tasks by themselves in their individual homes, larger social and economic efficiency was occluded. And, as we now know, scientific home management was an abysmal failure. Among the many reasons for its failure was that the discourse of scientific management from industry was grafted onto housework in the individual home without the actual division of manager and laborer positions as in industry.[54] The sole housewife could not be both.[55]

Class relations of work spread throughout the discourse and imagery on home technologies. The housewife was manager; she employed and supervised the devices that labored. Keywords and imagery for many household technologies included maids and servants. KitchenAid was the "electric maid"—"never sick—never tired—never takes a day off—never gives notice."[56] Edison Lamp Works of General Electric advertised "electric servants" for housework in the years of dwindling domestic service:

> Housework is hard work—and the problem of help in the home is growing more and more acute. But there's a way to simplify both the work and the problem—a way surprisingly easy and inexpensive. And here it is: Electric servants can be depended on—to do the muscle part of the washing, ironing, cleaning, and sewing.[56]

All that the housewife had to do was to turn on the servants and stand at the center of the technological system.

Racial discourses blurred with class discourses in the modern theme. Photos or drawings of the personal maid in crisp uniform or the frumpy washerwoman often completed the imagery of technology ads, and frequently the maids were black.[57] Two black maids talked, in a racist construction of black dialect, over a backyard fence in one 1934 Hoover ad:

"Mah Lady Gives Me Sundays Off"

"Yeah? Well, I Gets Part of Every Day Off. Mah Folk's Got the Hoover."[58]

The white employers owned the technology; the black maid labored with it. The black woman as servant typically had one of two relations to technology. The black woman either operated the technology that the white family or woman owned or the technology was said to replace the unreliable or inefficient black servant.[59] Technologies were embedded in social relations of class and race as well as gender.

Work and leisure were constructed very differently for man and woman as clearly represented by GE. In an ad titled "A Man's Castle is Woman's Factory," the visuals showed a woman in an apron washing dishes at the kitchen sink as a man smoking a pipe sat in the living room reading a book. The appeal began, "Men are judged by what they accomplish—women by the houses they keep . . . [so] modernize with the 10 best home servants."[60] In general, husbands and fathers would benefit from the housewife using labor-saving devices because she would have more time to beautify, develop herself as an interesting companion, and be a more devoted mother to the children, things asserted to be desirable by man.

Household technologies supposedly saved time and energy for woman so that she could have more leisure. Woman's leisure, in other words, was often constructed in the same text as her work. According to magazines and advertisements, leisure typically meant spending time with children, homemaking, beautifying, or shopping. Caretaking, home decoration, and shopping were things woman liked to do, not jobs or work. Technologies were seldom linked with "getting away from it all" for woman.

While male leisure was assumed or taken-for-granted in discourse about housework, it was explicitly articulated with broadcasting technologies, particularly the television set. Man wholeheartedly relaxed before the television, whether watching sports or ogling attractive women.[61] Images of leisure included smoking a pipe, lounging in a chair, laughing with friends, or simply watching a program.[62]

TV and man's leisure were even configured in advertisements for commodities other than television sets.[63] On the other hand, woman rarely watched television alone or purely for leisure. Instead, woman sat with a man and enjoyed tv as a couple or woman cared for children in front of the set.[64] In both cases, woman's leisure or caretaking with television was constructed through her courtship or marital relations.[65]

Radio and television were technologies of leisure for man, couples, and family, but were technologies linked with work for woman as mother and housewife. The development of radio soap operas in the 1930s might identify the basis of this linkage. Listening to daytime radio programming sponsored by soap manufacturers became part of woman's daily housework routine. She was the target of messages about particular products as she did the laundry, washed dishes, and mopped floors. Radio became a companion to the woman isolated in her individual or suburban home. Instead of being a source of leisure, radio in the home accompanied many housework jobs, such as sewing.[66] Daytime television later became linked to housework for women. The rhythms of soap opera narrative, game shows, and commercials match the fragmented, interruption-oriented, and never-ending patterns of housework and childcare. In an important sense, daytime "television and its so-called distractions, along with the particular forms they take, are intimately bound up with woman's work."[67]

The telephone, so long believed to be the technology of woman's idle gossip, was also part of woman's work. The housewife needed a phone for the kitchen, her "office at home." According to Bell Telephone Systems, a phone would be helpful "in the kitchen, where you do so much of your work."[68] The telephone work conducted by housewives and mothers included scheduling social occasions, making appointments with doctors, plumbers, and music teachers, and conveying birthday greetings.[69] In a contrasting Bell System ad, the telephone was linked with leisure:

> In many residences, the dressing room suggests itself as an appropriate location. A telephone here not only saves steps and time, but tends to prevent annoying delays when one is preparing for bridge, travel or the theatre.[70]

Similar to the personal maid and dressing room vanity, a telephone also signified luxury.

In modern discourse, home technologies did not have an essential function outside of social relations of work and leisure. Vacuum cleaners were technologies of work only for women and television was leisurely entertainment primarily for men. Technologies of work, communication, and entertainment fulfill such functions only in specific gender, work, and leisure relations.

## HIS OFFICE IN TOWN, HER OFFICE AT HOME

Public and private constituted large oppositions in modern discourse and each had significance in relation to the other. As discursive terms, public and private constructed two diverse, if not contradictory, messages. Either technology in the home was said to be analogous to technology in the industrial, political, or public world, or technology in the home was to be private and separated from the public.

A technological device often was said to have a place in the individual, private home based on a comparable device or counterpart in the factory, office, or political world. Technology and its use in a public space was said to be a model for technology and its use in the home. Distinct and separate spheres in discourse provided an opposition through which technology in the home could be meaningful and, in many cases, this clearly pertains to the fact that technologies were developed for commercial use and later adapted to the home.[71] For decades between the 1920s and 1960s, the Bell System argued that a woman needed a telephone for every room in her home as her husband had for every desk in his office. "Madam! Suppose you traded jobs with your husband?" asked a 1956 Bell Telephone System ad in which an image of a man in a business suit held a crying baby and a stack of dishes while a woman sat behind an office desk talking on the phone. The text continued:

> You can just bet the first thing he'd ask for would be a telephone in the kitchen. You wouldn't catch him dashing to another room every time the telephone rang or he had to make a call.
> He doesn't have to do it in his office in town. It would be mighty helpful if you didn't have to do it in your "office" at home.[72]

An "office in the kitchen" is analogous to an "office in town" where telephones were said to be widely available for use in answering and placing calls. A telephone in the house was assumed; the problem was how many phones, in which rooms, and in what colors. The office provided a space in discourse to construct meaningful places for more home phones.

In the late 1910s, General Electric ran a campaign in *Ladies' Home Journal* about the dependability of its motors for use in both public and private realms. One ad claimed that the housewife could depend on GE in her household just as the United States government, military, and business could depend on GE in Panama during that period of international crisis:

The same engineering pre-eminence that electrified the Panama Canal is concentrated in the little motor that makes WASHING MACHINES dependable.[73]

In another ad in the same campaign, GE claimed to be as dependable in the household as in the steel industry:

The same technical skill that has provided America's steel industry with electric motors of 10,000 horsepower has produced the motors that make HOUSEHOLD APPLIANCES dependable.[74]

Discourses about public and private were linked with keywords of "utility," "convenience," and "time and labor saving devices."

And articles about kitchen appliances used terms such as industrial and private to clarify which uses, practices, or values were being discussed. For example, the article, "The Electrical Kitchen for Private Houses," illustrated problems and disappointments about electrical appliances for use in the individual house which were not discussed by manufacturers' handbooks.[75]

On the other hand, ambivalence about technology's place in the home can be seen by another modern discourse and image. Technologies were also smuggled into the home as pieces of furniture. From refrigerators to television sets, technologies were envisioned as pieces of furniture rather than as mechanical and electrical apparati. Wires, tubes, and other components that made refrigeration or broadcasting possible were concealed inside of a cabinet. Advertisements often pictured interior home spaces such as kitchens or living rooms with the technology as a piece of furniture set in place. When displayed as fine furniture, phonographs, radio receivers, and television sets visually fit into the living room without calling attention to the technological apparatus which, in certain arguments, threatened privacy.[76] According to some interior design articles and advertisements, television sets even appeared to have no screen.[77] In other cases, images of refrigeration and broadcasting apparati accompanied images of the full appliance.[78]

Complex debates about radio and television broadcasting focused on relations between public and private. In its most basic form, debates about radio turned on broadcasting's new, modern dimension of "bringing the outside world into the individual home."[79] The place of the individual radio receiver was thoroughly inscribed with social tensions in the historical context of utopian and distopian arguments about such issues as commercialism, state regulation, moral uplift, and centralized transmission with mass or passive reception.[80] Later debates about television broadcasting also were set in terms of complex relations between public and private. Industry trade journals, interior design literature, and even television programming constructed a social space for the television set in conflicting relations between personal and communal, self-sufficiency and mobility, and domesticity and theatricality.[81]

In the discourses which displayed, advertised, and designed technologies for the home, public and private constituted meaningful relationships for creating technology's place in the  home.  Public and private were held in opposition or tension as relational terms in discourse, not as separate spheres.

## CONCLUSION: POSTMODERN DISCOURSE

Widely circulated discourses and images constructed ways of seeing, knowing, and valuing technology in terms of social relations. In discourses on gender, work and leisure, and public and private, modern messages articulated technologies with particular social practices and values. Repeated themes in the modern formation constituted everyday knowledges about who owned technologies, who operated them, for which purposes, and in what interests. Discourse and images that articulated technology's social places and uses demonstrate that "the most effective ground of ideology is not the domain officially defined as 'politics,' but rather the domain of everyday life [such as the home and the media.]"[82]

In 1983, the Bell System announced, "The Information Age is coming to your home. And your home will never be the same again."[83] The information age, also known as post-industrial or postmodern society, promises to bring both new technologies and discourse about them to the home. Modern linkages seem to be breaking down and emerging keywords include networking, programming, and interfacing. Cable television stations now urge viewers to "shop at home" and realtors advise sellers to put "your home on tv." Young black couples now display trash compactors; television sets with VCRs are promoted as kitchen necessities. The modern discourse theme or formation has not disappeared. Daytime television commercials still organize messages around the central image of the lone housewife with a cleaning device. Modern terms are even used to make sense of new communication technologies. The computer, for instance, has been called a labor-saving device.[84] Nonetheless, discourse on technology is changing the site of home in more ways than one. I can feel it as I finish these words composed on my personal computer. Engraved on my keyboard is a keyword: HOME.

## NOTES

1 Christopher Lasch, *Haven in a Heartless World* (New York: Basic Books, 1977).

2 Jean Baudrillard, "The Ecstasy of Communication," in *The Anti-Aesthetic: Essays on Postmodern Culture,* ed. Hal Foster (Port Townscend: Bay Press, 1983), 128.

3 Ruth Schwartz Cowan, "'The Industrial Revolution' in the Home: Household Technology and Social Change in the 20th Century," *Technology and Culture* 17 (1976): 1-23 and *More Work for Mother: The Ironies of Household Technology from the Open Hearth to the Microwave* (New York: Basic Books, 1983).

4 Marshall McLuhan, *Understanding Media: The Extensions of Man* (New York: Signet, 1966); Everett M. Rogers, *Communication Technology: The New Media in Society* (New York: The Free Press, 1986); and Raymond Williams, *Television: Technology and Cultural Form* (New York: Schocken, 1975).

5 Siegfried Giedion, *Mechanization Takes Command* (New York: Oxford University Press, 1948); Judith McGaw, "Women and the History of Technology," *Signs* 7 (1982): 798-828; Susan Strasser, *Never Done: A History of American Housework* (New York: Pantheon, 1982); and Delores Hayden, *The Grand Domestic Revolution* (Cambridge: MIT, 1981).

6 Cheris Kramarae and Paula A. Treichler, *A Feminist Dictionary* (New York: Pandora, 1986), 244.

7 Cheris Kramarae, "Gotta Go Myrtle, Technology's at the Door," in *Technology and Women's Voices: Keeping in Touch*, ed. Cheris Kramarae (New York: Routledge and Kegan Paul, 1988), 2-3.

8 See, for example, Joan Rothschild, ed., *Machina Ex Dea* (New York: Pergamon Press, 1983); Gena Corea, *The Mother Machine: Reproductive Technologies from Artificial Insemination to Artificial Wombs* (New York: Harper and Row, 1985); Teresa de Lauretis, *Technologies of Gender: Essays on Theory, Film and Fiction* (Bloomington: Indiana University Press, 1987); Donna Haraway, "A Manifesto for Cyborgs: Science, Technology, and Socialist Feminism in the 1980s," *Socialist Review* 30 (1985): 65-107; and Aihwa Ong, "Disassembling Gender in the Electronics Age," *Feminist Studies* 13 (1987): 609-262. Lana F. Rakow has provided a useful review of feminist analyses of technology and their implications for understanding gender and communication in "Gendered Technology, Gendered Practice," *Critical Studies in Mass Communication* 5 (1988): 57-70.

9 Kramarae, *Technology and Women's Voices*. Also see, Helen Baehr and Gillian Dyer, eds., *Boxed In: Women and Television* (New York: Pandora, 1987).

10 T. J. Jackson Lears, "The Concept of Cultural Hegemony: Problems and Possibilities," *American Historical Review* 90 (1985): 577. For different views on reification, see Roland Barthes, *Mythologies*, trans. Annette Lavers (New York: Hill and Wang, 1972) and Anthony Giddens, *Central Problems in Social Theory* (Berkeley: University of California Press, 1979).

11 Joan Kelly, "The Doubled Vision of Feminist Theory," *Feminist Studies* 5 (1979): 216-227; Michele Barrett, *Women's Oppression Today* (London: Verso, 1980); Nannerl O. Keohane, Michelle Z. Rosaldo, and Barbara C. Gelpi, eds., *Feminist Theory: A Critique of Ideology* (Chicago: University of Chicago Press, 1982); Teresa de Laurentis, ed., *Feminist Studies/Critical Studies* (Bloomington: Indiana University Press, 1986).

12 "Mass with Class," *HFD: The Weekly Home Furnishings Newspaper*, 23 January 1989, special supplement.

13 Jürgen Habermas, "Modernity—An Incomplete Project," in *The Anti-Aesthetic*, ed. Hal Foster, 3-15, and Andreas Huyssen, "Mapping the Postmodern," *New German Critique* (1984): 5-52.

14 Raymond Williams, *Culture and Society: 1780-1950* 2nd ed. (New York: Columbia University Press, 1983) and *Keyword: A Vocabulary of Culture and Society* (New York: Oxford University Press, 1983).

15 Michael McGee, "The Ideograph: A Link Between Rhetoric and Ideology," *Quarterly Journal of Speech* 66 (1980): 1-16.

16 Robert Atwan, Donald McQuade, and John W. Wright, *Edsels, Luckies, and Frigidaires* (New York: Dell Publishing, 1979), 124. "Reach out and touch someone" is a 1980s AT & T advertising campaign.

17 Ellis W. Hawley, *The Great War and the Search for a Modern Order* (New York: St. Martin's 1979).

18 Roland Marchand, *Advertising the American Dream: Making Way for Modernity, 1920-1940* (Berkeley: University of California Press, 1985).

19 Atwan, *et al.*, 22.

20 *House Beautiful* 92 (December 1950): 77.

21 Christine Frederick, *Selling Mrs. Consumer* (New York: Business Bourse, 1929).

22 Christine Frederick, *Household Engineering: Scientific Management in the Home* (Chicago: American School of Home Economics, 1920); Christine Frederick, *New Housekeeping* (Garden City, NY: Doubleday, Page, 1912); Lillian Gilbreth, *The Homemaker and Her Job* (New York: D. Appleton, 1927); Lillian Gilbreth, Orpha Mae Thomas, and Eleaner Clymer, *Management in the Home* (Chicago: American School of Home Economics, 1954); Barbara Ehrenreich and

Deidre English, *For Her Own Good: One Hundred Fifty Years of Experts Advice to Women* (Garden City: Anchor Press, 1978); and Alice Kessler-Harris, *Women Have Always Worked* (Old Westbury, New York: Feminist Press, 1981).

23 Marchand, 91.

24 William Boddy, "The Rhetoric and Economic Roots of the American Broadcasting Industry," *Cine-tracts* 2 (1979): 42.

25 Atwan, *et al.*, 132.

26 *Life*, 29 October 1951, 46-47.

27 *Life,* 15 October 1951, 31.

28 Marchand, 117-163.

29 Marchand, 273, 282 and *Good Housekeeping*, July 1928, 123.

30 Marchand, 124-125.

31 My argument should not be confused with another important claim often made about sociality and commodities in the twentieth century. It has been argued that advertising creates a social setting or function for products rather than focusing on the product itself. For example, people are scrutinzed by others for having dandruff, dirty floors, or aging skin in the attempt to sell shampoo, floor cleaners, or lotion. See Marchand; Stuart Ewen, *Captains of Consciousness* (New York: McGraw Hill, 1976); T. J. Jackson Lears, "Some Versions of Fantasy: Toward a Cultural History of American Advertising, 1880-1930," *Prospects 9*, ed. Jack Soltzman (New York: Bert Franklin, 1982): 349-342; and William Leiss, Stephen Kline, Sut Jhally, *Social Communication in Advertising* (New York: Methuen, 1986).

32 Catharine Beecher and Harriet Beecher Stowe, *The American Woman's Home* (New York: J. B. Ford, 1869) and Kathryn Kish Sklar, *Catharine Beecher: A Study in American Domesticity* (New York: Yale University Press, 1973). On women's history and the rhetoric of separate spheres, see Linda K. Kerber, "Separate Spheres, Female Worlds, Woman's Place: The Rhetoric of Women's History," *Journal of American History* 75 (1988): 9-39.

33 Kelly, 222.

34 Chantal Mouffe, "Hegemony and Ideology in Gramsci," in *Gramsci and Marxist Theory*, ed. Chantal Mouffe (Boston: Routledge and Kegan Paul, 1979): 168-204; Stuart Hall, "Signification, Representation, and Ideology: Althusser and the Post-Structuralist Debates," *Critical Studies in Mass Communication* 2 (1985): 91-114; and Lawrence Grossberg, ed., "On Postmodernism and Articulation: An Interview with Stuart Hall," *Journal of Communication Inquiry* 10 (1986): 45-60.

35 Hall cited in Grossberg, 53.

36 The distinction between woman as discursive construct and women as real historical subjects is drawn from Teresa de Lauretis, *Alice Doesn't: Feminism, Semiotics, Cinema* (Bloomington: Indiana University Press, 1984), 5-6.

37 For a study of women completing communication systems in the workplace, see Cheris Kramarae and Lana Rakow, "'After Awhile You Become Part of the Machine:' Women's Work as Communication Technology." Paper presented to the National Women's Studies Association, June 1986.

38 Atwan, *et al.*, 132.

39 Giedion, 547.

40 Martin Pumphrey, "The Flapper, the Housewife and the Making of Modernity," *Cultural Studies* 1 (1987): 179-194.

41 Marchand, 125 and *Good Housekeeping*, July 1928, 123.

42 Julian L. Watkins, *The Best Advertisements from Reader's Digest, 1955-1961* (New York: Random House, 1962), 114.

43 *House Beautiful* 92 (December 1950): 77.

44 Atwan, *et al.*, 15.

45 Marchand, 195.

46 Marchand, 252.

47 Marchand, 248-254.

48 *Better Homes and Gardens*, June 1948, 32-33; *Better Homes and Gardens*, October 1948, 140; *Saturday Evening Post* 27 May 1950, 5; and *Life*, 15 October 1951, 176.
49 Ann Gray, "Behind Closed Doors: Video Recorders in the Home," in *Boxed In*, eds. Baehr and Dyer, 42.
50 Lana F. Rakow, "Women and the Telephone: The Gendering of a Communications Technology," in *Technology and Women's Voices*, ed. Cheris Kramarae, 224-225.
51 Cowan, *More Work*, 42.
52 Kessler-Harris, 41.
53 Joann Vanek, "Time Spent in Housework," *Scientific American* 231 (November 1974): 116-120; Cowan, *More Work*; and Ewen.
54 Bettina Berch, "Scientific Management in the Home: The Empress's New Clothes," *Journal of American Culture* 3 (1980): 440-445.
55 For a discussion of the attempts and failure of alternatives to individualized housework (such as communal kitchens), see Hayden.
56 *Good Housekeeping*, June 1928, 143.
57 Strasser, 77.
58 Atwan, *et al.*, 96.
59 Atwan, *et al.*, 14.
60 Marchand, 188.
61 Lynn Spigel, "Installing the Television Set: Popular Discourses on Television and Domestic Space, 1948-1955," *Camera Obscura* 16: 47.
62 *Life*, 24 September 1951, 4.
63 *Newsweek*, 3 May 1948, 75 and Atwan, *et al.*, 78.
64 *Newsweek*, 12 December 1949, 20; *Life*, 15 October 1951, 176; and *Life*, 13 October 1952, 116.
65 For a study of discourses on gender and television, see Karen E. Altman, "Television as Gendered Technology: Advertising the American Television Set," *Journal of Popular Film and Television* 17 (1989): in press.
66 Cheris Kramarae, "Talk of Sewing Circles and Sweatshops," in *Technology and Women's Voices*, ed. Cheris Kramarae, 156-157.
67 Tania Modleski, "The Rhythms of Reception: Daytime Television and Women's Work," in *Regarding Television*, ed. E. Ann Kaplan (Frederick, Maryland: American Film Institute and University Publications of America, 1983): 74.
68 Atwan, *et al.*, 76.
69 Rakow, "Women and the Telephone," 217.
70 Marchand, 119.
71 Cowan, and Heidi I. Hartmann, "Capitalism and Women's Work in the Home, 1900-1930," Ph.D. Diss. Yale University, 1974.
72 Atwan, *et al.*, 70.
73 *Ladies' Home Journal*, July 1918, 2.
74 *Ladies' Home Journal*, August 1918, 2.
75 Giedion, 546.
76 Atwan, *et al.*, 131; Marchand, 91; *Better Homes and Gardens*, March 1953, 164-65.
77 W. W. Ward, "Is It Time to Buy Television?" *House Beautiful* 90 (October 1948): 169-173 and *Newsweek* 23 July 1951, 59.
78 Edgar R. Jones, *Those Were the Good Old Days* (New York: Simon and Schuster, 1959); Atwan, *et al.*, 137; *Life*, 13 October 1952, 50-51; and *Life*, 24 September 1951, 125.
79 Daniel Czitrom, *Media and the American Mind* (Chapel Hill: University of North Carolina Press, 1982), 60.
80 Czitrom and Mary S. Mander, "Utopian Dimensions in the Public Debate on Broadcasting in the Twenties," *Journal of Communication Inquiry* 12 (1988): 71-88.
81 Williams, *Television* and Spigel.
82 John B. Thompson, *Studies in the Theory of Ideology* (Berkeley: University of California Press, 1984), 83.
83 *National Geographic*, May 1983, n.p.
84 Joan Libman, "Why We Overwork," *Los Angeles Times*, 13 June 1988, V, 1-2.

Part Three

# Ideology and Language

# Experts, Rhetoric, and the Dilemmas of Medical Technology: Investigating a Problem of Progressive Ideology

*Michael J. Hyde*

The purpose of this essay is to address a question that is receiving much attention in the current literature of rhetorical scholarship. As recently stated by Walter Fisher: "What happens when 'experts' argue about moral issues in public?"[1] The raising of the question is not fortuitous. Rather, it reflects an attempt to further a critical understanding of a recurring symptom of our technological society that shows itself in the sociopolitical life-world. This symptom is the emergence of a "society of experts" whose acquired technical knowledge grants them authority and thus enables them to demonstrate more than a modicum of influence and power when arguing about matters that are perceived by the experts to be somehow related to their respective professional capacities. Before going on to state specifically how I intend to approach the question, it first will be helpful to offer some brief remarks concerning its historical underpinnings. These remarks will introduce various issues that are crucial to the subsequent analysis.

## A PROBLEM OF PROGRESSIVE IDEOLOGY

The "right" of people to question the legitimacy of authority is one that arose with the birth of classic democratic theory in the eighteenth century. Then, the right was used to challenge the power and influence of the King; now, the right is being used to make "experts" accountable for what they say and do.[2] Open discussion and debate are the means for achieving this goal; they define rhetorical practices that serve to concretize the theory of democratic politics. These practices help to insure that the private motivations and interests of individuals who claim authority will be disclosed to those who, according to classic democratic theory, constitute the sovereign collective known as "the public." Authority, so the theory goes, must always seek its justification by first offering a persuasive appeal to this collective. Such an appeal places a restriction on the subordinating tendencies of authority. For, as Hannah Arendt observes, persuasion "presupposes equality and works through a process of argumentation. Where arguments are used, authority is left in abeyance."[3] If authority is to play a role in the institutions of democracy, then those who profess to be authorities on matters that affect the life of these institutions must align their professions with the will of the people. Or, as C. Wright Mills states it: "the political structure of a democratic state requires the public; and, the democratic man, in his rhetoric, must assert that this public is the very seat of sovereignty."[4]

Like Mills, however, many observers of the historical fate of classic democratic theory are quick to point out that the theory no longer offers even an "approximate model" of the reality of political and economic systems founded on democratic principles. Consider, for example, Mills's 1956 assessment of the American situation:

> The issues that now shape man's fate are neither raised nor decided by the public at large. The idea of the community of publics is not a description of fact, but an assertion of a legitimation masquerading—as legitimations are now apt to do—as fact. For now the public of public opinion is recognized by all those who have considered it carefully as something less than it once was.[5]

In a summary of his consideration of the matter, Mills goes on to note:

> In economic and political institutions the corporate rich now wield enormous power, but they have never had to win the moral consent of those over whom they hold this power. Every such naked interest, every new, unsanctioned power of corporation, farm bloc, labor union, and governmental agency that has risen in the past two generations has been clothed with morally loaded slogans. For what is *not* done in the name of the public interest? As these slogans wear out, new ones are industriously made up, also to be banalized in due course. And all the while, recurrent economic and military crises spread fears, hesitations, and anxieties which give new urgency to the busy search for moral justifications and decorous excuses.[6]

What Mills saw in 1956 was an outcome of certain cultural and ideological changes that took form in the late nineteenth and early twentieth centuries when the scientific and technological advances of the industrial revolution began to transform the rural values and ways of life characterizing pioneer America. This transformation coincided with a Progressive, as opposed to a Populist, view of democracy. That is, with the intensification and growing sophistication of science and technology—a phenomenon which led American society to become more urbanized and its division of labor more pronounced—the general public found itself becoming more accustomed to relying on the specialized skills and educated judgments of professionals. Paul Starr identifies the rationale motivating this reliance when he notes: "The less one could believe one's own eyes—and the new world of science continually prompted this feeling—the more receptive one became to seeing the world through the eyes of those who claimed specialized, technical knowledge, validated by communities of their peers."[7]

Such receptivity performed a legitimating function with respect to the claims of the professions to competent authority. Allowing its expectations for a better and just society to be counseled by a Progressive ideology that promoted science as the means of moral and political reform, the public showed a willingness to make room for a "culture of expertise" and thus to share its sovereignty with an emerging professional class.[8] In this act of sharing, the public acknowledged its dependence on "elites" who purported to serve its interests. Professional authority, once considered to be antithetical to the traditional democratic values of self-reliance and individual common sense, was now given a license to achieve privilege, power, and prestige. These achievements were the rewards for those who could instigate progress through science and technology. The public need only act in accordance with this progress to have its expectations fulfilled. Perhaps the spirit of the times was stated best in the guidebook for the Chicago World's Fair of 1933: "Science finds—Industry applies—Man conforms."[9]

Robert Heilbroner once observed that "simplistic ideas of progress see only the near face of events when they look to the future. Hence such views of the future typically underrate its complexities. They do not consider that the solution of one problem is only the formulation of the next."[10] Was the Progressive ideology that established science and technology as the means for directing American democracy such a simplistic idea? Mills's assessment of the American situation in 1956 surely offers an affirmative answer to this query. For what he found then was a system of power weighted so heavily on the side of professional authority/sovereignty that "the public of public opinion" was present in name only. The technical knowledge of experts had advanced to such an extent that its relationship to the social knowledge of the laity was disintegrating. A problem of Progressive ideology had become apparent: the public was being cut off from the power of active decision making; with respect to the claims and requirements of technical knowledge, it no longer displayed the argumentative competence to engage experts in open discussion and debate and thus to make them accountable, in any real sense, for their actions.[11]

Major political, economic, and environmental developments (the Vietnam War, Watergate, the Equal Rights Movement, the oil crisis, Three Mile Island, the continuing threat of nuclear warfare) that arose throughout the 60s and 70s and that confronted the social consciousness of America with dire moral issues, contributed much to an awareness of the problem. The question directing the purpose of this essay—What happens when 'experts' argue about moral issues in public?—is one that reflects a critical interest in such developments, especially as they helped to refocus attention on the problem of Progressive ideology discussed above, or what Fisher describes as "the kind of system in which elites struggle to dominate and to use the people for their own ends or that makes the people blind subjects of technology."[12] This description echoes a common assessment offered by scholars who have discussed this question.[13] The discussions, to be sure, are illuminating; each, in its own way, dramatizes how modern societal rationalization can produce a society of experts who possess the technical knowledge but who lack the practical wisdom to shape a "good society." Furthermore, each discussion strongly motivates its readers to realize that in a democratic society conscientious rhetorical practices are necessary if professional authorities are to be made accountable to the general public such that its members' social knowledge and practical wisdom can play a role in the construction of public policy. I have no disagreements with these suggestions; I do not believe, however, that the assessment on which they are based allows for a sensitive enough understanding of the complexities that confront experts when they attempt to argue about moral issues in public.

In the following analysis I intend to clarify this belief by offering an examination of a professional community whose sovereign status in the twentieth century was made possible by its appropriation of the ways and means of Progressive ideology and, consequently, by its ability to use its scientific and technological expertise for the purpose of meeting the expectations of its clientele. The community I have in mind is that of medicine. Today, this community is being confronted by moral dilemmas that are threatening its reputation. These dilemmas, ironically, are due in great part to the scientific and technological capabilities of medicine, capabilities that have conditioned the members of the medical community to expect a willingness on the part of the public to grant them authority in the management of illness. Due to the dilemmas now confronting it, however, this authority is being challenged by the public; accountability is being demanded. To meet this demand in a democratic manner requires that the members of the medical community engage in public moral argument. Such argument is a "storytelling" activity, a form of rhetorical practice that has as its primary goal the creation of a narrative that can educate a general audience by taking into account its interests and social knowledge.[14]

The medical community, to be sure, is one whose very existence has always depended on the ability of its members to recognize and appreciate the importance of storytelling as a life-promoting activity.[15] Instructed by advances in science and technology, however, medicine's brand of storytelling has become an exercise in the creation and maintenance of technical knowledge, not social knowledge. That is, when telling their colleagues stories about the diseases that plague their patients' lives, physicians are obligated to construct a narrative whose primary purpose is that of making a medical diagnosis and prescribing treatments; the narrativity of these stories must therefore subscribe to the dictates of scientific reasoning.[16] What happens, then, when medical experts attempt to expand their appreciation of the scope and function of storytelling for the purpose of engaging in public moral argument?

By offering an answer to this question, I hope to show how the medical community may provide an illustration of why the problem of Progressive ideology noted above may not lend itself to an easy resolution by merely calling for what unfortunately may be today but another "simplistic idea": the institution of popular sovereignty. I begin the analysis with an examination of both the moral dilemmas that are confronting the medical community and the situation that has taken form because of this confrontation. I next comment on the way physicians typically use stories in the performance of their duties and then provide an examination of a story that was told by a physician who chose to engage in public moral argument. A concluding discussion is offered to suggest some of the implications that follow from my analysis.

## DILEMMAS OF MEDICAL TECHNOLOGY

Throughout the twentieth century, and in response to the public's increased desire for prolonged life, medical science has developed technologies that have greatly expanded the concept of health care. Respirators, bypass surgery, dialysis machines, CT scanners, pacemakers, chemotherapy, organ transplants, and artificial hearts are some of the better known technological advances that have enabled medicine to intensify its legendary "war" against life-threatening disease. Owing to these advances the expectations of both the medical community and those whom it serves have continued to grow.[17] But this success and its attending expectations have created dilemmas regarding life and death that the medical community is hard pressed to resolve. A story can help illustrate this point.[18]

At the age of fifty-eight, Mr. Howard F. Hubbs suffered complete kidney failure. For the rest of his life he would have to live with the help of a dialysis machine. His life was sustained but its quality was tragic: He lost his job, he lost his capacity to function sexually, he abandoned his religious faith, and he developed severe neurotic symptoms. Staring at his blood moving through a machine he continued to ask: "What is going on here? What is happening to my life?" Alone in his house he wrote in a diary that he kept to document personal responses to his deteriorating condition: "I can't die because I must stay to keep up with the 'Jones'. I can't die because I have traditions to carry on. I can't die because I am afraid to." In the next entry he wrote: "Everyone calls, the phone rings and rings with questions and professional advice. But its too, too bad that they can't help—because they don't know my pain—*nor do I*." Due to complications associated with his illness, Mr. Hubbs died four years after beginning his dialysis treatments.

A success of medical science had enabled Mr. Hubbs to live, but it was a living hell, evidencing how sophisticated medical care, despite the best of intentions, can increase levels of pain and suffering (morbidity). Thousands of patients whose lives are sustained by the technological capabilities of medicine experience this fate everyday. The irony here is discomforting. And the discomfort grows when one realizes that the success of medical science in developing its technologies is also related to skyrocketing health-care expenditures, accusations of impersonal medical care, and ethical questions concerning the proper and timely use of life-support systems. Dr. Willard Gaylin summarizes the irony of the situation quite well when he notes that "it is not the deficiencies of medicine that present the medical community with dilemmas, but the successes."[19] Clarification of the dilemmas, especially as they show themselves not as independent variables but rather as interrelated phenomena that together are promoting a serious problem for the health care system, follows below.

Today, for almost any life-threatening condition, some technical intervention is capable of delaying the moment of death. The problem of morbidity, as suggested above, results when the delay is successful, but due to the nature of the life-threatening condition, continual intervention is required to counter the patient's physiological pain and suffering. Then there is the tremendous economic burden that such intervention places on both the patient and the health care system. One hears about this burden, for example, as patients and family members describe their anxieties regarding insurance coverage and as medical personnel detail their efforts to cope with government regulations (i.e., "Diagnosed-Related Groups," or "DRGs") that are attempting to control costs by giving hospitals a financial incentive to limit hospital stays.[20]

Because the success of medical science has made possible a condition (morbidity) that results in pain, suffering, and financial stress, members of the medical community have been forced to scrutinize their own technical capabilities with the goal of judiciousness in mind. Dr. Ralph Crawshaw's findings are not atypical: "By a curious inversion of human values, scientific-therapeutic zeal has become a new, dangerous, iatrogenic disease, the technical fix. Caught up in the wonderment of our burgeoning technology, physicians become blind to the difference between physiology and life.[21] Dr. Lawrence D. Grouse adds to this when he notes: "The fact that the health care provided in the system may be improved as a result of the technology does not have as much impact as the subtle and hidden message that the machine has become the physician: the definitive adviser. The specialist-physician is metamorphosing into a technocrat and businessman. The physician retreats behind the machine and becomes an extension of the machine."[22]

In an age where consumers are becoming more aware of their social and political rights, such admissions of culpability on the part of physicians are eagerly welcomed by patients who feel that their expectations concerning the treatment of their illnesses are not being properly fulfilled. "Improper medical care" is the accusation; malpractice litigation is often the result. In an unsolicited letter sent to this author, one patient who was hospitalized in Chicago for cancer treatments states her case as follows: "Amazingly enough I am beginning to hear about others who have sued, not just for the money or even for the mistakes done to them, but because of the frustrations and anger created by thoughtless words or no words at all. It's amazing how much patients can forgive in deeds, but cannot forgive in words or lack of them."

Morbidity, rising health-care costs, and accusations leading to malpractice suits: that all of this could stem from medicine's success in developing technologies to combat death and disease is certainly ironic; yet, in the world of high-tech medicine the irony has become a reality. In this reality great expectations cannot always be met: Patients die or they linger in states of distress awaiting some words about their futures. And when some words are announced, they oftentimes are spoken by a physician who, aided by the success of high-tech medicine, must hurry on to counsel with other patients who remain alive because of this success. The scene is fast paced; and doctor-patient communication is easily influenced by this tempo of high-tech time. It is not uncommon to see this communication taking place between a patient who does not always know what to ask and a physician who does not always know what to say. Brief remarks or silence can result as the expectations of both interlocutors are transformed by the reality at hand.

Contributing to this transformation of expectations is a dilemma created by one of modern medicine's greatest technological successes—the development and use of sophisticated life-support systems. Medical personnel, patients, and family members who are involved in the decision to use these systems are forced to confront all of the dilemmas discussed so far: Morbidity is always a possible outcome, as is the burden of expense; for once the decision is made the system must be used until the patient recovers, dies, or consents to having the system terminated. But suppose the patient is incapable of making such a choice, suppose the patient is comatose, or too young or too senile to understand the circumstances of the case; and suppose the patient's family members are so distraught that they are incapable of reaching a rational decision at the present time. Here ethical and legal questions come into play: Should the attending physician intervene and risk the chance of some form of eventual prosecution? What set of criteria should the physician employ when determining a course of action? Are the empirical standards of medical science for gauging "survival rates" sufficient enough? Should "quality of life" considerations be given priority over these standards?

The technological capabilities of medicine, as well as the increased expectations that have been fostered by these capabilities, promote the asking of such questions. Medicine, however, is not alone in providing responses. Controversy reigns as religious, political, economic, and legal experts exert their influence and as public opinion polls exhibit contradictory data about whether or not physicians are obligated to confer "technological immortality on dehumanized patients."[23]

Dr. Norman Levinsky has argued that to meet the expectations of those whom they serve, "physicians are required to do everything that they believe may benefit each patient without regard to costs or other societal considerations. In caring for an individual patient, the doctor must act solely as that patient's advocate, against the apparent interests of society as a whole, if necessary."[24] Levinsky's position echoes the egalitarian attitude that has long contributed to both the credibility and the authority of medicine's elite community. Twenty-five years ago or so this attitude reigned supreme in the public mind. Today, however, this attitude is being challenged by dilemmas that are consequences of medicine's success in developing technologies to combat death and disease.[25] The medical community is being asked to resolve this ironic situation; its chief benefactor is demanding an immediate reply. And who is this benefactor? Dr. Lawrence Altman supplies his colleagues with the answer to this question: "Let me remind you that in this post-World War II era, the taxpayer, the public, has become the chief benefactor of medicine. Medicine is now a public institution because it is the taxpayer who is paying for the bulk of the costs of patient care, medical research, and medical education. Members of the medical profession have become public servants. And they are accountable to the public in the same way that other public servants are."[26]

Such accountability creates problems for the members of the medical community. As I will now attempt to show, this problem has much to do with the way in which the medical community uses stories in the performance of its duties.

## TELLING STORIES ABOUT LIFE AND DEATH

In being accountable to the public about the dilemmas fostered by its technological successes, the medical community is thrust into the world of public policy and must therefore engage in public moral argument. Such argument, according to Fisher,

> needs to be distinguished from reasoned discourse in interpersonal interactions and from arguments occurring in specialized communities, such as theological disputes, academic debates, and arguments before the Supreme Court. The features differentiating *public* moral argument from such encounters are: (1) it is publicized, made available for consumption and persuasion of the polity at large; and (2) it is aimed at what Aristotle called "untrained thinkers," or, to be effective, it should be. Most important *public* moral argument is a form of controversy that inherently crosses professional fields. It is not contained, in the way that legal, scientific, or theological arguments are, by subject matter, particular conceptions of argumentative competence, and well-recognized rules of advocacy.[27]

One reason why the medical community has a difficult time engaging in such argument can be illustrated by the following story.

A few years ago I had the opportunity to discuss with twelve, fourth-year medical students how they felt about the dilemmas under consideration here. I began the discussion by telling them the story of Mr. Hubbs. Neither the story nor the rationale for telling it were well received. Objections centered on the "subjective" and "unscientific" nature of the story. One response from a male student was especially revealing:

All of this research [on doctor-patient relationships] being done by Professors and PhDs is really biased . . . it's not science . . . it's subjective. . . . People doing this research are jealous of us. They don't make as much as we do; they don't have our status. And I believe that many of the Professors and PhDs are doing this research because they couldn't make it into medical school.

Why these defensive responses to a story that most certainly was true?

One must understand that medical students are not adverse to being told stories about cases that are relevant to their professional livelihood. As Dr. Eric Cassell points out, "the recitation of cases—telling stories—has been a way to teach medicine that has survived through the ages because nothing else does the job as well."[28] Oftentimes the cases that are recited by physicians are explications of "illness stories" that were told to a physician by a patient during a history-taking session. Such stories, according to Cassell, "are different from other stories because they almost always have at least *two* characters to whom things happen. They always have at least a *person* and that person's *body*."[29] Patients are under no professional obligation to make this distinction when telling the story of their illnesses, when trying to remember all that has been happening to them because of some bodily disorder that they believe is indicative of some disease. Physicians, on the other hand, do have a professional obligation to make the distinction; for only then can they properly diagnose and perhaps cure the diseases afflicting their patients. Thus, of the two characters that motivate the patient's illness story, it is the *body*, the place where a disease unfolds, that must assume priority as a matter of interest to the physician.

In explicating a patient's story to their students and colleagues, physicians are expected to offer a narrative that reflects an understanding of this priority and that does so with as much precision and accuracy as possible. The scientific and technological capabilities of medicine prove invaluable to the physician who must perform this task. By greatly enhancing the physician's ability to discover, diagnose, and then decide what to do about a patient's illness, these technological capabilities enable a physician to construct a story that provides a coherent account of the patient's disease (its pathophysiology) and thus enables the physician's peers to determine whether or not the story reflects an argument that is empirically verifiable and thus scientifically competent. In short, when physicians recite cases for the purposes of instructing their associates about the practice of medicine, they do so within a community whose members require from their teachers an ability to tell a certain kind of story—one whose narrative probability and narrative fidelity reflect an expert appreciation of how science and technology can disclose the workings of the body.

One perhaps can now understand why the medical students reacted defensively when asked to comment on the story of Mr. Hubbs. As told, the story was about a person, not about his body; it emphasized how the scientific and technological abilities of medicine adversely affected this person, not how they helped to sustain his bodily functions; it intimated that medical technology may prolong death, not life; it revealed moral dilemmas, not medical solutions. The story, in other words, did not coincide with the expectations and requirements that have long guided medicine in its telling of illness stories. Rather, it challenged the medical students with a difficult task: it asked them to step outside the boundaries of their training, their expertise, their traditional rationality, and to understand and accept as true a narrative that questioned the capabilities that make possible the students' professional existence. Who, one might ask, would not become somewhat defensive if their livelihood was so questioned, if after all their years of training and dedication their ability to serve the public was doubted?

Importantly, what is being suggested here should not be interpreted as an attempt to justify the medical students' behavior. Their story was told because it exemplifies the situation that the medical community presently finds itself in when it is forced to confront cases like those of Mr. Hubbs—cases that remind the medical community not only about the successes of its scientific and technological capabilities, but also about certain moral dilemmas that, as indicated earlier, are likewise made possible by these capabilities. Cognizant of this situation, Gaylin offers the following example as a way of indicating to his medical colleagues what the situation requires them to realize:

> With the development of advanced surgical techniques, antisepsis, procedures for controlling fluids and electrolytes, and other advances in surgery, we are now capable of performing operations so extensive that, while life-saving, they may be extraordinarily deforming. This factor destroys the comfort and convenience of having an empirical standard to measure outcome, such as survival days. Treatment A has 600 survival days. Treatment B has 300. Therefore Treatment A is twice as good as Treatment B. There was a comfort in the empiricism based on the scientific method. If, however, Treatment A is an operation that literally destroys or removes an entire face, then the empirical standard breaks down. Treatment B may be only two-thirds as effective in terms of survival rates, and yet it allows the patient to have a socially active, productive, and reasonably normal existence. Which treatment should you recommend? . . . What has happened is that technological developments have made us realize that many treatment decisions involve values and attitudes as well as medical judgments. But whose values and attitudes should prevail? The physician must exercise authority in questions of medicine, but not in moral matters disguised by a medical setting and language. We doctors have always assumed that we were not engaged in value judgments. That was never the case. We were always involved, but it was never quite as apparent. We must begin to share these moral decisions with our patients.[30]

What Gaylin is indicating here has important implications for how physicians must understand the issue of accountability when engaging in public moral argument. If physicians are to share moral decisions with their patients/ public, then they must expand their appreciation of the patients' narratives to include what these narratives are saying about "the person." Put another way, if physicians are to play a responsible role in dealing with the moral dilemmas that are currently plaguing both themselves and their patients, then they can no longer afford to be interested only in "the disease in the body in the bed."[31] These suggestions emphasize the need for physicians to establish a dialogical relationship with their patients. For such a relationship increases the possibility that patients will tell and physicians will hear illness stories wherein are contained those values and attitudes that patients consider to be important for their health care. Dr. Stanley Reiser puts the matter this way:

> That which is unique in a patient's illness can often be learned best by nontechnological inquiries based mainly on dialogue. Such inquiries deal with the patient's sensations and perceptions, with the values held by the patient that are relevant to treating the illness. An illness is not only a physical disturbance of the body's structure and function; it is an experience by the patient, who invests the illness with meaning. Knowing what this personal response involved is crucial information for therapy. Therapeutics is successful to the extent that it meets not only the needs of the disease but also the needs of the patient, who modifies and feels the effects of the disease in a unique way.[32]

Social scientists have made much of this point in their investigations of what Candice West describes as "troubles with talk between doctors and patients."[33] Owing to the moral dilemmas now confronting them, however, the troubles that physicians have to experience with their talk extends beyond the confines of doctor's offices and hospital examining rooms: Morbidity, rising health care costs, malpractice litigation, and ethical questions concerning the proper use of life-support systems have become public policy issues. As the chief benefactor of medicine, the public has the right to know about these issues. Accountability in the form of public moral argument thus becomes a responsibility for the medical community. And with this responsibility comes another one — that of being able to translate medical expertise into a narrative that the public of "untrained thinkers" can understand and appreciate.

Is all of this to say that when physicians engage in public moral argument they should tell stories wherein are contained values and attitudes that the public considers to be important for its health care? The narrative perspective offered by Fisher will be helpful, once again, in constructing a knowledgeable answer to this question.

## MEDICINE'S MOVE TO NARRATIVE RATIONALITY AND RESPONSES TO THIS MOVE

According to Fisher,

> the proper role of the expert in public moral argument is that of a counselor, which is . . . the true function of the storyteller. His or her contribution to public dialogue is to impart knowledge, like a teacher, or wisdom, like a sage. It is not to pronounce a story that ends all storytelling. An expert assumes the role of public counselor whenever she or he crosses the boundary of technical knowledge into the territory of life as it ought to be lived. Once this invasion is made, the public, which then includes the expert, has its own criteria for determining whose story is most coherent and reliable as a guide to belief and action. The expert, in other words, then becomes subject to the demands of narrative rationality.[34]

As used here, the term "narrative rationality" points to that which is at work in all forms of storytelling. As Fisher puts it, this ontological sense of rationality "is determined by the nature of persons as narrative beings—their inherent awareness of *narrative probability*, what constitutes a coherent story, and their constant habit of testing *narrative fidelity*, whether or not the stories they experience ring true with the stories they know to be true in their lives."[35] Narrative rationality thus makes room for those stories that may not meet all the requirements of argumentative competence dictated by the rational world paradigm but that still warrant serious attention because of the values they express—values that oftentimes offer people moral guidance when they are forced to confront the ultimate questions "of life and death, of how persons should be defined and treated, of preferred patterns of living."[36]

We have already seen that within the medical community the demands of narrative rationality require a physician to tell a story wherein the "body," and not the "person," assumes priority as a matter of interest. For the public, I would argue, the priority is different. Because of the way it has been influenced by the media to understand and appreciate the workings of medicine, the public expects physicians to tell it stories that are person-oriented.[37] Instructed by this orientation, the public can more readily determine whether the stories it is hearing from the medical community contain values and attitudes that the public deems important. This being the case, an affirmative answer to the question concerning what type of stories physicians ought to tell when engaged in public moral argument should be quite obvious.

In theory, the answer appears sound. But how are physicians supposed to apply this theoretical answer to their world of practice? Should physicians tell stories about people who, like Mr. Hubbs, have suffered from medicine's scientific and technological successes? Must they make sure that when telling such stories they at the same time promote those values (truth, harmony, equality, oneness with the Cosmos) that Fisher emphasizes as being "relevant to the good life"?[38] Which of these values, for example, would help solve the problem about whether or not physicians are obligated to confer technological immortality on dehumanized patients?

Dr. David Hilfiker recently attempted to answer such questions in an essay entitled "Making Medical Mistakes: How Doctors Harm Patients—and Themselves." The essay appeared in *Harper's* magazine (an earlier and very similar version, entitled "Facing Our Mistakes," was published in the *New England Journal of Medicine*).[39] Seven different cases punctuate the narrative, each detailing how the author contributed to the occurrence of iatrogenic disease, or doctor-influenced injuries.[40] All the cases are person-oriented; there is no attempt to mystify the moral dilemmas raised by the cases by disguising them in esoteric medical jargon. The matter of interest here is not the disease in the body in the bed. On the contrary, Hilfiker's goal is that of public moral argument. Thoughtful words are offered as Hilfiker transcends his community's boundary of technical knowledge, assumes the role of public counselor, and talks about medical mistakes that he knows are associated with occurrences of morbidity, rising health care costs, accusations of impersonal medical care, and ethical questions concerning the proper and timely use of life-support systems. But why did Hilfiker choose the sensitive and embarrassing topic of "mistakes" to address such dilemmas?

The topic of mistakes enables Hilfiker to place both the medical community and its critics on common moral grounds:

> Mistakes are an inevitable part of everyone's life. They happen; they hurt—ourselves and others. They demonstrate our fallibility. Shown our mistakes and forgiven them, we can grow, perhaps in some small way become better people. Mistakes, understood this way, are a process, a way we connect with one another and our deepest selves.[41]

Hilfiker does not use this understanding of mistakes as an excuse for medical malfeasance. He does, however, allow it to play an important role in his argument. He notes, for example, that

> because of its technological wonders and near-miraculous drugs, modern medicine has created for the physician an expectation of perfection. The technology seems so exact that error becomes almost unthinkable. We are not prepared for our mistakes and we don't know how to cope with them when they occur.[42]

And the situation becomes even more intolerable, he argues, when one realizes how the public also expects physicians to be perfect because of their scientific and technological capabilities.

> This perfection is a grand illusion, of course, a game of mirrors that everyone plays. Doctors hide their mistakes from patients, from other doctors, even from themselves. Open discussion of mistakes is banished from the consultation room, from the operating room, from physicians' meetings. Mistakes become gossip, and are spoken of openly only in court. Unable to admit our mistakes, we physicians are cut off from healing. We cannot ask forgiveness, and we get none. We are thwarted, stunted; we do not grow.[43]

This last point is especially important to Hilfiker's understanding of mistakes. For what is thwarted when physicians are unable to admit mistakes is their capacity to deal with their emotions in a healthy manner. "Guilt," "anger," and "anxiety" receive the most attention as Hilfiker shares with his readers what he did not share with the persons whose illness stories are being told. One hears how these suppressed emotions inspire feelings of incompetence and self-doubt and how such feelings are oftentimes disguised by an arrogant and all-knowing demeanor. "Little wonder," Hilfiker remarks, "that physicians are accused of playing God. Little wonder that we are defensive about our judgments, that we blame the patient or the previous physician when things go wrong, that we yell at nurses for their mistakes, that we have such high rates of alcoholism, drug addiction, and suicide."[44] Determined to put an end to these debilitating problems that adversely affect the doctor-patient relationship, Hilfiker concluded his story with the following prescription:

> At some point we must all bring medical mistakes out of the closet. This will be difficult as long as both the profession and society continue to project their desires for perfection onto the doctor. Physicians need permission to admit errors. They need permission to share them with their patients. The practice of medicine is difficult enough without having to bear the yoke of perfection.[45]

In telling his story, Hilfiker did not use his expertise "to pronounce a story that ends all storytelling" about the dilemmas arising from medicine's scientific and technological successes. Rather, he engaged in public moral argument by telling an illness story that allowed "persons" to be the primary matter of interest. Guided by this interest, the story offered an authentic response to the "demands of narrative rationality": it asked its readers (both the public and members of the medical community) to acknowledge that they are fallible, that they thus will make mistakes, and that they must be willing to admit and accept such errors if they expect to maintain some semblance of well-being in their personal and professional lives. The value of this acknowledgement, it seems to me, is undeniable. But is this value strong enough to bring together the medical community and the public in such a way that their common interests are recognized and that neither will be plagued any longer by the successes of modern medicine?

The case of medicine presented in this discussion, and the way this case was conceptualized by turning to the rhetorical theory of public moral argument, requires one to ask this question. The goals of public moral argument—its emphasis on how stories that speak to the highest human values can lead to the "good life"—would suggest that Dr. Hilfiker's story should warrant serious consideration from its audience. I hope that this suggestion is correct, for then one could be somewhat confident that a forthcoming answer to the above question would be: "yes." At the present time, however, I am afraid that such confidence may be misleading. I say this for the following reasons, all of which point to certain consequences made possible by our appropriation of the ways and means of Progressive ideology.

The dilemmas confronting the medical community and its public today are part of a dense social and technological reality; to separate the dilemmas from this reality is to obscure their real meaning: The scientific-technical expertise of the medical community must be juxtaposed to the psychological orientation of the patient (public); moral and legal questions raised by death cannot be separated from historical, social, and economic issues; and in a democratic society none of these concerns can be separated from the political reality in which they occur. In this reality many stories are being told and competing to be heard. Advocates of public moral argument tell us that this competition is necessary if the "truth" is to prevail. But in the case of medicine presented here there are many truths at work: those of doctors, patients, lawyers, theologians, philosophers, politicians, and so on. Not all of these truths coincide, despite the fact that they are all attempting to promote the good life. How is the public to decide about which of these stories deserves the most attention? Is the public or any field of experts "competent" enough to make such a decision? And what criteria should inform a working and acceptable definition of competence?

From the theoretical standpoint of public moral argument, Dr. Hilfiker should be judged competent. For, in addition to the reasons noted above, he told what Fisher designates as a "good" story: He constructed a narrative that sought *"to justify . . . decisions or actions already made or performed and to determine future decisions or actions."*[46] Reactions to Dr. Hilfiker's story that members of both the medical community and the general public have shared with me through a questionnaire suggest, however, that this judgment is too narrow in scope to resolve the conflict that presently exists between these two communities.[47] For example: Some respondents from the medical community praised Dr. Hilfiker for "going public" about medical mistakes; most condemned him of "negligence" for making the mistakes; and, as exemplified in the following response, all expressed reservations about how his story would affect the public's ability to appreciate fully the complexities of the medical profession:

> I think it's one more assault on their senses. It will do what most information does, help some and hurt others. I'm all for the public being informed but this article has a negativity about it that will stand to convince plenty of people that they were "right all along" about the medical profession. I also believe that part of the problem is the public's lack of information and that they are easily intimidated by their uncertainty. The public should realize that medical attention is a service oriented field and they should quickly go elsewhere when they are not satisfied. Too few people trust their own feelings and instincts about how their care is being handled. I don't hear too many people voicing sympathy for physicians, most people have an illusion that it is a wonderful existence.

Perhaps the skepticism of this response could easily be met with a popular indictment of Progressive ideology: Here, for example, is but another indication of how "elites" respond when their professional authority is challenged, when they are exposed for using "the people for their own ends" or for making "the people blind subjects of technology."[48] This indictment does find support in nearly all of the responses from the public—responses that favored Dr. Hilfiker's rhetoric because "frankness has been lacking in the Medical Community. Medical Doctors have *not* been monitored and have had *carte blanche* freedoms"; or because "We don't have the experience and expertise to challenge the doctor. Some roles are exempted from our normal curiosity and suspicion. These role players then get too much decision power over our lives. Articles such as Hilfiker's provide knowledge and incentive to allow us to reclaim that power"; or because "If the Surgeon General can warn us of [the] potential health hazards of cigarettes, why not publicize the risks of making gods out of our physicians?" But perhaps the skepticism heard from the medical community is justifiable: it could be a rejoinder to a not uncommon cynicism that finds expression in the following two responses from members of the public: "I would like to know why [Dr. Hilfiker] told the public, maybe so we could forgive him for his sins, maybe to get more money, double payments for the same mistakes"; and "Pushing aside all of the sympathetic feelings his article might evoke, wouldn't you love to sue the bastard?" Then, again, the skepticism may echo a warranted caution against a documented tendency of the public to be extremely ambivalent when asked to make a "competent" judgment about matters of high-tech health care.[49] A form of this ambivalence showed itself in twenty-three of the thirty questionnaires submitted by members of the public. Although these individuals acknowledged *their* support of Dr. Hilfiker, they expressed doubt that *others* would react favorably to his narrative. As noted in one of the questionnaires, the reason for such doubt is this: "I think the general public would rather not deal with the realities that Dr. Hilfiker discusses. Because of the human fear element involved with illness, they would prefer to hold on to the illusion of doctors plus technology equals perfection."

Is this illusion a barrier to the power of the word, to the path of virtue and good sense that public moral argument not only sees as its task to set forth but also encourages experts to follow? I believe so. But I also believe that the barrier is more than an illusion. For its presence today is rooted in historical developments that, as discussed in the beginning of this essay, allowed the ideology of Progressivism to become a reality. This reality continues to play a decisive role in our current sociopolitical system: For we, the people, still expect science, technology, and expertise to be vehicles of progress. And if this expectation is to be met, then we, the people, must be willing to dwell in a world where specialized, technical knowledge is granted the liberty to differentiate itself from the social knowledge of the laity. "Experts" and "the public" are, respectively, the sovereign keepers of these epistemic realms. Democracy allows for such privileges. This essay has dealt with a rhetorical problem that shows itself when these privileges find it difficult to co-exist in a harmonious way. My concluding remarks speak to this difficulty and to an issue it poses to advocates of public moral argument.

## CONCLUSION

In a democracy, accountability is the price that must be paid by those who enact the liberty and thus the privilege of developing and maintaining communities where technical knowledge and expertise can be cultivated. Such communities are known by different professional names: medicine, law, the military, government, academe, and so on. But the accountability that the members of these communities must demonstrate is not limited to the general public and its social knowledge; the members also are accountable to colleagues who call upon them to promote the technical knowledge of their professions. The claims of the professions to competent authority are dependent on this call being answered: only then can the expert hope to fulfill the expectation of progress (and perhaps perfection) that the will of the people desires; only then can the expert's colleagues offer an educated and good faith judgment about whether or not "work in progress" is competent enough to be made public.[50]

Of course, the power of this judgment can inhibit its assigned purpose. In a society such as ours, where great expectations about the wonders of science and technology can easily fuel the flames of public discontent when these wonders are not immediately forthcoming, professional communities may opt for a form of reticence. That is, they may educate their members to believe that public visibility is a genuine liability to expert respectability.[51] This belief functions to limit accountability to the world of technical knowledge; and it can be readily criticized for fostering an elite mentality, one that is taking advantage of a liberty that permits it to exist. But the belief also can be said to perform another, more noble function. For it bespeaks an awareness of the limitations of technical knowledge and how such limitations may prevent experts from currently identifying and thus talking about all of the risks associated with a given technical endeavor. And when the belief performs this function, it also does something more: it raises in the minds of experts a complex issue that neither they nor the members of the general public have been able to resolve. As stated by Mary Douglas and Aaron Wildavsky, the issue is one that has much to do with the "path" that public moral argument sees as its task to set forth:

> Where does the path of virtue and good sense lie—in announcing every possible risk as soon as it arises, or in waiting until there is more conclusive evidence or safer alternatives? One side says, "Do not start unless you're sure it's safe." The other side says, "Do not stop until you've got something better." Since it is not possible to say everything at once, or with equal emphasis, something has to be slighted. Some sort of risk has to be hidden. Is it better to hide the risks of action or of inaction?[52]

Another way of stating the issue is this: Owing to the great expectations that influence it, the path of virtue and good sense may at any time find itself having to respond to one of two audiences: either a community of experts or the general public. Given the factors of the risk involved, as well as those individuals who may have to suffer because of this risk, which of these audiences should warrant priority? When is the right time to make such a choice? And who is to decide?

The situation that the medical community finds itself in today provides an illustration of how experts are attempting to confront such a complex issue. The difficulty of dealing with it by choosing to engage in public moral argument is evidenced in the reported reactions to Dr. Hilfiker's narrative. Both the issue and the difficulty it poses are consequences of what has been defined as a problem of Progressive ideology; that is, they presuppose the existence of a society whose lay members 1) expect progress in the form of science and technology, 2) allow experts to create and maintain communities of technical knowledge that can instigate this progress, and then 3) condemn the ways and means of these communities because they do not always succeed in meeting the material and rhetorical needs of those who still expect these ways and means to produce progress.

Again, I must ask: Can this problem be solved by having experts engage in the rhetorical practice of public moral argument? If what this essay sought to accomplish warrants any consideration, then those who feel that the question deserves *only* an affirmative response may want to reconsider their position. This is not to say that public moral argument should simply be dismissed as something that has no value for experts who choose to argue about moral issues in public. On the contrary, the theory of public moral argument forces one to take seriously how rhetoric may be performed in a noble and authentic way.[53] Thus, as Fisher suggests, the theory can be used as "a basis for critique" because it implies "a praxis constant with an ideal egalitarian society."[54] But we, the people, do not live in such a society today. (Has such a society ever really existed?) For nearly a century ago we awoke from the dream (the myth?) of popular sovereignty and opted for a reality built on a Progressive ideology. We thus showed a willingness to live in a world marked by a tension between technical knowledge and social knowledge. We believed that this tension would produce progress. It did. But it also produced problems. Today, we still desire this progress; we do not, however, desire its attending problems. The theory of public moral argument allows us to dream about how such problems may be resolved: Truth founded on the will of the people is what it seeks. Surely, it is better to dream than to have recurring nightmares. But the dream must not be allowed to blind us to the reality at hand. The people expect progress in the form of scientific and technological advances; they want experts to fulfill their expectations. The history of such expectations has made possible a complex state of current, and most probably future, affairs. Before critiquing any group of "experts" that must necessarily conduct its rhetorical transactions within this complex state, I believe it would be wise to acquire first a sensitive understanding of how these transactions are influenced by and thus must confront that which makes them possible: Great Expectations.

## NOTES

*The earliest version of this essay was presented at the Fourth Speech Communication Association/American Forensic Association Conference on Argumentation (1985) and was published in its proceedings. Portions of that early version are contained here with the permission of the publisher. Later versions have been presented at various universities in the U.S. Grateful acknowledgment for research support of this essay is due to The Institute for Modern Communication, School of Speech, Northwestern University.*

1 Walter R. Fisher, *Human Communication As Narration: Toward a Philosophy of Reason, Value, and Action* (Columbia: University of South Carolina Press, 1987), 71.

2 See C. Wright Mills, *The Power Elite* (1956; rpt. New York: Oxford University Press, 1971-2); John Dewey, *The Public and Its Problems* (1927; rpt. Chicago: Swallow Press, 1954); Walter Lippman, *The Public Philosophy* (New York: The New American Library, 1955); Michael Calvin McGee and Martha Anne Martin, "Public Knowledge and Ideological Argumentation," *Communication Monographs* 50 (1983): 47-65; and Magali Sarfatti Larson, "The Production of

Expertise and the Constitution of Expert Power," in *The Authority of Experts: Studies in History and Theory*, ed. Thomas L. Haskell (Bloomington: Indiana University Press, 1984), 28-80.

3 Hannah Arendt, *Between Past and Future* (New York: Viking, 1961), 93.

4 Mills, 323.

5 Mills, 300.

6 Mills, 344-345.

7 Paul Starr, *The Social Transformation of American Medicine: The Rise of a Sovereign Profession and the Making of a Vast Industry* (New York: Basic Books, 1982), 19.

8 Starr, 3-19. See also Arnold Pacey, *The Culture of Technology* (1983; rpt. Cambridge, MA.: The MIT Press, 1984), 13-54.

9 Quoted in Pacey, 25.

10 Robert L. Heilbroner, *The Future as History: The Historic Currents of Our Time and the Direction in which They are Taking America* (1959-60; rpt. New York: Harper & Row, 1968), 202.

11 From a philosophical/sociological standpoint, this problem receives its most systematic treatment in the ongoing writings of Jürgen Habermas. See, for example, his *Toward a Rational Society: Student Protest, Science, and Politics,* trans. Jeremy J. Shapiro (1970; rpt. Boston: Beacon Press, 1971); *Communication and the Evolution of Society,* trans. Thomas McCarthy (Boston: Beacon, 1979); and *The Theory of Communicative Action, Vol. 1: Reason and the Rationalization of Society,* trans. Thomas McCarthy (Boston: Beacon Press, 1984). See also Manfred Stanley, *The Technological Conscience: Survival and Dignity in an Age of Expertise* (New York: The Free Press, 1978); and Hans-Georg Gadamer, *Reason in the Age of Science,* trans. Frederick G. Lawrence (1981; rpt. Cambridge: The MIT Press, 1983).

12 Walter R. Fisher, "Narration as a Human Communication Paradigm: The Case of Public Moral Argument," *Communication Monographs* 51 (1984): 9.

13 I am speaking here especially of *rhetorical* scholars. See, for example, Lloyd F. Bitzer, "Rhetoric and Public Knowledge," in *Rhetoric, Philosophy, and Literature: An Exploration*, ed. Don M. Burks (West Lafayette, Idiana: Purdue University Press, 1978), 67-93; Thomas B. Farrell and G. Thomas Goodnight, "Accidental Rhetoric: The Root Metaphor of Three Mile Island," *Communication Monographs* 48 (1982): 271-300; McGee and Martin; and Walter R. Fisher's analysis of "current nuclear war controversy" in his *Human Communication As Narration,* 69-71.

14 See Fisher, *Human Communication As Narration.* Fisher's position owes much, as he admits, to Alastair McIntyre's *After Virtue: A Study in Moral Theory* (Notre Dame: University of Notre Dame Press, 1981).

15 For a superlative discussion of the rhetorical nature of this activity, especially as it dates back to ancient Greece, see Pedro Lain Entralgo, *The Therapy of the Word in Classical Antiquity*, ed. and trans. L. J. Rather and John M. Sharp (New Haven: Yale University Press, 1970).

16 Research for this essay included spending much time in hospital settings where I could observe and document the practice of such stories during "rounds" in Intensive Care Units and during "grand rounds" presentations to the hospitals' medical community. Further discussion of the function of these stories is offered later in this essay.

17 See, for example, Stanley Joel Reiser, *Medicine and the Reign of Technology* (New York: Cambridge University Press, 1978); Starr; Charles Mangel and Allen B. Weisse, *Medicine: The State of the Art* (New York: The Dial Press, 1984); Samuel Vaistrub, *Medicine's Metaphors: Messages and Menaces* (Oradell, New Jersey: Medical Economics Company, 1977).

18 The patient's name in this story has been changed to provide anonymity for his family.

19 Willard Gaylin, "Modern Medicine and the Price of Success," *Bulletin of the American College of Surgeons* 68 (1983): 4. See also Samuel Gorovitz, *Doctors' Dilemmas: Moral Conflict and Medical Care* (New York: Oxford University Press, 1982); and Melanie Phillips and John Dawson, *Doctors' Dilemmas: Medical Ethics and Contemporary Science* (New York: Methuen, 1985).

20 See, for example, Joan O'C. Hamilton, Emily T. Smith, and Susan Garland, "High-Tech Health Care: Who Will Pay?," *Business Week*, February 6 (1989): 74-78; Victor Fuchs, "The 'Rationing' of Medical Care," *New England Journal of Medicine* 311 (1984): 1572-1573; Warren Butt and

Duncan Neuhauser, "The Machine and the Marketplace: Economic Considerations in Applying Health Care Technology," in *The Machine at the Bedside: Strategies for Using Technology in Patient Care*, ed. Stanley Joel Reiser and Michael Anbar (New York: Cambridge University Press, 1984), 135-151; Carol M. McCarthy, "DRGs—Five Years Later," *New England Journal of Medicine* 318 (1988): 1683-1686; Edward L. Schneider, "Options to Control the Rising Health Care Costs of Older Americans," *Journal of the American Medical Association* 261 (1989): 907-908. For an example of the kind of thinking and analysis that helped to initiate DRG regulations, see Louise B. Russell, *Technology in Hospitals: Medical Advances and Their Diffusion* (Washington, D.C.: The Brookings Institution, 1979).

21 Ralph Crawshaw, "Technical Zeal or Therapeutic Purpose—How to Decide?," *Journal of the American Medical Association* 250 (1983): 1859.

22 Lawrence D. Grouse, "Has the Machine Become the Physician?," *Journal of the American Medical Association* 250 (1983) 1891; also see E. Richard Brown, *Rockefeller Medicine Men: Medicine and Capitalism in America* (Berkeley: University of California Press, 1979).

23 Norman G. Levinsky, "The Doctor's Master," *New England Journal of Medicine* 311 (1984): 1573-1575; Richard D. Lamm, "Copernican Politics: It's Time to Ask Heretical Questions," *The Futurist* 17 (1983): 5-11; Robert J. Blesdon and Drew E. Altman, "Public Attitudes about Health-Care Costs: A Lesson in National Schizophrenia," *New England Journal of Medicine* 311 (1984): 613-616; Daniel Callahan, *Setting Limits: Medical Goals in an Aging Society* (New York: Simon and Schuster, 1987); Tom Tomlinson and Howard Brody, "Ethics and Communication in Do-Not-Resuscitate Orders," *New England Journal of Medicine* 318 (1988): 43-46; William Winkenwerder and John R. Ball, "Transformation of American Health Care: The Role of the Medical Profession," *New England Journal of Medicine* 318 (1988): 317-319; Barry M. Mannel, "A Contemporary Physician's Oath," *New England Journal of Medicine* 318 (1988): 521-522; and Maurice McGregor, "Technology and the Allocation of Resources," *New England Journal of Medicine* 320 (1989): 118-120.

24 Levinsky, 1573.

25 For one of the first major works that brought attention to the situation, see René Dubos, *Mirage of Health: Utopias, Progress, and Biological Change* (New York: Harper & Row, 1959).

26 Lawrence K. Altman, "After Barney Clark: Reflections of a Reporter on Unresolved Issues," in *After Barney Clark: Reflections on the Utah Artificial Heart Program*, ed. Margery W. Shaw (Austin: University of Texas Press, 1984), 119.

27 Fisher, *Human Communication As Narration*, 71.

28 Eric J. Cassell, *Talking with Patients, Vol. 1: The Theory of Doctor-Patient Communication* (Cambridge: The MIT Press, 1985), 8.

29 Eric J. Cassell, *Talking with Patients, Vol. 2: Clinical Technique* (Cambridge: The MIT Press, 1985), 15.

30 Gaylin, 6-7. The medical community, it must be stressed, has become well aware of what is being suggested here. See, for example, John Edward Ruark, Thomas Alfred Raffin, and the Stanford University Medical Center Committee on Ethics, "Initiating and Withdrawing Life Support: Principles and Practice in Adult Medicine," *New England Journal of Medicine* 318 (1988): 25-30; David C. Lowance, "Withdrawal from Dialysis: An Ethical Perspective," *Kidney International* 34 (1988): 124-135; and Jerry A. Green, "Minimizing Malpractice Risks by Role Clarification: The Confusing Transition from Tort to Contract," *Annals of Internal Medicine* 109 (1988): 234-241. Empirical research on "quality of life" issues is also being conducted. See, for example, Marion Danis, Donald L. Patrick, Leslie I. Southerland, and Michael L. Green, "Patients' and Families' Preferences for Medical Intensive Care," *Journal of the American Medical Association* 260 (1988): 797-802. In this study quality of life was found to be a poor predictor of willingness to repeat intensive care. As indicated by the authors, what makes this finding especially relevant for the purposes of the present essay is that it poses "a dilemma regarding decisions about intensive care treatment. If the majority of patients and families unconditionally desire intensive care even toward the end of life, as they did in this study, these wishes may conflict with the need to limit medical care costs in the United States" (802).

31 Jay Katz, *The Silent World of Doctor and Patient* (New York: The Free Press, 1984), xix.

32 Stanley Joel Reiser, "The Machine at the Bedside: Technological Transformations of Practices and Values," in Reiser and Anbar, p. 17. See also William J. Donnelly, "Righting the Medical Record: Transforming Chronicle Into Story," *Journal of the American Medical Association* 260 (1988): 823-825.

33 Candice West, *Routine Complications: Troubles with Talk between Doctors and Patients* (Bloomington: Indiana University Press, 1984).

34 Fisher, *Human Communication As Narration*, 73.

35 Fisher, *Human Communication As Narration*, 64.

36 Fisher, *Human Communication As Narration*, 71-72.

37 Cf. Altman; Maryann Haggerty and Charles D. Bankhead, "Medicine and the Media," *Medical World News-Psychiatry Edition*, (June 1984): 28-38; Donald F. Phillips, "Physicians, Journalists, Ethicists Explore Their Adversarial, Interdependent Relationship," *Journal of the American Medical Association* 260 (1988): 751-752,757.

38 Fisher, *Human Communication As Narration*, 180-190.

39 David Hilfiker, "Making Medical Mistakes: How Doctors Harm Patients—and Themselves," *Harper's* 268 (May 1984): 59-65; David Hilfiker, "Facing Our Mistakes," *New England Journal of Medicine* 310 (1984): 118-122.

40 Hilfiker's admissions—which include how he once aborted a healthy fetus after misdiagnosing it as dead—are especially relevant for the purpose of this essay since they were not confined to the pages of one of the most prestigious medical journals in the world. Dr. Arnold Relman, editor of the *New England Journal of Medicine*, is reported to have called Hilfiker "to give him a chance to withdraw the article from publication" and thereby avoid doing further and "serious damage to his career." See Mark Frankel, "The Healing of a Healer: A doctor decides that he just can't play God," *Chicago Tribune*, October 5, 1986, Section 2 (Tempo), 1, 4. Hilfiker also published a book, *Healing The Wounds* (New York: Pantheon Books, 1985), detailing his professional and personal experiences and his observations about American medical practice. With his *Harper's* essay, however, Hilfiker first broke (in a very "public" way) the so-called "code of silence" about medical mistakes. For an excellent study of how "medical mistakes are an intrinsic feature of medical work," see Marianne A. Paget, *The Unity of Mistakes: A Phenomenological Investigation of Medical Work* (Philadelphia, PA.: Temple University Press, 1988).

41 Hilfiker, 60.

42 Hilfiker, 62.

43 Hilfiker, 62.

44 Hilfiker, 65.

45 Hilfiker, 65.

46 Fisher, *Human Communication As Narration*, 187.

47 I developed the questionnaire in an attempt to document responses to Dr. Hilfiker's article. The questionnaire contained four parts: 1) Do you favor Dr. Hilfiker's decision to go public with his experiences and views? If yes, why? If no, why? 2) How do you think the general public will react to Dr. Hilfiker's article? Please explain. 3) Is there additional information that you would have included in the article to corroborate, correct, and/or extend Dr. Hilfiker's assessments concerning "medical mistakes" and their effects? Please explain. 4) Other responses that you wish to share about Dr. Hilfiker's article. The questionnaire was attached to photocopies of the article. A cover sheet indicated that a similar version of the article first appeared in the *New England Journal of Medicine* and that respondents were not to sign their names on the completed questionnaires. Thirty-two questionnaires were distributed to the members of an adult continuing education class located at a midwestern university. All but two of these questionnaires were filled out and returned. Fifty additional questionnaires were distributed to members of medical staffs from two university hospitals located in the Midwest. Eighteen of these questionnaires were filled out and returned. Clearly, this way of conducting the questionnaire research is far from being scientifically rigorous. My objective, however, was only that of obtaining some feedback concerning Hilfiker's article that could be used for the purpose of this essay.

136 *Communication and the Culture of Technology*

48 See note 12 and accompanying text.

49 Cf. Blesdon and Altman.

50 The point I am making here receives support from such notable works as Michael Polanyi, *Personal Knowledge: Towards a Post-Critical Philosophy* (New York: Harper & Row, 1964); John Ziman, *Public Knowledge: The Social Dimension of Science* (Cambridge: Cambridge University Press, 1968); and Thomas Kuhn, *The Structure of Scientific Revolutions* (Chicago: University of Chicago Press, 1970). Lewis Thomas states a version of this point when he notes: "It is when physicians are bogged down by their incomplete technologies, by the innumerable things they are obliged to do in medicine when they lack a clear understanding of disease mechanisms, that the deficiencies of the health-care system are most conspicuous. If I were a policymaker, interested in saving money for health care over the long haul, I would regard it as an act of high prudence to give high priority to a lot more basic research in biologic science. This is the only way to get the full mileage that biology owes to the science of medicine, even though it seems, as used to be said in the days when the phrase still had some meaning, like asking for the moon" (36). See "The Technology of Medicine," in his *The Lives of a Cell: Notes of a Biology Watcher* (1974; rpt. New York: Penguin Books, 1982), 31-36. See also Arnold S. Relman, "More on the Ingelfinger Rule," *New England Journal of Medicine* 318 (1988): 1125-1126, wherein the author (and editor of the *Journal*) makes it clear how the point is especially relevant for medicine.

51 Cf. Cullen Murphy, "In Darkest Academia," *Harper's* 257 (October, 1978): 24-28.

52 Mary Douglas and Aaron Wildavsky, *Risk and Culture: An Essay on the Selection of Technical and Environmental Dangers* (Berkeley: University of California Press, 1982), 27.

53 I have committed myself to this view of rhetoric in the following essays: "Hermeneutics and Rhetoric: A Seen But Unobserved Relationship" (with Craig R. Smith), *Quarterly Journal of Speech* 65 (1979): 347-63; "Rhetorically, Man Dwells: On the Making-Known Function of Discourse," *Communication* 7 (1983): 201-220; "The Hermeneutic Phenomenon and the Authenticity of Discourse," *Visible Language* 17 (1983): 146-162; "Treating the Patient as a Person," *Quarterly Journal of Speech* 72 (1986): 456-469; and "A Rhetoric of Risk: Medical Science and the Question of Wrongful Life," in *Argument and Critical Practices: Proceedings of the Fifth SCA/AFA Conference on Argumentation*, ed. Joseph W. Wenzel (Annandale, VA.: Speech Communication Association, 1987), 129-136.

54 Fisher, *Human Communication As Narration*, 66. For Fisher's most recent defense of his theory—which I believe still avoids the problem being raised here—see his "Clarifying the Narrative Paradigm," *Communication Monographs* 56 (1989): 55-58.

# Rhetorical Maintenance of Technological Society: Commercial Nuclear Power and Social Orthodoxy

*Micheal R. Vickery*

One of the motives for studying technology and communication is the need to understand more fully the ethos of what many have called technological society.[1] Lloyd Bitzer has argued that this need is imperative:

> The exigencies of our time are dangerous almost beyond the power of the imagination to conceive: war, famine, threat of accidental destruction by instruments of technology, ruin of an irreplaceable environment—these and other exigencies threaten to destroy individuals and societies.[2]

Such prospects produce intense and increasingly public debate. These debates reveal much about how both experts and non-experts view the general topic of technology. To study the discourse that is "about" technology is to study the fundamental *work* that people do to interpret the possibilities and the problematics of social life in our times. By examining discourse concerning the commercial development of nuclear fission technology in the electric utility industry,[3] I will describe the doctrinaire rhetorical methods through which industrial, governmental, and academic agents of "technological society" interpret the "facts" of life in that society and thereby maintain an orthodox vision of social order.

## PROBLEMATICS OF TECHNOLOGY AND SOCIETY

Although scholars have sketched provocative images of the nature, value, and problems of technological society, much of this analysis begins by defining technology in its most material and instrumental form and celebrating technology as a functional necessity.[4] Often, analyses of the relationship between communication and technology, per se, have focused on specific episodes of technological controversy and/or failure, thereby dramatizing the dialectical tensions between technology and society.[5]

Although a dialectical view of technological controversies is useful, it has also been suggested that a strict concern with social conflict may blur the deeper evolutionary questions about the relationship between technology and society or may obscure the rhetorical and ideological similarities of seemingly disparate technological problems.[6] Yet, comparatively little has been done to reveal the ways in which technology functions as a premise in the communicative process through which a social order is constituted and maintained.[7]

The materialistic and dialectic presumptions regarding the nature of technology and the character of technological controversies have been called into serious question in the writings of Jacques Ellul, Martin Heidegger, and Arnold Pacey.[8] Ellul's entire purpose in *The Technological Society* and *The Technological System* is to implicate technological instrumentalities as manifestations of technological consciousness. Similarly, Heidegger argues that the most important questions regarding the nature of technology have almost nothing to do with technological *instruments* themselves.[9] Rather, these technological "things," Heidegger argues, are manifestations of a way of thinking. Since they are the manifest form of technological thinking, technological instruments also *reveal* technological thought and it is this fact that demands that technology be intellectually engaged as meaning something more than instrumental means.

Pacey also argues that discussions of technology often treat the term incorrectly, limiting it to its restricted sense as machine, technical skill, or the activity of making things work.[10] Limiting the meaning of technology only to this *technical aspect*, Pacey argues, obscures the *organizational* and *cultural* aspects of technology-practice. Pacey defines technology-practice as "the application of scientific and other knowledge to practical tasks by ordered systems that involve people and organization, living things and machines."[11] The organizational aspect of technology involves administration, public policy, and the range of people and jobs through which technology gets "done" as well as the complex of persons who use and consume the products of technological activity. The cultural aspect of technology includes the values, habits of thinking, and beliefs that underpin the activities characteristic of technology and science.

I argue that these perspectives on technology as cultural practice, world view, and mode of consciousness constitute a speculative hypothesis about how people make sense to one another in a technological society. If, as Ellul, Heidegger, and others have postulated, technology really embodies cultural consciousness, then technological discourse should reveal that consciousness. Discourse that is alleged to make sense when deliberating technological facts, benefits, and risks must also embody technological predicates of thought and action as preeminent grounds for "making sense." Consequently, talk about technological systems may be said to enact the technological culture and a technological system such as the commercial nuclear power industry may therefore be thought of as "a product of successful sense-making."[12] To study the rhetoric of nuclear technology is, then, to study some of the most important sense-making habits of modern culture.

Carroll Arnold argues that rhetors construct solutions to social problems by weaving together three themes: (a) the pragmatic details constituting a point of decision, (b) the doctrinaire formulae by which a rhetorical community has traditionally addressed problems, and (c) the transcendental, intuitive vision the audience has of its place and its future in the "real" world.[13] In this essay I will argue that after more than four decades of explaining and arguing for commercial nuclear power plants, the meaning of nuclear energy is revealed through the rhetoric of its advocates to be a variety of technology-practice. Indeed, the discourse itself is a variety of technology-practice, embodying technical, organizational, and cultural predicates. Analysis of the discourse through which development of the commercial nuclear power industry has been advocated reveals the inner workings of an orthodox social philosophy that subsumes cultural problems of knowledge, value, and power under the covering ideology of technological progress.

The rhetoric of commercial nuclear power offers conventional solutions to the problematics of any technology-practice by: (1) celebrating the technical skills and knowledge required to assess, amass, and apply information needed to act usefully in an empirically defined human reality, (2) invoking doctrinaire materialistic assumptions about the quality of life, and (3) ritualizing the transcendental dream of a homocentric natural order. The rhetoric of commercial nuclear power weaves a utilitarian vision of "facts," a materialistic vision of "progress," and a homocentric vision of "natural order" into a web of orthodox presumptions regarding right life in the technological society. The remainder of this essay briefly describes each of these themes in the rhetoric of nuclear power and then suggests how this particular body of discourse may characterize a more general rhetoric of technology; a rhetoric endemic to technocratic society and through which the technological society *maintains* itself.

## The Utilitarian Vision of Empirical Reality

Technology depends for both its meaning and its effectiveness on the possession and manipulation of "facts" about "how things work." Yet, technologies often fail and "facts" are revealed as transitory. One of the most striking features of discourse advocating the development of commercial nuclear power over the past four decades is that, even when technical expertise has gone wrong and "facts" have turned out to be less than factual, a utilitarian faith in the ability of an idealized "technical man" to manipulate and manage reality has not wavered. Even when empirical claims regarding the technical and organizational "facts" about nuclear power and the nuclear industry have been directly contradicted—when nuclear energy turned out to be expensive enough to meter and when people did die because of a nuclear accident—advocates have designed new "facts," thereby reconstituting the utilitarian faith in an empirically knowable and socially useful reality.

It is not surprising that nuclear power is often interpreted in distinctly utilitarian terms. Americans venerate "useful knowledge" and "experts" who possess and "know how" to use information.[14] The rhetoric of nuclear energy affirms this veneration. Marking the keel-laying of the *USS Nautilus*, President Truman paid homage to the ability of "technical man" to "design" and "work out" the problems of the "whole complicated mechanism" comprising the world's first nuclear-powered submarine.[15] Similar homage was paid by an electric utility executive who toasted "technical man's" focus on facts rather than opinion and reverence for rationality rather than emotionality.[16] Moreover, technical man is revered as the social and intellectual superior of "the public at large—the young, housewives, artists, teachers, social scientists, and small businessmen" who "have a basic fear of all advanced technology that they don't understand and a distrust of people who do understand."[17]

Advocates of commercial nuclear power have consistently portrayed technical data as the antidote to corrosive and dangerous influences posed by opinion and, especially, emotion. For its advocates, the creation of the entire nuclear industry is warranted by a "correct" reading of "accurate" information about both the natural and technical worlds. People and ideas that contradict these warrants for nuclear energy are therefore dismissed as ignorant of the "real" facts, responding emotionally rather than "rationally," or both. The clash between those who know the "facts" and those who do not—between the clear-headed technical person who understands the technical beauty of nuclear power and the muddle-headed emotional person who does not understand the "realities" of social life and the "necessity" of nuclear power—is one of the most enduring themes in the rhetoric of nuclear energy.[18]

Nuclear energy has remained, even to the present day, firmly enmeshed in a cultural "web of significance"[19] that captures the "meaning" of commercial nuclear power in terms that sanctify "rational" thought, define social problems in empirical terms, treat technical information as truth, and venerate instrumental skills for solving not only mechanical but social problems as well. Discourse engendered by and expressing these issues reveal both the content and style of thought through which the technological society interprets its utilitarian need for nuclear energy. The rhetoric's empirical focus not only denigrates opponents as uninformed but it equates engineering with science[20] and confuses technical data and hypothetical knowledge.[21]

The rhetorical depiction of the risks of nuclear power has also manifested the utilitarian reverence for objective data. Risks are computed by counting—frequencies and amounts of radioactive leaks at nuclear power plants, the number of cases of cancer in a community, the number of dead bodies resulting from nuclear accidents or radiation—and comparing one pile of facts to another. Of course, until the Chernobyl accident, using a body count as a "real" measure of technological risk was a powerful token in the rhetoric used to justify nuclear energy. Until Chernobyl the bumper sticker that read, "More people have died in Ted Kennedy's car than have died because of nuclear power," revealed in popular culture what advocates of nuclear power expressed in quite serious terms.[22]

Even when nuclear technology has yielded destruction and bodies could be counted, the utilitarian vision has remained clear. For example, a former president of General Dynamics was able to reconstitute the atomic bombing of Japanese cities into a great boon for humankind. "So great is the expansion of uses of radioactivity in the field of medicine," said he, "it is estimated that already more lives have been saved by the atom than were lost in the Hiroshima and Nagasaki explosions."[23] Similarly, the plant failure at Three Mile Island produced the optimistic assessment that the event was not so much an accident or a demonstration of the failure of technological reasoning as a learning experience; an experience that was certainly "frightening" but also "useful and educational," yielding new facts to be computed in the technical design and operation of nuclear power plants.[24] Not even the disaster at Chernobyl in the Soviet Ukraine can daunt the utilitarian reverence for empirical "data." Timothy Luke's analysis of the symbolic "packaging" of that disaster in both Western and Eastern Bloc media reveals that the event has been "explained" as just another technical malfunction that, although serious, could have been—if you count piles of money and bodies and compare them to other industrial accidents— worse.[25] This perspective on the event allows it to be recast as just another laboratory event, like Three Mile Island, from which to learn more useful facts about the technical performance of nuclear machines and their human components.

Obviously, the rhetoric employed by advocates of nuclear technology has relied on an explicitly stated reverence and quest for technical "facts"—physical and social in nature—about nuclear power and nuclear power plants. However, this discourse also reveals in its utilitarian faith in *relevant* "facts" a tacit imperative to manipulate the meaning of facts in order to satisfy utilitarian social ends. The rhetoric affirms the potential of nuclear power to satisfy social goals because social problems are defined in empirical terms and a nuclear power plant is, therefore, a fitting *empirical* answer to problems so defined. The focus of the discourse has remained, even through the last decade of diminished activity in the nuclear industry, consistent with its functional quest: *how much*, compared to other means of generating electricity, does nuclear power cost; *where* and *how* should reactors be built and waste be stored; and *when* will the non-technical public recognize the "facts" and get on with building nuclear reactors?

The presumption of the legitimate social role of those who "know how" to amass empirical data and turn it into something socially desirable is a first premise in discourse used by advocates and opponents of nuclear power to construct the "correct" perception of "information." This cultural presumption legitimizes indictments of those who oppose one's own perspective in a technological controversy because their claims to technical knowledge are "inadequate" and their perceptions of the "realities" of social life are "incorrect." This presumption underpins the construction of claims that are *culturally* valid. These claims assert, in their simplest forms, that human experience is a product of either a correct or incorrect reading of the composite of measurable, empirical facts of the natural and social worlds; that the perception and use of relevant facts is best left to experts; and that all important questions about nuclear technology are technical and organizational in nature.

From its rhetorical inception as a social premise and as a commercial product, nuclear energy has been viewed through a thick utilitarian lens. Advocates of nuclear energy captured the fearsome and esoteric complexities of nuclear technology from the very first in the familiar language and ordinary images of transportation, medicine, agriculture, and manufacturing that depicted "normal" life in an industrially and economically burgeoning postwar society. By objectifying the idea of nuclear power in such familiar contexts and rendering it in some sense "visible" to ordinary people, the technology could be treated in the conventional ways in which automobiles, airplanes, farming, steel mills, and family doctoring would be treated. That is, nuclear energy was made useful by turning the abstract idea into the concrete "things" that populate everyday "normal" life. Of course, this also obscured the empirical differences between *nuclear* things and mechanical things. Nonetheless, by rendering nuclear technology into the comfortable and preferred empirical images of life in an industrially productive postwar America, the meaning of nuclear energy was shifted away from its military, destructive images and embedded within the quintessential American images of a materially prosperous social order.

## The Materialistic Vision of Progress

If utilitarian presumptions have been used to construct the meaning of nuclear energy as a knowable and therefore useful *means*, it can also be said that materialistic presumptions have been used to interpret those means into culturally desirable ends. Indeed, one of the most ubiquitous features of the rhetoric of nuclear technology is the presumption of inevitable, forward, linear, and, especially, material progress of "industrial democracy."[26] From its rhetorical inception, nuclear power has employed the symbology of the expanding industrial economy to explain the transcendent social value of commercial nuclear power. In its rhetorical applications this theme has encapsulated the technical entities of nuclear plants and the electricity they produce within a schema of materialistic values. The effect has been to transform the technical *thing* into a cultural *sign*. Nuclear power plants have come to signify economic progress and material prosperity.

Since nuclear energy and nuclear power plants have been consistently defined as social as well as technical marvels, we can argue, using Pacey's terminology, that they must be understood in organizational terms such as professions, jobs, consumers, and administrative policies as well as in terms of cultural values such as wealth, status, and security. This broader meaning of nuclear power as technology-practice is dramatically illustrated in the advertising campaign by the United States Committee for Energy Awareness. This campaign explicitly links nuclear plants and electricity to the cultural symbologies of political independence and economic growth.[27] USCEA literature and print advertorials tap into patriotic and emotional themes[28] by characterizing nuclear power plants as sources of "homegrown" electricity, linking electricity growth to growth in GNP, and linking concerns about "adequate supplies of electricity" to "standards" of comfort, convenience, and mobility.[29]

The presumption of material progress is legitimized by the organizational and cultural aspects of the nuclear industry. Institutional and cultural habits of industrial democracy have made it possible for "energy" to be translated from its meaning as a natural potential to that of a socially useful, institutionally managed commodity. From the outset, those who advertised the wonders of the coming nuclear age employed rhetorical visions of better ships, better planes, better health, and more everyday conveniences. A utility company president expressed it succinctly:

> In a nuclear plant, the heat is produced by burning nuclear fuel in a reactor. From there on, the heat makes steam, the energy in the steam is converted into mechanical energy, the mechanical energy into electric energy, the electric energy into better living.[30]

The embedding of nuclear energy within a larger socio-cultural framework of industry and economics is, if not as explicitly stated as the above, just as certainly implied in the rhetoric of nuclear energy. The images of capitalist industrial economics permeate the justifications of nuclear technology. Technical questions about the operations and safety of nuclear plants are yoked to the "economic necessity" of electricity and the material marvels of an industrially expanding social order. Once the "need" for electricity is linked to a positive vision of the material signs of having *made* progress—jobs, money, automobiles, electrical conveniences—and to a negative vision of not being able to make progress without *enough* electricity, it is easy to conclude that the "quality of life" depends on the generation of electricity by any and all available technical means.

The effect of symbolically constituting "energy" in terms of industrial and economic activity has slowly but inexorably driven out uses of the term to mean the inherent potentials and forces in nature. Captains of the nuclear industry and apologists for industrial democracy have successfully redefined energy as *manufactured power*. The argument has proceeded from the presumption that the material rewards of physical comfort and convenience flow from the centralized industrial manufacture and commercial distribution of electricity.[31]

Centralized bureaucracies of government and industry are seen in this rhetoric as essential to the public good. Rhetorical agents of these institutions have consistently advocated a reactive social system that, it is argued, simply responds to the needs of citizens. Moreover, what is good for "industry" is seen *de facto* as good for the citizens. For example, former U.S. Secretary of Energy, Donald Hodel, argued that nuclear energy is needed by the citizenry because industry needs to "be able to acquire the energy they need to produce the products, ship the goods, or maintain their business activities . . . and remain competitive."[32] Indeed, to be against commercial nuclear power is characterized by one executive vice president of an electric utility company as being against "business," against "growth," and against "progress."[33] Such arguments tacitly define the debate over nuclear power as an ideological fight between those either for or against progress, which is defined in terms of industrial growth and material comfort. Consequently, the personal and social *desire* for comfort and convenience are transformed into *needs*. Such rhetoric sets up a paradoxical relationship between individuals who "need" convenience and the institutions that "need" for the public to want more electricity. One major function of discourse by advocates, therefore, has been to sell the goodness of fit between nuclear technology, the utility industry's social role of electricity producer, and the public's role as a component in the technological system that produces both electricity and the lifestyle that demands it.

The public's role is sometimes inelegantly stated as that of a "ratepayer"[34] but the role is more complex than that. The public must want what "industrial democracy" offers. If the material products of industrial democracy are *needed* then the means of generating those products are less important than the needs they fulfill. The linkage of suppliers and consumers of electricity as partners in this materialistic vision of progress underpins an argument for progress that has cultural validity; that is, having achieved electric lighting and its attendant social benefits, the technological society cannot go backwards and is constrained to maintain both the forward trajectory of progress and the means of producing the material signs of progress. Indeed, the discourse legitimizes a preferred lifestyle and propagates a vision of right living. For example, an executive of Rockwell International's electronics division asserted his preference for

> infinite energy and continuous expansion of our (and the world's) economy . . . with all the comforts and luxuries that go with the progressive society we have created. I have no appetite for austerity, nor do I feel any glow of nostalgia for the so-called "good old days" of simple living—of ice boxes, coal furnaces, and horse drawn carriages.[35]

Such talk exemplifies the tactic of denigrating "dated" technology and illustrates the acceptance of material wealth and physical comfort as sufficient justifications for new technologies. These tactics have been used by utility executives, U.S. Senators, publishers, engineers, and energy consultants for more than three decades to construct symbolic relationships between energy, industrial expansion, national welfare, and quality of life.[36]

The rhetoric of nuclear power, then, celebrates and advocates as necessary a fundamentally orthodox economic relationship between cultural, organizational, and technical components of the nuclear industry as a technology-practice.

### The Homocentric Vision of Natural Order

The rhetoric of nuclear energy embodies not only orthodoxies of knowledge and value but potency as well. This third element in the rhetoric of nuclear energy is reflected by a distinctive deterministic theme in the discourse. This theme is manifest in a paradoxical vision of the "inevitable" imposition of nuclear technology by irresistible forces of nature on a human community characterized by its irresistible forces of free will and technical skill. The human will-to-control is rhetorically envisioned as manifesting two forms. One form is the instrumental manipulation of physical elements and processes of a preeminent natural order. The second is the control of social order through the manipulation of the laws, regulations, and institutions of the technological social order. The will-to-control and the instrumental product of that will—nuclear power plants—are depicted, in the first case, as a preeminent force and, in the second case, as a naturally logical extension of the preeminent order of things.

Rhetorical depictions of the inevitability of commercial nuclear power and of the homocentric control of nature form one of the nuclei around which the rhetoric of nuclear energy turns. It took little time following the bombing of Nagasaki and Hiroshima for advocacy of "peaceful" uses of nuclear energy to become common refrains in public discourse. The editor of *Debater's Digest* told a Business and Professional Women's Club that "the controlled release of atomic energy is the most momentous fact ever to confront the human race."[37] It was not only a momentous fact but its successful commercial development was envisioned as a virtual certainty. The same speaker said that "the first development will undoubtedly be the production of power; probably to supply electricity for lighting, heating, and industrial power purposes."[38]

Commercial development of nuclear energy was subsequently described as "absolutely inevitable,"[39] and "as certain to come one day soon as anything can be certain in life."[40] The absence of doubt about the inevitable commercial development of nuclear energy continues to mark the discourse, even in the most recent years in which the development of nuclear energy has slowed. In fact, even after the Three Mile Island accident a utility company executive assured his audience that efforts to obstruct the construction of more nuclear power plants would only delay the inevitable.[41]

Thomas Farrell and Mark Pollock have argued that there are many different kinds of power in the world, including the power of possibility, the power of accident, the power of relationship, and the power of "harnessing" or control of nature.[42] It has been argued that the need to control nature is a virtual commonplace in contemporary definitions of the human condition. Historian of science Lynn White argues that control of nature undergirds the entire Judeo-Christian philosophical tradition and is the foundation of modern science.[43] At the very least, control is a god-term in the lexicon of modernity.

If harnessing nature is the dominant sense of power in human affairs, and if technology is an instrumental means of exercising such control, then technological discourse must both celebrate control and account for the other forms of power in the world. The rhetoric of nuclear power certainly construes and celebrates power as the control of nature. Nuclear technology is seen as an embodiment of technical and organizational skills through which nature and human behavior may be harnessed. The celebration of technical and organizational skill at controlling reality is invoked as the counterforce to accidents and errors; forms of power that belie the possibility of the control of either nature or human behavior. To be able to control nature it is necessary to hold in check potential negations of that power: errors—the misapplication by human agents of the means of control—and accidents—the intrusion of forces not adequately enframed by a technological means of control.

The definition and celebration of power as harnessing and controlling nature is used, at least by the high priests of the nuclear power industry, to construct a kind of transcendental confidence[44] in their ability to control nuclear energy and, therefore, control the future of society. The idealization of the human capacity to control nature through technical skill and organizational procedures means that problems that arise—nuclear accidents, for example—are rhetorically containable within the conceptual boundaries of potent human agents and a controllable natural world.

Such idealization has practical consequences in the ways in which the problematics of nuclear-generated electricity are treated. Faith in rational means of controlling was expressed early and often. One year after the bombing of Hiroshima, a delegate to the United Nations Atomic Energy Commission said that "the United States has proposed an international authority with unequivocal power to exercise full and effective control over atomic energy from birth to death."[45] Not even the facts of a disaster like Chernobyl daunts the human spirit of control. In his televised speech following the Chernobyl accident, Mikhail Gorbachev assured the Soviet people that:

> All aspects of the problem . . . are under the close scrutiny of the government commission. . . . It goes without saying that when the investigation of the causes of the accident is completed, all the necessary conclusions will be drawn and measures will be taken to rule out a repetition of anything of the sort.[46]

Technological phenomena of all types—whether nuclear plants or the bureaucratic regulations that govern them—are grounded in a kind of totalitarian confidence that no part of nature—physical or human—lies outside the bounds of human technological expertise.

None of the three visions described here is manifest in isolation from the others. They work in concert to envision the realities of life in technological society. One illustration is the way in which nuclear power advocates have defined nuclear plants as technical, empirical entities but also as social, axiological signs of progress. The argument for nuclear power is in some sense an argument from signs. The argument has been that if nuclear power plants are built, they would not only produce electricity effectively and economically, but also signify the will of an industrial, materialistic society to meet the increasing demands for electricity, demands that are virtually created by the socio-political engines of the industrial democracy. Even in the face of lethal events such as the Chernobyl accident, nuclear power is defined by its advocates as safe and as a social necessity.[47] Nuclear plants and the nuclear industry are self-justified in this discourse as proof from signs that a knowable, preferable, and *controllable* social order has been maintained. That is, they are seen to constitute a normal organizational and technical means of maintaining orthodox social values.

Similarly, the rhetoric of nuclear energy has been fraught with images similar to President Truman's vision of harnessing technology like a farm animal—a kind of socio-economic mule to be "hitched up" in order to do the chores of the industrial democracy.[48] The opposite form of this linkage of the technical and cultural dimensions of nuclear power envisions the industrial society as riding the nuclear mule to the "fantastic future" guaranteed by the "atomic revolution."[49] In more contemporary times, the rhetoric of nuclear power is still employed to weave together images of a dark and dangerous past in which people labored in drudgery to eke out a low standard of living and to contrast this with images of a bright, prosperous, indeed electrified, future.[50] Questions of value are presupposed in a quest to acquire the material products of industrial democracy, products which themselves are the artifacts of "making progress." The principal values espoused are couched in images of humans in control of the physical environment (as masters of the dark) and as materially prosperous. Through such images of control the rhetoric celebrates material power, harnesses the natural world to human whim, need, and desire, and does not differentiate between human well-being and materialistic lifestyle.

Each of the three primary themes are important in understanding the cultural meaning of commercial nuclear power. Each reveals a part of an orthodox vision of the technological society. Cyril Welch argues that one of the primary conditions governing the relevance of an act of speaking is that discourse must be perceived as a response to some interrogative.[51] In that sense, each of the three themes of commercial nuclear power reveal a culturally significant question, an arena of uncertainty that must be addressed in order for the culture to see itself.

This discourse, like all rhetoric, is a mirror. It reflects an implicit vision of a culture that warrants some kinds of questions and sanctions some kinds of answers. Each of the themes in the discourse reveals, indeed, enacts its own significant interrogative condition in which that theme will work to *make sense* of things. For example, the utilitarian theme essentially enacts a condition of uncertainty that is defined as a deficit of empirical information. Consequently, the questions about nuclear power are empirical questions and the resolution of those questions is possible because there are experts who "know how" to find out how things work. The materialistic vision of progress sees the problematics of modern life in terms of the available amounts of material convenience and physical comfort. Questions about nuclear power are, therefore, seen in terms of the potential for nuclear plants to sustain the style of life to which the society—

at least those members who have made "progress"—has become accustomed. The homocentric vision of the nature of the real world sees the problematics of technological society in terms of implacable natural and social forces that threaten preferred images of normal life. These forces are to be met with force. The elemental force in this view of the natural order of things is the human will to control. Within these interdependent visions of technological society lies the seed germ for technological optimism and the intellectual architecture of beliefs that justify and maintain the technological society.

## Nuclear Power, Technology and Social Orthodoxy

Heidegger has argued that understanding technology is impossible if we see it *only* as the instrumental means to ends.[52] He observes that the essence of technology is the "way" in which "the instrumental" reveals the experience of human existence. Only when this essential quality of technology is grasped can humans become truly responsible for (and to) the relationship between means and ends that instrumental consciousness entails. Hyde interprets this to mean that human communities may be forever consigned to chasing after the *consequences* of instrumental thinking and reacting to the aftermath of the social applications of technological instruments unless they begin to respond to "the 'way' and not merely to the 'what' of technology."[53] The confusion of the *what* of technology (the instrumental forms and their material effects) and the *way* of technology (the symbolic forms and their emotional and perceptual effects) confounds a full understanding of the rhetorical *construction* of technological society by use of an orthodox language of objective reality, materialistic values, and a potent homocentric natural order.

No sense of technology or technological discourse could be made without presupposing the purposeful control and use of an empirically eminent world. For example, advocacy of the commercial nuclear industry presupposes the possibility of mastering elemental properties of nature and of managing the mechanical and social instruments constituting a technological system; that is, the technological system is predicated upon the assumption that both the facts of nature and the instrumental artifacts of human intervention into nature are at least *potentially* controllable. Therefore, nature in both its physical and its artifactual forms is defined as a resource to be managed. As resources, neither nature nor the elements of the technological system are potent facts in and of themselves. In Heideggerian terms, everything relevant to the technological enterprise is meaningful only as a standing reserve, waiting to be used.[54]

Yet a technological system is composed of more than the mechanical mastery of the physical components of technological instruments. Indeed, rhetoric during the first two decades of the nuclear age focused significantly on the question of organizational mastery of bureaucratic and political components of nuclear technology. The debate focused explicitly and intensely on whether government or commercial bureaucracies would be given the task of mastering the peaceful applications of nuclear energy.[55] Mastery of the social nature of human beings through bureaucratic manipulation in the form of laws, regulations, and procedural guidelines is an enduring premise in the rhetoric of nuclear energy. Rhetors advocating nuclear energy imply that such must be so, for the technological system is, after all, designed by and for human agents. Moreover, humans are themselves components of the technological system and must, therefore, be controlled as well.

The mastery of human nature is rhetorically constructed as technically feasible via a faith in "correct" organizational techniques. This is a transcultural phenomenon. As we have seen, Soviet President Mikhail Gorbachev responded to the Chernobyl accident by asserting the coercive power of bureaucratic procedures to protect citizens from future nuclear accidents. The Boy Scouts of America agree, attesting that in the United States "we have laws to protect us" from too much radiation.[56] It would seem that the principles of bureaucracy that explain how to control human actions are as universally shared by industrial societies—both capitalist and communist—as are the principles of physics that allow them to control the splitting of atoms.

Despite claims by Mr. Gorbachev and the Boy Scouts, it is clear that procedural guidelines in the technological system do not actually affect the physical properties of technological machines. Bureaucratic rules are a class of "facts" different from the facts of splitting atoms, radiation, or nuclear reactors. Envisioning the possibility and the necessity of mastering these bureaucratic facts, however, may be the single most important element of the discourse that is intended to manage and maintain the technological system. Certainly, the *rhetorical* facticity of nuclear energy must be accounted for if one is to understand adequately, appreciate, and become responsible for technology.

The social control of nuclear technology is, per force, a function of rhetorical behavior. The language of "law," "order," "profit," "procedure," "risk," "well-being," and so on is assembled into the *symbolic* structure of nuclear power as surely as concrete, steel, and graphite are assembled into the physical structure of a nuclear plant.

It is clear that, given the mechanical and social components of technological systems, the effectiveness of the commercial nuclear industry in selling its product has required that not just nature and nuclear power plants be masterfully managed, but that the meaning of "nuclear power" be managed just as skillfully. Three Mile Island and Chernobyl have laid to rest the absoluteness of the former kind of control and made quite clear that it is the latter that is essential to the success of a technological enterprise.[57] To master any technological system certainly requires that empirical facts and technical instruments be controlled. Yet it is equally important to recognize that a means of control would not be sought, would not be used, indeed, would make no sense were the patterns of perception in human communities not made habitual through rhetorical practices. It is for this reason that the nature and use of the orthodox rhetorical facts that constitute technology-practices must be better understood.

The rhetoric examined here is about what nuclear technology *means*. In this sense, the three primary themes in this rhetoric reveal the orthodox attitudes, visions, and consciousness embodied in orthopraxic cultural behaviors such as building commercial nuclear power plants. That is, the rhetoric discussed here is a rhetoric of the *way of technology*, of the habits of thought and action revealed in discourse about technological change.

An important implication of treating the rhetoric of nuclear technology as a kind of social orthodoxy is that one's perspective on large-scale technical problems or accidents is informed by that treatment. Simply put, technical failures will always be excusable because they will be defined as sources of new information that will lead to technical improvements and more progress. Large technological systems are thereby made technically and organizationally sustainable because the culturally embedded first premises of the technological thesis are nonfalsifiable.

Indeed, despite the difficulties of the nuclear industry since the late 1970s, proponents of commercial nuclear power maintain a powerful rhetorical position. This is because they have, from the beginning, bespoken a doctrinaire vision of right-life in an industrialized, consumer-driven society. Such proponents have spoken the familiar language of communal life in a materialistic, utilitarian society. They invoke instrumental symbols such as jobs, electricity, and dollars as signs of the consummatory values of a technologically advanced society—comfort, convenience, wealth, and power. These familiar symbols embody a materialistic definition of human progress and are, in a paradoxical fashion, granted status as natural resources to be husbanded like any other. The products of the expanding, technological, industrial society come to be defined as pre-existing resources that the technological society is designed to protect. In this strange loop of logic, the products of technological society become divorced from the processes that produce them. For example, electricity gets rhetorically constructed as a need and the orthodox materialism of the technological society encourages whatever is necessary to protect the supply of electricity. Both the technical and social consequences of creating that electricity using nuclear means become symbolically embedded in this rhetorical vision of the necessities of life. In this way the debate over such technologies as nuclear power plants becomes deliberations over the technical means of producing what is already defined as a necessary social end.

## CONCLUSION

Through its orthodox utilitarianism, the rhetoric of nuclear energy creates the raw materials of social control in a technological society. By creating the symbolic capital of "objective information" and "empirical facts" about nuclear energy and the "realities" of social life, the discourse is able to construct the deterministic theme of "mastering" that kind of reality. Orthodox visions of material well-being and images of economic progress constitute telic images of a mastered world; a world that feeds desires and yields material rewards. Finally, orthodox visions of the irresistible natural and social forces provide the politically galvanizing legitimacy for taking control of both natural resources and human nature. Taken together, these themes constitute a secular faith in the absolute potency of human will-to-control and the legitimacy of technological agencies through which that will is exercised.

This technological discourse is therefore didactic. It asserts, informs, instructs, and, in its highest form, makes demands of its audience relative to what must be done in the present to maintain a technological society that in the future will be the same only better. The rhetoric is also rationalistic; that is, it uses the imagery of rational deliberations grounded in objective information to justify its driving motive of control. An idealized rationalistic view of correct human behavior is used to define the "non-expert" human as ignorant, emotional, and politically devious *unless* he or she accepts the "necessity" for the advocated technology-practice. In that case, the human is defined as eminently rational indeed.

In effect, the natural framework within which nuclear power plants exist as designed, constructed, fallible, technical instruments has been rhetorically subsumed within a social framework of organizational goals and materialistic cultural values.[58] Within this framework a nuclear plant "means" the power *to* produce and maintain a preferred way of life. The argument for nuclear power rests not upon the viability of nuclear instruments but upon the acceptability of the technological social order and the techniques by which it is maintained. The argument spins around interlocked social techniques: (a) an orthodox vision of materialistically defined quality of life is used to justify nuclear power, (b) the construction and maintenance of nuclear plants to sustain that quality of life requires that orthodox political and economic techniques be employed to attribute value to and distribute "necessary" resources, and (c) since political and economic techniques are *public*, the entire system rests upon the application of orthodox rhetorical visions and norms to interpret nuclear power into an acceptable framework of meaning. In essence, nuclear power is as much a social and rhetorical phenomenon as a natural and empirical one. Nuclear power exists as a reified piece of the symbolic mosaic that explains and gives meaning to human experience in modern technological society. The argument for nuclear power is a minor premise in the cultural argument for a vision of social life in which human techniques control reality and therefore make life good.

Through the rhetoric of commercial nuclear power the meaning of human community is also subsumed under the functional needs of an instrumentally defined social order. Human beings and the natural world alike are defined as resources to be used by the mechanical and organizational systems that "operate" modern life. In a very direct sense, the rhetorical orthodoxies of technological society help to create a vision of social order in which human beings are responsible *to* rather than *for* their technological actions. The direction in which responsibility in a technological society lies is cogently articulated by engineer J. A. Haddad:

> All of us have work to do, whether we are legal experts, or technologists, or engineers, or scientists, or sociologists, or philosophers, or political scientists, or theologians. Together we have to create new systems to allow us to live with the new [technological] facts. We, each, need the other's wisdom, and knowledge, and creativity. We have a new world to design.[59]

A seminal vision of the rhetoric constituting technological society is revealed here. By the use of the term "orthodoxy" in this essay, I have meant to suggest that within the rhetoric of commercial nuclear power there is evidence of something quite conventional in the way in which symbolic means are employed to *design the perceptions* of "facts," "necessities," and "actions" that permit a technologically advanced social order to be enacted.

## NOTES

1 Among the more interesting treatments of what "technological society" means to the conduct of everyday life are Jacques Ellul, *The Technological Society* trans. John Wilkinson (New York: Alfred A. Knopf, 1964) and *The Technological System* (New York: Continuum Publishing, 1980); Theodore Rozak, *The Making of a Counter Culture: Reflections on Technocratic Society and its Youthful Opposition* (Garden City: Anchor Books, 1969); Martin Heidegger, "The Question Concerning Technology," *Martin Heidegger: Basic Writings*, ed. D.F. Krell (New York: Harper and Row, 1977), 287-317; Carolyn Miller, "Technology as a Form of Consciousness: A Study of Contemporary Ethos," *Central States Speech Journal* 29 (1978): 228-236; Neil Postman, "The Technical Thesis," *Vital Speeches of the Day* 46 (1979), 180-185.

2 Lloyd Bitzer, "Rhetoric and Public Knowledge," in *Rhetoric, Philosophy, and Literature*, ed. Donald Burks (West Lafayette, Indiana: Purdue University Press, 1978), 91.

3 This essay draws from a body of speeches, broadcast interviews, and pro-nuclear partisan literature that spans the years 1945 to 1989. This body of data was the focus of my dissertation. See Micheal R. Vickery, "The Rhetoric of Commercial Nuclear Power: A Study of Technique." Ph.D. Diss., The University of Texas at Austin, 1988, 384-399.

4 Apologists for technology, per se, generally take the position that technological problems are the necessary cost of human development and are always solvable with new, more, and better technology. See, for example, Buckminster Fuller, "An Operating Manual for Spaceship Earth," in *Environment and Change: The Next Fifty Years*, ed. William R. Ewald, Jr. (Bloomington: Indiana University Press, 1968); Charles Susskind, *Understanding Technology* (Baltimore: Johns Hopkins University Press, 1973); Gerhardt Mensch, *Stalemate in Technology: Innovations Overcome the Depression* (Cambridge, Massachusetts: Ballenger, 1975).

5 See, for example, Thomas B. Farrell and G. Thomas Goodnight, "Accidental Rhetoric: Root Metaphors of Three Mile Island," *Communication Monographs* 48 (1981): 271-300; Alan G. Gross, "Public Debates as Failed Social Dramas: The Recombinant DNA Controversy," *Quarterly Journal of Speech* 70 (1984): 397-409; Timothy W. Luke, "Chernobyl: The Packaging of Transnational Ecological Disaster," *Critical Studies in Mass Communication* 4 (1987): 351-375.

6 The concern for overemphasis on dialectics at the expense of understanding evolutionary and developmental qualities of technological deliberations is expressed by Kenneth Boulding, *A*

*Primer on Social Dynamics* (New York: Free Press, 1970). An interesting treatment of the rhetorical and ideological similarities between very different topics of controversy is found in Allan Mazur, *The Dynamics of Technical Controversy* (Washington, D.C.: Communications Press, 1981).

7 It has been argued that critical scholarship tends, in general, to focus on the clang and clatter of value conflicts to the detriment of a broader understanding of the discourse through which a community maintains itself. See, Roderick P. Hart, "The Functions of Human Communication in the Maintenance of Human Values," in *Handbook of Rhetorical and Communication Theory* ed. Carroll C. Arnold and John Waite Bowers (Boston: Allyn and Bacon, 1984), 749-760.

8 Particularly, Ellul, in *The Technological Society* on pages 3-22 and 61-64 and in *The Technological System* on pages 17-20. Also see Heidegger, "The Question Concerning Technology," and Arnold Pacey, *The Culture of Technology* (Cambridge: MIT Press, 1983), 1-12.

9 Heidegger, 287-288.

10 Pacey, 5.

11 Pacey, 6.

12 Karl Weick, *The Social Psychology of Organizing* (Reading, Massachusetts: Addison-Wesley, 1979), 131.

13 Carroll C. Arnold, "Reflections on American Public Discourse," *Central States Speech Journal* 28 (1977): 73-85.

14 Studies have consistently shown that utility and practical knowledge are central values in American cultural life. See Edward Steele and Charles Redding, "The American Value System: Premises for Persuasion," *Western Speech Journal* 26 (1962): 38-42; Wayne Minnick, *The Art of Persuasion*, 2d ed. (Boston: Houghton-Mifflin, 1968).

15 Harry Truman, "The Threshold of a New Age," *Vital Speeches of the Day* 18 (1952): 559.

16 Harry Slater, "Bridging the Gap: Atomic Power," *Vital Speeches of the Day* 36 (1970): 399-402.

17 Slater, 400.

18 Opponents of commercial nuclear power plants have been characterized as, among other things, misinformed, misguided, victims of propaganda, unfamiliar with technology, naive, uninterested, childlike, weak, precocious, irrational, slavishly devoted to political agendas, virulently anti-technology, and intentionally mischievous. While such indictments are found throughout the body of speeches I have reviewed, excellent exemplars include Francesco Costagliola, "Nuclear Power: The Information Communication Predicament," *Vital Speeches of the Day* 37 (1971): 436-441; William R. Gould, "Energy Options for the Future: Acceptability of Risks," *Vital Speeches of the Day* 39 (1973): 605-607; Endicott Peabody, "Toward an Energy Vision of America's Future: The Hard Choices Are Just Ahead," *Vital Speeches of the Day* 42 (1975): 136-140; Llewellyn King, "Nuclear Power in Crisis: The New Class Assault," *Vital Speeches of the Day* 44 (1978): 713-716; Russell C. Youngdahl, "Energy in the 80s: What If? And Yes If," *Vital Speeches of the Day* 46 (1980): 112-114.

19 This phrase is used to describe the matrices of symbolic relationships through which the deluge of everyday human experience is filtered, selected, and ordered into a pattern of meaning. Clifford Geertz, *The Interpretation of Culture* (New York: Basic Books, 1973), 5.

20 For example, Gould, 607.

21 See W. Donham Crawford, "Electricity: A Necessity for the People," *Vital Speeches of the Day* 43 (1977): 294; Frank G. Zarb, "Nuclear Power: A Time for Decision," *Vital Speeches of the Day* 41 (1975): 669; William R. Foulkes, "The Atom: Death—Or Life Abundant," *Vital Speeches of the Day* 14 (1948): 346.

22 Peabody, 138. In his speech, Peabody argues that the proof of the nuclear industry's "excellent safety record" was, at that time in 1975, "zero deaths and zero injuries."

23 John Jay Hopkins, "The Atomic Revolution: Creation Rather Than Destruction," *Vital Speeches of the Day* 29 (1954): 490.

24 Charles Robbins, "The Changing Picture of Energy: Need for Perspective," *Vital Speeches of the Day* 46 (1980): 297.

25 Luke, 367-370.

26 This characterization of industrial and technological social order reveals the interwoven relationships between cultural, social, economic, and political dimensions of what commercial nuclear power "means" to its proponents. See, for example, Hopkins, 489.

27 USCEA's vice president for advertising asserts that independence and economic growth are the two strategic pillars upon which rests the campaign to sell nuclear power to the public. Edward L. Aduss, "Why and How We Use Advertising to Communicate." Paper presented at the annual meeting of the Speech Communication Association, Chicago, November, 1986.

28 USCEA's advertising is handled by Ogilvy and Mather Advertising. Matthew C. Ross, vice president in charge of the USCEA account, asserts that energy independence, jobs, and economic prosperity form the emotional core of USCEA's appeals. See "Research and Execution of the USCEA Message: The Role of Ogilvy and Mather." Paper presented at the annual meeting of the Speech Communication Association, Chicago, November, 1986.

29 Pamphlets, booklets, and advertorials are the "curriculum" of USCEA's campaign according to Ross, "Research and Execution," 5. Although the topical focus of advertorials vary they are unified by the themes of independence and economic growth. Full page advertorials have appeared since 1983 in numerous issues of popular magazines including *Time*, *Reader's Digest*, *National Geographic*, *Newsweek*, *Psychology Today*, *Smithsonian*, and *U.S. News and World Report*.

30 Elmer I. Lindseth, "1958: Year of Decision in Atomic Power," *Vital Speeches of the Day* 24 (1958): 316.

31 For example, Crawford, 294-296; Brooke Hartman, "The People Want Electricity: Power Shortages and Nuclear Plants," *Vital Speeches of the Day* 39 (1973): 373-376; L. Manning Muntzing, "To Protect the Public Interest: A Regulator's Quest," *Vital Speeches of the Day* 41 (1975): 262-266; Robert W. Scherer, "Industry's Prospects for the Future of Nuclear Power: The Needs of the Public," *Vital Speeches of the Day* 42 (1975): 27-30.

32 Donald P. Hodel, "National Energy Policy Plan: A Preview," *Vital Speeches of the Day* 50 (1983): 131.

33 Youngdahl, 113.

34 This is not a particularly rare term in the lexicon of electric utility spokespersons. Representative is its repeated use by Linn Draper, Senior Vice President of Gulf States Utilities, on *The Phil Donahue Show*, October, 1982.

35 Donn L. Williams, "World Power: A New Meaning," *Vital Speeches of the Day* 40 (1974): 453.

36 James A. McClure, "Nuclear Energy: The Moral Issue," *Vital Speeches of the Day* 45 (1979): 327; King, 1978, 716; Costagliola, 1971, 441; John Ray Dunning, "Electrical Power Potentialities," *Vital Speeches of the Day* 20 (1954): 452.

37 Foulkes, 349.

38 Foulkes, 346-347.

39 Hopkins, 491.

40 Leland I. Doan, "How Will Atomic Energy Affect Industry," *Vital Speeches of the Day* 20 (1954): 451.

41 Youngdahl, 114.

42 Thomas Farrell and Mark Pollock, "Against the Purity of Power: Rhetoric Once Removed." Paper presented at the annual meeting of the Speech Communication Association, Chicago, November, 1986.

43 Lynn White, "The Historical Roots of Our Ecological Crisis," *Science*, 10 March, 1967, 1203-1207.

44 Arnold uses this phrase in "Reflections on American Public Discourse" to describe the conditions that a successful American orator must meet. Speakers must, says he, be "transcendentally confident in some visible ideal, systematically, doctrinally cogent, and at least apparently practical," p. 84.

45 John J. Hancock, "Atomic Development Authority," *Vital Speeches of the Day* 12 (1946): 734.

46 Mikhail Gorbachev, "The Chernobyl Accident: Nuclear Energy and Radiation," *Vital Speeches of the Day* 52 (1986): 514.

47 USCEA began within a week after the Chernobyl accident to run full page advertorials in newpapers and magazines throughout the United States. The ad carried the banner, "Energy Update" and purported to explain, "Why what happened at Chernobyl didn't happen at Three Mile Island." Chernobyl is explained as a result of a faulty design that did not provide, as American reactors do provide, adequate "defense in depth" against Chernobyl-like accidents. This ad appeared in May 26 issues of *Newsweek*, *Time*, and *U.S. News and World Report*. The necessity of choosing the nuclear option was maintained by technical experts and popular media in both Western and Eastern countries after Chernobyl. See, especially, Luke, 370-375.

48 Truman, 557-558.

49 Buell G. Gallagher, "Prometheus Rampant: Responsibilities of the Atomic Age," *Vital Speeches of the Day* 13 (1946): 142.

50 Williams, 453.

51 Cyril Welch, "Speaking and Bespeaking," in *New Essays in Phenomenology*, ed. James M. Eadie (Chicago: Quadrangle Books, 1969), 72-82.

52 Heidegger, 1977.

53 Michael Hyde, "The Human Component in Technology," in *Phenomenology and the Understanding of Human Destiny*, ed. S. Sousgaard (Washington, D.C.: Center for Advanced Research in Phenomenology and the University Press of America, 1981), 36.

54 Heidegger, "Question Concerning Technology," argues that one of the elemental features of technological consciousness is that the ontological status of everything is subservient to what it may be used for. This conception corroborates Richard Weaver's argument that modernity has come dangerously close to deconstructing the social order because concern for the function of things has obliterated the philosophical question of ontological status. See Richard Weaver, "Status and Function," in his *Visions of Order: The Cultural Crisis of Our Time* (Baton Rouge: Louisiana State University Press, 1964), 22-39.

55 The debate concerning public versus private control of nuclear energy has been well documented. See, especially, the discussion in Richard G. Hewlett and Jack M. Holl, *Atoms For Peace and War, 1953-1961: Eisenhower and the Atomic Energy Commission* (Berkeley: University of California Press, 1989).

56 *Atomic Energy* (Irving, Texas: Boy Scouts of America, 1983), 36.

57 See discussions by Farrell and Goodnight and Luke concerning the rhetorical consequences of these nuclear accidents.

58 See Erving Goffman, "Primary Frameworks" in his *Frame Analysis* (Cambridge: Harvard University Press, 1975), 21-39, for a discussion of natural versus social frameworks for constructing the meaning of human action.

59 J. A. Haddad, "Technology and Human Values," *Vital Speeches of the Day* 52 (1985), 241.

# The Technological Priesthood:
# A Case Study of Scientists,
# Engineers, and Physicians for
# Johnson-Humphrey

*David Henry*

The explosion of the atomic bomb changed radically and permanently the role of science and technology in American culture. Not least noteworthy was the scientist's emergence from the laboratory and subsequent entry into the fray of political decision making. The change was attributable partly to the prominence of men like J. Robert Oppenheimer, wartime director of the Los Alamos laboratory, site of the initial nuclear blast. Oppenheimer's appeal resided in the absence from his demeanor of the "popularly held image of the scientist as cold, objective, rational and therefore above human frailty, an image that scientists themselves fostered by underplaying their personal histories and the disorder that precedes the neat scientific conclusion."[1] Oppenheimer and his co-creators of the bomb came to be "looked upon by the public with almost reverent awe."[2] Concurrent with their new-found adulation, scientists like Niels Bohr, James Franck, and Leo Szilard began to experience a sense of responsibility for their discovery. Who better to decide the future of nuclear policy than those who best knew the bomb's devastating potential? Scientists saw, and met, two responsibilities: They assumed the role of the experts they were and began to speak out on public issues, and they stimulated discussion by encouraging non-experts to study and to contribute informed commentary to the emerging dialogue on nuclear policy.[3] The scientist as public advocate thus began to surface.

A major difficulty in the surfacing process quickly became apparent: the public's tendency to view the scientific community as monolithic. Marked divisions within the scientific elite characterized the often-heated debates in the 1940s over such issues as international cooperation and military vs. civilian control of the atom.[4] Yet two decades later, the public too often failed to recognize that scientists' judgments in political and social matters were as susceptible to non-scientific considerations as those of any other disputant. Hans Bethe and Edward Teller, for example, both possessed impeccable scientific credentials. But on the question of curtailed tests of nuclear weapons in the early 1960s, what Robert Gilpin terms a "problem of conflicting expertise" arose. Both scientists worked from the same data base. Bethe concluded that the evidence proved it *"technically feasible* to devise a system of detection stations and inspections which give reasonable assurance against clandestine testing." Teller, however, surmised that without doubt there *"are no technical methods* to police a test ban." Further investigation of the conflicting conclusions revealed that each scientist had interpreted the "scientific facts" according to his own political predispositions.[5] Teller again exercised his predilection for authoritative omnipotence in 1962, when he informed a group of NATO military and civilian leaders that the "scientific world" would soon produce a "class of weapons that the Russians could never counter." Lord Solly Zuckerman of the British contingent rose at the conclusion of Teller's remarks. Teller, said Zuckerman, was "presuming when he said that he was speaking for the world of science." Teller's non-laboratory-based retort after the meeting—"I will never forgive you for that"[6]—revealed the very human level at which arguments about scientific "fact" transpired.

Perhaps even more telling for immediate purposes were the sojourns of the technological elite into debates about public policy on issues outside their areas of expertise. Given that political argument is much too important to be left to the politicians, there is surely nothing inherently wrong with the scientists' entry into debate over public policy. There is, however, a fair question to be raised about the assumption that expertise in any field, even one as vital as nuclear physics, establishes authority on all issues. That, nonetheless, proved one consequence of the scientists' entry into public affairs. Thus, during the 1940s and 1950s scientists increasingly held forth on virtually all public issues even distantly related to science and technology, even though their conclusions were often not grounded in hard data. At best, wrote sociologist George Lundberg, it was not "always clear when a scientist is expressing a scientific conclusion and when he is expressing a personal preference." At worst, he continued, it was "simple fraud" for scientists to "pose as disinterested scientists announcing scientific conclusions" when in fact they "are merely expressing personal or group preference."[7] Stephen Toulmin warned of the ultimate harm of the emerging cult of expertise. When the public tries "to read in the scientist's palm the solutions of difficult problems in other fields," Toulmin argued, "we are asking of him things that he is in no position to give, and converting his conceptions into myths."[8] The present case study traces the construction and perpetuation of that myth by the scientists themselves, in association with partisan politicians.

For despite the clear absence of a unified scientific community,[9] it is my central thesis that a rhetorically and politically potent "technological priesthood" ultimately evolved from the scientist-citizen's activism. After nearly two decades of public advocacy, the technological elite had achieved a rhetorical stature deemed potentially powerful in mainstream electoral campaigning. Recognizing that potential, election strategists for Lyndon Johnson integrated into the campaign framework an organization identified as Scientists, Engineers, and Physicians for Johnson-Humphrey (SEP). More than an endorsement group, SEP members worked at all levels of the campaign, from grassroots tasks such as walking precincts and stuffing envelopes, to public rhetoric inclusive of writing articles and making personal and televised appearances on behalf of the Democratic ticket. Examination of two major print documents and a 30-minute televised commercial from SEP's public rhetoric illumines the unique discursive functions of the scientist as political advocate. The analysis begins with a delineation of the rhetorical dimensions of the technological priesthood, proceeds to an explication of the impact of the consequent political rhetoric on the 1964 campaign, and concludes with observations about the phenomenon's potential influence in future political discourse.

## THE TECHNOLOGICAL PRIESTHOOD AS RHETORICAL CONSTRUCT

The notion of a technological priesthood derives from a merging of Thomas Lessl's concept of the priestly voice with the theory of bardic rhetoric advanced by John Fiske and John Hartley. Priestly rhetoric is didactic, working vertically from the rhetor down to the audience. The language of the priestly form originates in nature; it is "filled with extra-human authority." Priestly rhetors address two audiences, other scientists and the public at large. This priestly voice is "*extensive*, the speech of one culture to another." Bardic rhetoric, in contrast, works horizontally, rhetors communicating as ordinary members of the community in conversation with other members of the community. The bard's language derives from the culture's shared symbols, as the audience addressed consists of the public. Bardic rhetoric is "*reflexive* . . . a culture speaking about itself to itself."[10]

Priestly rhetoric presumes centrality of the source, placing a "premium on its own specialized literacy." Similarly, scientists in the public realm combine technical terminology with the public's vernacular, "a habit that does more to evoke a desirable ethos than to disseminate knowledge." In addition to language, the "locus of priestly authority" originates in "institutional circles from which the audience is largely excluded." Placement of the priest—whether religious or secular—at the center of the culture results in a "synecdochic rhetoric." Just as a "priest interprets all experience" in terms of a "specialized subculture," so the scientist "depicts history as the unfolding of deterministic processes that lead naturally to modern science." Ethos and world view fuse, resulting in a "hierarchical rendering of history with science at its apex—the subtle suggestion being that the superiority of the scientific ethos is firmly reflected in the evolutionary character of nature." Critically, however, that ethos is used to emphasize the reciprocity between the scientific and non-scientific worlds:

> [A]s in the mediational discourse of the religious priest, the synecdoche we find in the rhetoric of public science manifests two directions of transference: (1) Elements of the scientific that overlap with ordinary experience are drawn upon in an effort to humanize the otherwise alien world of scientific objects, and; (2) elements of ordinary human experience that coincide with specialized scientific norms and values are reconstructed so that they achieve scientific meaning.

The consequent consubstantiality says, "Look at me. See how I think and act as a scientist. You also think and act in these ways. These scientific aspects define your humanness, making you a scientist as well." Despite the invitation to the public to "think of themselves as scientists," though, the scientist's world view is "decidedly that of the elite." Scientists may choose to speak on religious and political issues, "but they habitually resent the efforts of outsiders to speak on scientific ones." Public efforts to discuss science impinge too closely on the scientist's specialized, sacred domain.

Bardic rhetoric, on the other hand, attends to the "negotiated central concerns of its culture," de-emphasizing the particularized interests of individual groups.[11] Common language, the first of four relevant traits of bardic communication,[12] is central to the expression of those concerns. Where technical expression infuses priestly rhetoric, the bard works with the "available linguistic resources of the culture" to produce for members of the audience a "confirming, reinforcing version of themselves." Second, authority for bardic communication resides in the audience or public, rather than in the source. Hence, messages are structured "according to the needs of the *culture* for whose eyes and ears they are intended," rather than for the demands of the text or source. Third, though authority resides in the public or audience, bardic communicators are at the center of culture because they speak "to all members of [a] highly fragmented society." That centrality is more a matter of communicative function as a conduit for messages, however, than as a source of influence or power. Finally, bardic rhetoric is "*oral*, not literate." In combination, these characteristics enhance communication by a culture "with its collective self."

The activism of Scientists, Engineers, and Physicians for Johnson-Humphrey evidences the working in tandem of the bardic and priestly voices to create a scientific ethos unique to the technological priesthood. In public policy debate, the priestly voice enhances technological rhetors' credibility and lends force to their pronouncements, while their adaptation to the demands of bardic communication promotes sufficient audience attention to allow them to take advantage of that credibility.

## THE JOHNSON-HUMPHREY CAMPAIGN

"While the fundamental purpose of the campaign is to re-elect [sic] President Johnson," stated an internal memorandum circulated at the president's advertising agency in May, "it is our additional goal to create a landslide, a mandate." William Bernbach of the agency, Doyle Dane Bernbach, reiterated that goal three months later in a letter to Bill Moyers, Johnson's press secretary and political adviser: "No one knows better than you why we took on the Presidential campaign. There is only one reason. We are ardent Democrats who are deadly afraid of Goldwater and feel that the world must be handed a Johnson landslide."[13] The landslide objective helps to establish the campaign context and to explain the attention given groups like Scientists, Engineers, and Physicians in 1964.

Despite Johnson's obvious strength as an incumbent running against an opponent of dubious threat to his election, campaign strategists were determined to move toward a January inaugural under the force of a popular mandate. Essential to such a mandate was wide support among a broad spectrum of the electorate. Two facets of the overall campaign strategy were designed to stimulate such support. Johnson-Humphrey (1) developed an array of citizens committees intended to exploit the perceived credibility and expertise of their members, and (2) fitted mass media use carefully to the strengths of each group.

Though such planning might easily define any campaign, strategists engaged a unique challenge in integrating scientists and technologists into the overall scheme. On one hand, planners understood the esteem in which a broad range of the public held the scientific community. On the other, the scientists' inexperience in transferring that credibility to campaign messages transmitted through the electronic media might hinder optimum exploitation of that esteem. For despite their increased activism in the post-war era, scientists were far more comfortable with the print culture than with the alien world of television.[14] The rhetors' competence thus endowed them with a priestly voice, but television's pre-eminence placed them in a bardic culture. Resolution of this potentially problematic dilemma came in two steps. First, Scientists, Engineers, and Physicians for Johnson-Humphrey (SEP) formed under the auspices of an overarching citizens' committees program, wherein the group received a clearly defined target audience and task. Second, rather than allowing the requirements of television to dictate what the group said and how, SEP first established its campaign credibility through a thorough print campaign rooted in the scientific ethos, then imposed that ethos on electronically mediated messages, particularly television.

## Organizing Scientist-Citizens

Preliminary discussions between Doyle Dane Bernbach and the Democratic National Committee revealed three groups of voters that comprised the campaign's target audiences: (1) the "middle class, middle-income, suburban dwelling blue and white collar family"; (2) ethnic groups of "specific old-world nationalistic origins," a target whose "complete definition . . . is still to be determined"; and (3) "Intellectual groups. These are the academic and culture-oriented groups among whom it is felt Mr. Johnson's image is diffuse and his background within their environment is unknown." Interacting with these demographic groups were the geographic requirements that the president focus on 12 states, where victory would provide overwhelming Democratic support in the Congress, thus ensuring the implementation of the electoral mandate.[15]

The scientists' involvement in the campaign targeted two of the three groups. Most obviously, they aimed to appeal directly to "intellectuals." At this level, advocates directed their message to an elite audience whose members presumably shared the rhetors' assumptions, analytical approach to problems, and faith in a logical-positivist philosophy.[16] Less obvious but equally significant were the scientists' messages directed at the campaign's primary target, the middle class. Here the scientist campaign's bardic dimensions surfaced. On television, engaging a cross section of the public, the scientists became part of the culture "addressing itself," as but one of over two dozen citizens' committees.

Organized under the direction of 30-year political veteran James Rowe, the citizens' committees targeted the interests of women, rural America, youth, the arts, entertainers, labor, district attorneys, military veterans, and business, among others.[17] Unlike similar organizations in past campaigns, the citizens' committees were to function as action groups rather than as campaign auxiliaries or think tanks. Rodney Nichols, who helped organize Scientists, Engineers, and Physicians, recalled that the group "decided at the outset" that its work "would only be meaningful if it was a vote-getting mechanism, not another brain trust."[18]

With Donald MacArthur its head, SEP was at the forefront of the citizens' committees program.[19] Indeed, because of MacArthur the scientists led the way. A physical chemist by training, the 33-year old MacArthur had virtually no background in politics, but he feared the possibility of Barry Goldwater's election. A week after the Senator's nomination, MacArthur was at the Pentagon to discuss with Harold Brown, Johnson's Defense Director of Research and Engineering, the feasibility of organizing scientists into a political force. With Brown's support, MacArthur spoke with the President's science advisor, Donald Hornig, who in turn encouraged MacArthur to contact numerous scientists from government, education, and industry who had expressed an interest in just such an organization. Though Rowe advised caution, urging MacArthur to await the central campaign's instructions on forming and launching the scientists' organization, MacArthur used his own money to secure a headquarters, initiate a communication network, and recruit a staff, including Nichols. Recognizing the limitations imposed by their own political inexperience, SEP quickly sought professional advice, hiring David Garth, a young but experienced campaign adviser and Peabody Award winner for his work on television.

The organization issued guidelines for action, affirming the group's commitment to a meaningful contribution to the campaign. The overarching objective was to "influence opinion on science-related campaign issues through public pro-Administration statements from spokesmen representing the scientific and engineering communities" and to "stimulate and support scientists and engineers across the country to organize 'get out the vote for Johnson' groups."[20] In pursuit of that objective, specific guidelines emphasized that: (1) the aim was not to lobby on behalf of science, but to "enlist these communities in behalf of Johnson and to turn their prestige against Goldwater"; (2) Scientists and Engineers for Johnson would "not simply serve as a brain trust or as window dressing for Johnson's candidacy" but would be a "political action" outlet; and, (3) the organization was to be bi-partisan and financially independent of central Democratic Party fundraising. In essence, these guidelines identified the rhetorical strategy the organization would employ through November in weaving the scientific ethos derived from the technological priesthood into the fabric of the campaign. That ethos consisted in the rhetors' objectivity, authority, and placement of their roles as citizens above their narrower professional concerns.

On August 13, 1964, the press release announcing the organization's formation and goals made explicit the centrality of that strategy.[21] "Forty-two of America's leading men and women," the press release declared, "today announce formation of SCIENTISTS and ENGINEERS for JOHNSON, a bi-partisan committee pledged to the election" of the president.[22] Indented and set off from the text were five representative members, selected to reinforce the organization's bi-partisanship, varied service in government and industry, and excellence in scientific achievement. "Among the organizing group," the statement continued, were:

> *Dr. George Kistiakowsky,* Science Advisor to President Eisenhower, now Professor of Chemistry at Harvard University.
>
> *Dr. Jerome Wiesner,* Science Advisor to the late President Kennedy, now Dean of Science at M.I.T.
>
> *Dr. Paul Dudley White,* internationally known heart specialist.
>
> *Dr. Kelly Johnson,* Vice President, Lockheed Aircraft Corporation and noted designer of the A-11 aircraft.
>
> *Dr. Katharine McBride,* President, Bryn Mawr College.

The committee intended to "support the work of scientists and engineers of both parties in a strong pro-Johnson vote drive at the grass-roots level" and to "stimulate the involvement in active politics of scientists and engineers across the country." Most important, members of the committee would "comment through national media on science-related issues throughout the campaign." Those issues included a mix of topics, some clearly suited for technological expertise—a "balanced and rational" defense policy, "arms limitations consistent with the national security," and continued progress in "medical and scientific research"—and others with a less-obvious scientific orientation, such as "unqualified enforcement of the Civil Rights Act" and "rejection of extremism under any guise." A list of over three dozen prominent physicians, research scientists, and engineers comprised the announcement's second page.

In the body of its "Statement of Principles" issued three days later, SEP urged exploration of outer space, research of the oceans and of the earth's crust, and the application of new technologies to improve health care,[23] all topics where their expertise might be fairly presumed. At the same time, though, the statement used that authority to lend force to claims clearly beyond the edge of predominantly scientific investigation. "As citizens and members of the scientific and engineering community," stated the opening lines, "we have the right and the obligation to speak out in this critical campaign." The text then moved beyond this clearly unobjectionable claim to enumerate the group's positions on space, ocean research, and health care. But in a foreshadowing of the tactics that would define the campaign to follow, emphasis then shifted from technical matters to issues of public policy.

On two counts in particular, the document intimated the scientists' authority in areas about which they held, at best, marginal expertise. On both, the groundwork was being laid for subsequent print and electronically mediated messages. The target, though only implied at this stage of the campaign, was Barry Goldwater. First, SEP "reject[ed] 'extremist' solutions to any of the problems that confront us. We urge our colleagues not to surrender to the myths and distortions which are seductively easy and simple." Here, ironically, the scientists at once warned against deception and myth while employing the myth of technological superiority to strengthen their appeal. Second, the statement specified one such distortion by indicting—albeit still not by name—Goldwater's misleading representation of the complexity of nuclear arms control. Focus was on the Senator's alleged contention that under some circumstances it might be unwise to limit control of the nation's atomic capacities to the president. The scientists affirmed their "unqualified support of the time-tested policy of *exclusive* Presidential determination of the use of nuclear arms, whether strategic or tactical. The use of nuclear weapons can be contemplated only when the gravest national interests are at stake, which is a decision only the President can make."

MacArthur, Nichols, and Garth—working in conjunction with Moyers and others at the White House, as well as with Doyle Dane Bernbach—predicated the SEP strategy on the supposition that the public attributed to scientific leaders special skills and knowledge. Once again taking advantage of the myth of technological expertise, the campaign stressed the classic dimensions of ethos: the rhetors' knowledge, respect, and goodwill. In a list of talking points recommended for emphasis by central SEP headquarters to state and regional offices, John Rubel, Vice President of Litton Industries and an original member of the organization's founding committee, urged the group's advocates to note the potential persuasiveness of messages built on their acknowledged expertise. What should be pushed, Rubel maintained, was the "excellence and integrity of American science and technology . . . vital to our continued economic development, as well as to national security. Scientists and engineers are as deeply concerned as other citizens with all political questions, but they are particularly concerned where their knowledge, and indeed, their work are involved in the major issues of the day."[24] In response to critics who charged that scientists overstepped their authority upon entering political debate, one writer contended that "science has entered so far into the public domain, and with such prestige and influence, as to be useful to nonscientist politicians; then is it not better that the scientist act rather than leave totally in the hands of others actions involving themselves and the accumulated prestige of their fields?"[25] To reassure the public that their faith in the scientists' knowledge and prestige was well-placed, advocates stressed the motivation of goodwill that undergirded their actions. Implicit in the technological community's involvement in the campaign was the conviction that such action was essential for the public good. The country's future, for instance, moved Nobel prize-winning chemist Harold Urey to work for Johnson-Humphrey, the first campaign in which he had "ever worn a campaign button and ever had a bumper sticker on my car." Similarly, George Kistiakowsky, science advisor to President Eisenhower, told a gathering in Washington that he had been "non-political" until the contest between Johnson and Goldwater. But because the "issues to be decided are so vital" to the nation, he said, "no one can afford to be neutral."[26] With the organization in place and a scientific ethos established, Scientists, Engineers, and Physicians for Johnson-Humphrey moved forward to the fall campaign.

## Mass Media and the Scientist-Citizen

Integral to the campaign, was a multimedia plan that built carefully toward optimum exploitation of television. From the inception, Johnson's strategists plotted a well-integrated advertising and public communication scheme involving speakers' bureaus, network radio, radio spots, and newspapers.[27] And to some extent Scientists, Engineers, and Physicians for Johnson-Humphrey campaigned in all of these media.

What accommodated the scientists' facility for fulfilling the rhetorical expectations of the technological priesthood, however, was the intertwining of the group's print and television messages. The campaign proceeded serially. First, print messages—an introductory pamphlet and a campaign book, in particular—put into operation the plan laid out in the organizational phase: group members' scientific expertise was stressed as a foundation for warranting acceptance of their claims about political issues. Second, the credibility engendered by the print campaign set the stage for television appearances, culminating with the 30-minute television special, "Sorry, Senator Goldwater." Despite the closing telecast's poor use of the medium, consideration is due the program's context as an extension of the scientists' credibility established through print messages. Such consideration warrants the claim that what may have been "bad television" by spot commercial standards, constituted sound strategy for persuasion dependent on scientific ethos and linear argument.

*Constructing the Print Campaign*

Media use in the 1964 campaign may be accurately defined as in transition. Not only was television's dominance emerging, but the roles of other message sources were undergoing re-definition as well. Books and pamphlets, for example, "achieved unusual prominence . . . in the shaping of perceptions of preferences, issues, and candidate images in both parties."[28] SEP produced two such texts that demand critical examination: a short pamphlet issued at the campaign's outset, and a 40-page monograph, "The Alternative is Frightening." Though the graphic or non-linear facets of these works deserve attention, it is their conventional print features that establish and reinforce both the advocates' credentials and the lines of argument that eventually permeated their televised communication.

Entitled simply, "Scientists and Engineers for Johnson," the pamphlet introduced key elements of the effort to generate support for the president based on arguments to which, it was believed, the technological community would adhere.[29] Printed in two slightly different shades of beige, the pamphlet's cover carried the title in the upper left-hand corner. In the lower right-hand corner appeared the words, on four lines, "bi-partisan campaigners pledged to the election of Lyndon Baines Johnson." Eight mathematical symbols, in large print, bordered the right-hand margin and dominated the cover. The president's speech to the National Rocket Club on March 23 of the previous year provided the opening statement on the first page. The "future capabilities of our country," the excerpt stated in part, "depend upon mutual understanding and mutual trust . . . between the community of technology and the community of public affairs." Scientists and Engineers was then identified as a "bi-partisan, self-supporting committee pledged to the election of Lyndon Johnson," whose members "will comment through national media on science-related issues." Opening the pamphlet to its center pages revealed SEP's relative emphasis on issues and credibility in the campaign's initial stages. The top half of the left-hand page listed eight issues the committee "specifically supports," while a roster of over 40 organizing committee members and their credentials covered the remainder of that page and all of the next. A second Johnson quotation concluded the pamphlet, the president noting that "it is neither too much to ask— nor too much to hope—that responsible men of the technological community and the political can and will find their way to achieving the goals essential to our freedom."

In microcosm, the pamphlet foreshadowed the scientists' subsequent print and television campaign. Note first that among the "science-related issues" about which their expertise invited audience adherence were, as in the press release announcing the group's origin and the statement of principles that followed, "rejection of extremism under any guise," enforcement of the Civil Rights Act, and enhanced educational opportunity. Though surely the SEP members' rights as citizens guaranteed the privilege to express themselves on these issues, it was clearly intended that their perceived credibility as scientists warrant acceptance of their testimony on topics only vaguely linked to their expertise. A second feature worthy of critical attention is the emphasis on numbers of group members and their credentials. Drawn broadly from government service, the academic world, and private industry, SEP members cut across party lines and ranked in the highest echelons of their fields. Receipt of a Nobel prize in chemistry, ran the message's reasoning, certified an advocate's credibility on all issues even marginally identifiable as "science-related." Finally, the pamphlet's language hinted at the interrelationship between scientific ethos and the rhetors' expression of their ideas. Consider the call for "a balanced and rational defense policy," for instance, or their "rejection of extremism under any guise." In either case, the perceived authority of ostensibly objective scientific testimony imbues rather nebulous phrases with the force of assumed neutrality. In reality, the introduction of subjective yet powerful terms like "extremism," "rational," and "balanced" paved the way for a severe indictment of Barry Goldwater's character and judgment, and thus of his fitness for office.

Scientists, Engineers, and Physicians for Johnson-Humphrey explicated that indictment most fully in "The Alternative is Frightening." Graphics again enhanced the argument, but given the work's monograph form, the emphasis was on the text's apparent structural integrity. The cover nevertheless combined color and imagery complementary to the argument.[30] Below the title, which was printed in red, the cover declared in blue print, "Johnson-Humphrey Must Be Elected." Serious-faced photographs of President Johnson and Senator Humphrey accompanied the declaration. In contrast, a dated photograph of Barry Goldwater's face, topped by a navigator's cap to reinforce the SEP's portrayal of the Senator as a dangerous warmonger, appeared above the words "alternative" and "frightening."[31] The mathematical symbols that dominated the short pamphlet's cover remained as part of the logo, though their placement at the bottom and in smaller print de-emphasized their importance in relation to the message imparted by the text.

The title page established not only the book's contents, but the structure of the SEP argument as well. That argument had four parts: the preface explained, "Why we have decided to participate"; part one detailed the "Black and White World of Mr. Goldwater"; "The Case for JOHNSON-HUMPHREY" followed; and, "The Ways You Can Help" closed. The text's structure thus moved logically from establishment of the organization's credibility, to an interpretation of Goldwater's record that gained credence from its linkage to the rhetors' ethos, to delineation of the Johnson-Humphrey alternative, to the action phase calling for monetary and volunteer support.

"Scientists and engineers are not, as a rule," stated the first line, "politically minded. The public generally thinks of them as brainy, bespectacled fellows, sitting in their laboratories or before their drawing boards, contemplating lofty questions." But in recent years, there has been a "growing awareness among scientists of their political responsibilities, as citizens and as members of the technical professions." While it may have been possible in the past for scientists to "do their work and let politics take its course," the demands of modern society dictated a new role. Because "[s]cientists have come to recognize that they have very special competence and understanding in fields that weigh heavily on the decisions that the leaders of this nation must make," it had become their "duty to contribute to the resolution of many basic questions facing the nation." Responding to that duty, SEP's "*bi-partisan* and *self-supporting*" membership aspired to only "one aim: electing to the highest office in the land in 1964 the most competent, most responsible men available." In remarkably succinct fashion, the characterization depicts men and women possessed of the noblest qualities: reluctance to exercise undue influence on others, special expertise that would benefit others, and movement to action out of a sense of duty.

In fulfilling that duty, the scientists addressed their message to peer and public alike[32] in a fashion reflective of priestly rhetoric's synecdochic form, evincing reciprocity between source and auditor. The "discussion [was] intended to be of special interest to the technical world," the campaign book stated. But "because science and technology today are inseparable from considerations of public policy, it is by *necessity* devoted to topics which affect the future of *every citizen.*" SEP hoped to show that "the American people, and the science and technology which work for them, will best be served by the election of Lyndon B. Johnson and Hubert H. Humphrey."

Derogation of Barry Goldwater constituted the essential first step in demonstrating the desirability of a Johnson-Humphrey victory. Having established a basis on which readers might infer the scientists' credibility and the common bond between reader and writer, SEP built on the language-use tactic only hinted at in the earlier pamphlet. Employing connotative words as if the writers' credibility endowed them with denotative meaning, "The Alternative is Frightening" portrayed Goldwater as simplistic, ill-informed, and dangerous. Stressing the importance of electoral participation, the indictment of the Senator began, "In some elections, some of us may feel the decision doesn't matter. In this election, many of us feel that it matters very much." Goldwater himself provided the cause for concern, for

> his views are almost always *out of step* with every trend in American life. He sees the world as simply *black or white*, generally takes a *negative*, a *destructive* stand. His *little world* is revealed clearly through his statements.
>
> Decide whether this man, attended by advisers attracted to the *extreme minority views* he mouths, is equipped with the *understanding*, *patience*, *information*, and *crisis decision-making qualities* required to lead us as President of the United States.
>
> Goldwater's *confusion*, his *shallowness* in responding to the large and subtle problems facing this nation at home and abroad, are displayed—and displayed very *frighteningly*—in his own words on the following pages (emphasis added).

The explicitly negative language of the first and third paragraphs makes enthymematic completion of the more implicit, rhetorically framed second paragraph a matter of course. Moreover, packaging the litany of adjectives between the recitation of the scientists' benevolence and the cleverly constructed review of Goldwater's record that followed, intimated the force of evidence without the bother of developing a sound argument.

To complete the picture of Goldwater as inept, the Senator's record, political philosophy, and concept of government were then presented in his own words and those of his presumed allies. Fully one-fourth of the text offers excerpts of the following sort:

sister doubts if he ever read a book . . .

reads sports page, but "reluctantly thumb(s) to the front of the paper"

"I've been exposed to problems and I don't have to stop and think in detail about them . . ."

"The child has no right to an education. In most cases the children will get along very well without it."

"The Supreme commander of the North Atlantic Treaty Organization . . . should have direct command over a NATO nuclear force. . . . A nuclear NATO could meet local invasions on the spot, with local tactical nuclear weapons."

Concludes the SEP book, "Have you *ever* heard or read of a President of the United States capable of the kind of rash, irresponsible comments which Goldwater makes?"

Even when attention shifted to why President Johnson deserved the people's support, Goldwater's shortcomings occupied center stage. "We believe," the scientists concluded, "that President Johnson possesses the perception, restraint and understanding to be granted this authority (control of nuclear arms) by the American people." The conclusion evolved from the following data: Senator Goldwater "recklessly . . . would allow field commanders to exercise this authority in many instances"; the Senator was quoted as saying he would "drop a low-yield atomic bomb on the Chinese supply lines in North Viet Nam, or maybe shell 'em with the Seventh Fleet"; Goldwater consistently exhibited that he "lacks the judgment to hold the fate of the world in his hands." Though construction of "The Case for Johnson-Humphrey" faithfully described the incumbent's record and goals, on the driving issues of science and technology "The Alternative is Frightening" continually returned to "the case against Goldwater." The book urged support of the case by concluding with a series of "action suggestions," ranging from inviting SEP speakers to address local groups, to soliciting funds, to circulating petitions. Consistent with the organization's founding principles, Scientists, Engineers, and Physicians sought direct participation, not backroom wisdom.

At the national level and in already active local branches, such rhetoric enhanced the scientists' ethos so painstakingly constructed over the late summer and early fall. Their credentials as members of an elite whose expertise justified the assumption of a didactic role in relation to both peers and public reflected clearly the functions of priestly rhetoric. Their discourse was "*extensive*, the speech of one culture to another." The shift from print to television in October and November therefore presented a dilemma of considerable import. A bardic medium, television ostensibly succeeded best when rhetors employed the common symbols and language of the culture, thus promoting a *"reflexive"* rhetoric—"a culture speaking about itself to itself." In "Sorry, Senator Goldwater," Scientists, Engineers, and Physicians balanced bardic and priestly functions in using television to extend a solidly constructed print campaign.

## Adapting Print to Television

Goldwater's statements and record occupied much of the Johnson television planning. Extending the themes of the print campaign to the electronic medium, Johnson's advisers intended to make the "anti-Goldwater schedule . . . the first commercials aired. . . . [B]ecause of the importance of the anti-Goldwater effort, we should consider devoting 40% of network volume and 60% of local volume to these commercials." To emphasize the threat posed by a Goldwater presidency, "All commercials are to be tagged 'Vote for President Johnson on November 3. The stakes are too high for you to stay home.'"[33] A particular concern was Goldwater's effort to modify his perceived extremism. William Bernbach warned Bill Moyers, for example, of Goldwater's intention to "adjust his extreme position to one more acceptable. Knowing the short memory span of the average person, it is entirely possible he might succeed in creating a new character for himself if we are unable to remind people of the truth about this man."[34] Television, Lloyd Wright added in another memorandum to Moyers, allowed the Johnson campaign to "keep Goldwater on the defensive," and prevented the Republican candidate from "covering up his extremist stands, his irresponsibility, etc."[35] Coupled with implicit and explicit claims about the qualifications of the incumbent and those who campaigned on his behalf, concentration on Goldwater's alleged deficiencies characterized both the spot commercials and the longer telecasts.

In contrast to detailed analyses of the spot commerical campaign,[36] the 30- and 60-minute programs have occupied far less critical attention,[37] often being dismissed as inappropriate for a medium aimed at the auditor's limited attention span. Such dismissal overlooks their contextual significance in the Johnson election campaign, however, specifically in light of their role in exploiting the scientific ethos characteristic of the technological priesthood. Consider the importance the campaign planners attached to "Sorry, Senator Goldwater."

According to Democratic Party officials, it was Johnson's idea to sponsor a panel discussion in which members of the scientific elite would denigrate the Goldwater challenge and extol Johnson's virtues.[38] Working independently of Doyle Dane Bernbach, David Garth took charge of producing the program which was taped under the sponsorship of Scientists, Engineers, and Physicians on October 18, two weeks before the telecast.[39] After reviewing the tape and showing it to Bill Moyers, Lloyd Wright, and others at the White House, Donald MacArthur described the program to the president as "a tremendously strong and frank political endorsement of President Johnson from the point of view of peace, national preparedness, continuity of leadership, and a range of domestic issues including education." To be "educated" was to come to know the secrets [secret knowledge] of the technological priesthood, since scientists worked with the "hard data" of "real knowledge." MacArthur outlined three distribution choices, concluding with the "third proposition [which] is for a 128-major station network hook-up covering 93% of the U.S. television homes. The cost is $34,297. I strongly recommend option 3."[40] The most expensive of the options, the third proposition also had the advantage of reaching the largest market, thus prompting Johnson's approval of MacArthur's recommendation.[41]

In a series of letters following the taping, SEP staff director Rodney Nichols thanked participants and detailed the contributions they had made to the campaign's effort to align the credibility of the technological and scientific spheres with Johnson's candidacy. To Benjamin Spock, for example, Nichols wrote, "Your participation sharpened and strengthened the image we wanted to convey: a broadly based technical community supporting Johnson and Humphrey." The people who had seen the show, "including Moyers of the White House Staff, have been most impressed with your forceful and effective 'performance,' particularly on the education issue." George Kistiakowsky received high praise for his "remark concerning 'my former boss, President Eisenhower' supporting the Test Ban Treaty. All those who have seen the tape . . . have singled this comment out as a potent political endorsement." Nichols apologized to Kistiakowsky for the participants' having had to work from a script, but noted that "this was required to insure a fast-moving and complete discussion." Though not all of the program adhered strictly to the plan, Nichols told Herb York his work as "the moderator was excellent, holds the show together in a way that could not have been duplicated by any professional interviewer, and in fact deepened the sense of conviction in all exchanges." He attributed that in part to York's "inserting 'personal experience' and informal moving of commentary" which "made the show come alive." And with two weeks remaining before the election, Nichols wrote to Harold Urey that the media planners had been moved to consider ways to integrate Urey's "eloquent reminder" of the parallels between Teddy Roosevelt and the president into the ongoing spot commercial campaign.[42] In providing expertise, bi-partisanship, eloquence, and evidence based in current data and historical examples, the scientists' participation had anointed the Johnson election effort with the blessing of the technological priesthood.

### "Sorry, Senator Goldwater"

The merging of the priestly and bardic forms attendant to the anointing process culminated in the telecasting of "Sorry, Senator Goldwater" on November 1. At the conclusion of that Sunday afternoon's American Football League game, an announcer for the American Broadcasting Company informed viewers that a political program would replace the network's regularly scheduled show. After a brief pause, the screen revealed a man familiar to millions, seated on the edge of a desk. "My name is Henry Fonda," he began. "The following program is a paid political telecast sponsored by the Scientists, Engineers, and Physicians for Johnson-Humphrey. This organization is bi-partisan. It's been working hard across the country for the election of President Johnson and Senator Humphrey."

Fonda continued to speak as he moved casually toward the camera: "In the past several weeks, and in fact in past presidential campaigns, we've heard and seen politicians, famous personalities, even movie stars endorse candidates. What is unique about this campaign is that for the first time men and women from every level of science, medicine, and engineering have joined behind one candidate, President Johnson." Six of those "thinkers, leaders, and planners in fields ranging from medicine to missiles, from national defense to the life sciences," were about to give the nation "their thoughts on the candidates, based on their knowledge and years of experience in areas that you and I simply read or have heard about." These men "are not here to entertain us," said Fonda in concluding his brief introduction, "but to discuss the decision that you and I must make this Tuesday—a decision that will most certainly affect our lives for the next four years, and quite probably the lives of every man, woman, and child on earth."[43]

From the confident and accomplished Fonda, the camera shifted to a logo. Imposed over a black and white drawing of a mushroom cloud was the program's title, "Sorry, Senator Goldwater, The Country Just Can't Risk It." The program's moderator uttered these words from off camera, adding, "The country just can't risk your election." As he spoke, the camera panned away from the visual to a full screen shot of six distinguished-looking, but not terribly comfortable, participants. The focus shifted to the moderator. "My name is Herb York," he said. "During the Eisenhower administration, I was Director of Defense Research and Engineering. Here with me are men from science, engineering, and medicine. Some of us are Republicans and some Democrats. None of us are politicians. And most certainly none are television performers." York proceeded to identify each panelist, and as he spoke the camera moved to close-up range:

> Dr. Benjamin Spock, internationally known child care expert.
>
> Dr. Jerome Wiesner, Dean of Science at MIT and former science advisor to President Kennedy.
>
> Admiral W.F. Raborn, U.S. Navy—Retired, Vice President of the Aerojet General Corporation.
>
> Dr. George Kistiakowsky, Professor of Chemistry at Harvard University, and former science advisor to President Eisenhower.
>
> Dr. Harold Urey, Professor of Chemistry, University of California, and Nobel Prize winner.

The introductions yielded a logical segue to the program's first major purpose, establishment of the scientist-citizen's ethos.

## Scientific Ethos

York immediately addressed the technological community's reluctance to assume a visible political role, but then echoed the claim expressed in "The Alternative is Frightening," that duty and responsibility necessitated such action. As the telecast progressed, the participants' authority and objectivity would be added to this foundation tenet of scientific ethos. "Traditionally," York observed, "we scientists and engineers are supposed to represent the so-called non-political community in America. . . . We usually stand in the background, giving advice whether we're asked or not. But this time we're involved in grassroots political activity, because this time it's different." What made it different were the roles several panelists had played in ushering in the nuclear era: "Most of us have spent our lives working to build, to create, to improve. While at the same time, we've helped develop a power that could destroy mankind. We have worked to build this power to ensure national security. National security in every sense is our deepest concern." Contrast the scientists' commitment to "building, creating, and improving" with the earlier depiction of Goldwater as irresponsible and dangerous. In both language and substance, the opening reinforced the print campaign's message and established the groundwork for encouraging a positive perception of the scientists' authority.

## Technological Authority

York frequently attached positively charged labels to the enumeration of the panelists' achievements as he introduced them. In addressing Admiral Raborn, for instance, York observed, "Red, there's [*sic*] been a lot of accusations thrown around lately about our supposed lack of preparedness. You were responsible for the Polaris program from its very inception to the time of the deployment of the first ships. You were responsible for the submarine itself and for the missile. As a *real expert* in the field of military preparedness, what's your feeling?"

Another tactic was to integrate references to the pariticipants' shared experiences with the public into the flow of the dialogue. The technique is clearly bardic, for it exploits the culture's common symbols or language to invite rhetor-auditor identification. In contrast to priestly rhetoric's synecdochic form with science at the apex, such bardic identification is reflexive, a culture speaking to itself about itself. Thus the tenor of York's introduction of Benjamin Spock: "Dr. Spock, you work with what is probably the world's most precious resource, our children. Millions of Americans, *my wife and I included*, have come to rely on your advice. What do you feel would be Goldwater's effect on our children?" The identifying label, "Child Care Expert," accompanied the closeup of Dr. Spock that then filled the screen.

Affording panelists the opportunity to work their own credentials into their answers constituted a third tactic. After his own extended discourse on arms policy, York asked Aerojet General Corporation Vice President Raborn if he could "comment on this issue of nuclear weapons and their control." Not surprisingly, Raborn could, "from several points of view—my long service in the country's military forces, knowledge of engineering, and corporate experience."

Finally, mention of a panelist's distinction or honors might accompany the introduction. When York turned to Nobel-recipient Harold Urey at about the half-way point of the telecast, for example, he began, "Harold, the Nobel prize is an international award. You've been concerned all your long and illustrious career with technology and with international understanding. What is your reaction to Goldwater's stated attitudes?" The joining of the participants' expertise with the properties they shared with the television audience thus served to construct and reinforce the scientist-citizen's authority.

### Priestly Objectivity

Bipartisanship constituted the third dimension of the advocates' credibility. York's technique for drawing George Kistiakowsky into the discussion illustrates: "George," said York, "you and I worked very hard and faithfully for President Eisenhower during his administration. Can you tell me why so many Republicans are involved in behalf of President Johnson in this campaign?" The camera shifted to Kistiakowsky, below whose face was printed, "Science Advisor to Eisenhower." A similar technique was used when York asked Jerome Wiesner to comment generally on the scientist's place in contemporary culture. Dean of MIT's School of Science, Wiesner had worked in the Kennedy White House, but had not limited his service to Democratic presidents. As York noted, "Jerry, you've also worked closely with Eisenhower, Kennedy, and Johnson. What, uh, what are your observations of the way President Johnson understands and takes into account the role of science and engineering and technology in the modern world?" The technological ethos evolving from the combination of bipartisanship, expertise, and the panelists' sense of duty provided a perception of competence which lent force to the participants' argumentative claims.

### The Argument

Demonstration of the threat that Barry Goldwater posed to the community proceeded in two steps: (1) indictment of his character, leading to (2) serious questions about the wisdom of his policies. Extremism comprised the focus of allegations about the Republican candidate's character. Former Eisenhower advisor Kistiakowsky contended that it was that extremism which had put the *Saturday Evening Post* in Johnson's camp. Kistiakowsky observed that he had

> read the other day in the editorial in the *Saturday Evening Post*, that hasn't endorsed one Democratic candidate since the beginning of the twentieth century, which has endorsed the election of President Johnson and urged the defeat of Barry Goldwater. They said, and I quote, "Goldwater is a grotesque burlesque of the conservative he pretends to be. He is a wild man, a stray, an unprincipled and ruthless political jujitsu artist."

Thus, concluded Kistiakowsky, "he is outside the mainstream of responsible American thinking, and is clearly unqualified to be trusted with the great powers of the presidency." Harold Urey concurred, depicting Goldwater as a "blustery, threatening man, who talks often without thinking—shoots from the hip, as they say. This frightens me. I am quite certain that it frightens our friends in Europe, and I fear that it might threaten . . . uh, frighten, the U.S.S.R." The portrayal of Goldwater as an irresponsible extremist in turn guided the analysis of Goldwater's record and proposals.

On domestic policy, for instance, these leading men of science explicated education's vital role if the nation was to keep pace in a technological world. Dr. Spock addressed the education issue by first citing recent pledges by President Johnson to pursue accelerated funding for education. "Now," said Spock, "contrast this with Senator Goldwater's statement: 'The child has no right to an education. In most cases, children get along very well without it.'" And Goldwater's votes against aid to elementary education, to vocational schools, to college education, and to dental and medical students "back up that kind of statement." As a result, Spock concluded, "I just don't think that he understands the vital importance of education to our children or to our nation, and I don't see how parents can support him under these circumstances." Note, first, that the initial quotation from Goldwater is the same one published in "The Alternative is Frightening." Second, the subsequent method of analysis illustrates the Johnson campaign's adapation of linear argument to an image-oriented medium.

Virtually every dimension of the scientists' persuasive strategy came together in their examination of Goldwater's nuclear weapons proposals and statements, the topic of nearly the entire second half of the telecast. The Senator's most basic problem, York charged, is his lack of fundamental knowledge: "He's said that such weapons [so-called small conventional nuclear weapons] are no more powerful than the firepower we've faced on other battlefields. As a weapons scientist, I'm appalled that anyone who is an Air Force Reserve General could be so wrong on the basic facts of our weaponry." Goldwater's alleged proposal that NATO commanders should have control of certain tactical weapons came under Raborn's close scrutiny. In the quotation under examination, Goldwater contended that in some instances time constraints made problematic sole control by the president. Again replicating a citation in "The Alternative is Frightening," Raborn quoted Goldwater as saying, "A NATO commander should not be required to wait, while a White House calls a conference to decide whether or not these tactical weapons should be used." The reason he "disagree[d] firmly on this," Raborn offered in rebuttal, was "the simple word 'escalation.'" Granting of nuclear control to more than one person would simply increase the chances that at some point someone might commit an "irrational act and . . . use nuclear weapons," setting off a mutually destructive retaliatory chain. "So I just can't believe anything except that a mature, stable, responsible person in the presidency of the United States should be granted sole control of the use of nuclear weapons." Finally, Harold Urey moved the discussion from the realm of Goldwater's proposals to his Senatorial actions, with a specific focus on the Nuclear Test Ban Treaty. Disputing the contention that Goldwater's proposals were "bold," Urey argued that "his views are simply rash and punitive." Goldwater argued and voted against the Test Ban, said Urey, even though such strong advocates of military preparedness as, "My former boss, President Eisenhower, supported the treaty. And almost everyone else, in all walks of life, saw this treaty as a first and limited, but very important step toward reducing the possibility of nuclear war. The choice which Senator Goldwater advocated was horror and destruction."

York closed the discussion by integrating the spot commercial tag line into his concluding sentences. "I hope we've been able to make our point convincingly. For us, the stakes are too high. Ladies and gentlemen, we all here agree with the statement, 'Sorry, Senator Goldwater, the country just can't risk it. The country just can't risk your election.'" As it had at the start, the logo bearing these words re-appeared on the screen inscribed over the drawing of a mushroom cloud.

## CONCLUSION

The technical community's entry into political debate during the Cold War eventuated in a scientific ethos that endowed its advocates with the status of secular priests. In exercising that status in disputes about technological issues, rhetors addressed their peers and the public authoritatively, employed the language of nature, and spoke extensively from one culture to another. As the Scientists, Engineers, and Physicians for Johnson-Humphrey discovered, however, success in addressing the political culture on broader matters necessitated adoption of a more reflexive rhetoric. Hence, the scientists' campaign messages in 1964 evinced appeals to a repository of shared symbols in the promotion of bardic identification. The resultant merging of the bardic and priestly forms produced a technological priesthood whose members grounded their persuasion in being *of* the community as well as *in* the community.

Recent analyses of argument about technical policy making, the scientist as public advocate, and the link between scientific discourse and social drama,[44] reflect increasing scholarly attention to communication's vital role in shaping as well as explaining a technological culture. Examination of the scientist-citizen's activism in the 1964 presidential campaign suggests the wisdom of such attention. It advises as well a course of scholarly and public vigilance, aimed to guard against unwitting acceptance of apparent expertise and presumed authority. For in a society troublingly defined as a democracy without citizens,[45] it becomes rapidly evident that a technological world is too important to be left to the technologists alone.

# NOTES

1 *Robert Oppenheimer: Letters and Recollections*, ed. Alice Kimball Smith and Charles Weiner (Cambridge: Harvard University Press, 1980), vii.

2 Ralph E. Lapp, *The New Priesthood* (New York: Harper and Row, 1965), 114.

3 Lapp, *The New Priesthood*, 153 and 226-227.

4 Two excellent accounts of the conflicts among scientists appear in Alice Kimball Smith, *A Peril and A Hope: The Scientists' Movement in America, 1945-47* (Chicago and London: University of Chicago Press, 1965), and Donald A. Strickland, *Scientists in Politics: The Atomic Scientists Movement, 1945-46* (W. Lafayette, IN: Purdue University Studies, 1968).

5 Robert Gilpin, *American Scientists and Nuclear Weapons Policy* (Princeton, NJ: Princeton University Press, 1962), 262-263.

6 Richard H. Ullman reports this exchange in his review of Solly Zuckerman's *Monkeys, Men and Missiles*: "A Scientist Among Warriors," *Los Angeles Times*, Book Review section, 28 May 1989, 7.

7 Cited in Marcel Evelyn Chotkowski La Follette, "Authority, Promise, and Expectation: The Images of Science and Scientists in American Popular Magazines, 1910-1955" (Ph.D. dissertation, Indiana University, 1979), 266-267.

8 Stephen Toulmin, *The Return to Cosmology* (Berkeley: University of California Press, 1982), 82.

9 Lapp, *The New Priesthood*, 218.

10 Thomas M. Lessl, "The Priestly Voice," *Quarterly Journal of Speech* 75 (1989): 184-186. All subseqent citations about the priestly voice are from this source, 183-197.

11 The interpretation of bardic rhetoric used by Lessl, and therefore employed here, is in John Fiske and John Hartley, "Bardic Television," in *Television Criticism: The Critical View*, ed. Horace Newcomb, 4th edition (New York: Oxford University Press, 1987), 603. All subsequent citations about bardic rhetoric are from this reference, 600-612.

12 Seven characteristics of bardic television are established by Fiske and Hartley, "Bardic Television," 601-603.

13 Memorandum, A. Petcavage to J. Graham, 21 May 1964 (p. 2), Democratic National Committee Papers, Container 224, Doyle Dane Bernbach file; Letter, William Bernbach to Bill Moyers, 17 August 1964 (p. 1), Democratic National Committee Papers, Container 224, Doyle Dane Bernbach file, Lyndon Baines Johnson Library, Austin, TX (hereafter cited as LBJ Library).

14 LaFollette details the scientists' image in popular periodicals in "Authority, Promise, and Expectation." David Henry examined the scientists' movement's use of print in "Idealism vs. Realism: Rhetorical Form in the Scientists' Movement, 1945-1965," paper presented at the Western Speech Communication Association convention, 1988. The conflict the scientists experienced because of the shift from print to electronic messages may be explained in terms of Marshall McLuhan's concept of source-medium match. See: Bruce E. Gronbeck, "Rhetoric and/ of McLuhan," in *Contemporary Theories of Rhetoric*, ed. Richard L. Johannesen (New York: Harper and Row, 1971), 295-297; Stephanie McLuhan's excellent documentary on her father's life and work, "McLuhan: The Man and His Message," Public Broadcasting System, 1985; Marshall McLuhan, *Understanding Media* (New York: McGraw-Hill, 1964); Bruce E. Gronbeck, "McLuhan as Rhetorical Theorist," *Journal of Communication* 31 (1981): 117-128; *McLuhan: Pro & Con*, ed. Raymond Rosenthal (New York: Funk & Wagnalls, 1968); and *McLuhan: Hot & Cool*, ed. Gerald E. Stearn (New York: The Dial Press, Inc., 1967).

15 Memorandum, Petcavage to Graham, 2. The 12 states were California, Illinois, Indiana, Maryland, Massachusetts, Michigan, Minnesota, New Jersey, New York, Ohio, Pennsylvania, and Wisconsin.

16 Scientist-to-scientist rhetoric and scientist-to-public communication reflect the concepts of "elite" and "universal" audiences as advanced in Chaim Perelman and L. Olbrechts-Tyteca, *The New Rhetoric: A Treatise on Argumentation*, trans. by John Wilkinson and Purcell Weaver (Notre Dame and London: University of Notre Dame Press, 1969), 33-35.

17 Theodore H. White, *The Making of the President 1964* (New York: Atheneum Publishers, 1965), 351-352; John H. Kessel, *The Goldwater Coalition: Republican Strategies in 1964* (Indianapolis and New York: Bobbs-Merrill, 1968), 230.

18 Cited in "Professionals Emerge as Political Force," *Aviation Week and Space Technology*, 12 October 1964: 26.

19 Background material in the following two paragraphs is from two very thorough pieces by D.S. Greenberg: "Venture into Politics: Scientists and Engineers in the Election Campaign." Part I appeared in *Science*, 11 December 1964, 1440-1444, and Part II followed on 18 December 1964, 1561-1563.

20 Cited in Greenberg, "Venture into Politics (I)" 1442.

21 Press Release, "Scientists and Engineers Organize for Johnson," 13 August 1964, Democratic National Committee Papers, Container 290, Organize for President Johnson file, LBJ Library.

22 Despite the presence of a prominent heart specialist, Dr. Paul Dudley White, on the founding committee, the organization's initial name was "Scientists and Engineers for Johnson." As physicians like Benjamin Spock became more active, and once Johnson selected Hubert Humphrey as his running mate, the longer name used throughout this paper was adopted.

23 "Statement of Principles," 16 August 1964 (pp. 1-2), Diana T. MacArthur Papers (AC 69-19), Container 2, National Organizing Committee file, LBJ Library.

24 Letter, John H. Rubel to Donald M. MacArthur, 14 August 1964 (p. 1), Diana T. MacArthur Papers (AC 69-19), Container 1, National Organizing Committee file, LBJ Library.

25 Richard L. Kenyon, "Scientists in Partisan Politics," *Chemical and Engineering News*, 12 October 1964, 7.

26 Elliott Maraniss, "Urey Says Politics Must Solve Peace," *The (Madison) Capital Times*, 14 October 1964, 27; and Marquis Childs, "Scientists Join Politicians in Battle for the Presidency," *Salt Lake Tribune*, 8 October 1964.

27 Memorandum, Petcavage to Graham, 11-12; Greenberg, "Venture into Politics (II)", 1563; Night Letter, Donald MacArthur to All State Headquarters, 13 October 1964, Diana T. MacArthur Papers (AC 69-19), Container 1, National Organizing Committee file, LBJ Library; and, Memorandum, Donald MacArthur to All States, 15 October 1964, Diana T. MacArthur Papers (AC 69-19), Container 1, National Organizing Committee file, LBJ Library.

28 Charles A.H. Thomson, "Mass Media Performance," in *The National Election of 1964*, ed. Milton C. Cummings, Jr. (Washington, D.C.: The Brookings Institution, 1966), 111.

29 "SCIENTISTS and ENGINEERS for JOHNSON," campaign pamphlet, Diana T. MacArthur Papers (AC 69-19), Container 2, Publicity file, LBJ Library.

30 Kessel, *The Goldwater Coalition*, 231-232; Greenberg, "Venture into Politics (II)," 1561.

31 "The Alternative is Frightening: Johnson and Humphrey Must Be Elected," Diana T. MacArthur Papers (AC 69-19), Container 2, Publicity file, LBJ Library. Subsequent references to this document are from this source.

32 Despite the explicit claim that the discussion would be of "special interest to the technical world," it is reasonable to interpret the scientists' peers as the book's ostensible audience, while the actual target was the public. Richard P. Fulkerson delineates the general strategy in "The Public Letter as a Rhetorical Form: Structure, Logic, and Style in King's 'Letter From Birmingham Jail'," *Quarterly Journal of Speech* 65 (1979): 121-136.

33 Memorandum, J. Graham to Selected Democratic National Committee Members and Doyle Dane Bernbach Staff, 15 July 1964 (p. 3), Democratic National Committee Papers, Container 224, Doyle Dane Bernbach file, LBJ Library.

34 Letter, Bernbach to Moyers, 2.

35 Memorandum, Lloyd Wright to Bill Moyers, 13 August 1964, Democratic National Committee Papers, Container 224, Doyle Dane Bernbach file, LBJ Library.

36 Johnson's spot commerical campaign has received thorough treatment in multiple sources: Kathleen Hall Jamieson, *Packaging the Presidency* (New York: Oxford University Press, 1984), 173-205; Edwin Diamond and Stephen Bates, *The Spot: The Rise of Political Advertising on Television* (Cambridge: The MIT Press, 1984), 121-147; Thomson, "Mass Media Performance,"

127; and, Kessel, *The Goldwater Coalition*, 237-239. Audiovisual references include, "The 30-Second President," *A Walk Through the Twentieth Century with Bill Moyers*, Public Broadcasting System, 1984, and a 30-minute film of 21 Johnson commercials, available through the Audiovisual Archive, LBJ Library.

37 Diamond and Bates, for instance, describe "Sorry, Senator Goldwater" in 10 lines. See: *The Spot*, 136.

38 Greenberg, "Venture into Politics (II)," 1562-1563.

39 Memorandum, Scientists and Engineers for Johnson/Humphrey to James Rowe, 16 October 1964, Diana T. MacArthur Papers (AC 69-19), Container 1, National Organizing Committee file, LBJ Library.

40 Memorandum, Donald M. MacArthur to the President, 21 October 1964 (pp. 1-2), Diana T. MacArthur Papers (AC 69-19), Container 1, National Organizing Committee file, LBJ Library.

41 Greenberg, "Venture into Politics (II)," 1561.

42 Letters, Rodney Nichols to Dr. Benjamin Spock (21 October 1964), Dr. George B. Kistiakowsky (21 October 1964), Dr. Herbert York (21 October 1964), and Dr. Harold Urey (20 October 1964), Diana T. MacArthur Papers (AC 69-19), Container 1, National Organizing Committee file, LBJ Library.

43 "Sorry, Senator Goldwater . . . The Country Just Can't Risk It," 30-Minute Telecast (Scientists, Engineers, and Physicians for Johnson-Humphrey; David Garth, Producer, 1964). Quotations from the telecast are drawn from the author's transcription of a videotape procured through the Audiovisual Archive, LBJ Library. Philip Scott, Audiovisual Archivist at the Library, provided indispensable assistance in the collection of data for this paper.

44 Representative studies include: Janice Hocker Rushing, "Ronald Reagan's 'Star Wars' Address: Mythic Containment of Technical Reasoning," *Quarterly Journal of Speech* 72 (1986): 415-433; Thomas M. Lessl, "Science and the Sacred Cosmos," *Quarterly Journal of Speech* 71 (1985): 175-184, and "Heresy, Orthodoxy, and the Politics of Science," *Quarterly Journal of Speech* 74 (1988): 18-34; and Alan G. Gross, "Public Debates as Failed Social Dramas: The Recombinant DNA Controversy," *Quarterly Journal of Speech* 70 (1984): 397-409.

45 Robert M. Entman, *Democracy Without Citizens: Media and the Decay of American Politics* (New York: Oxford University Press, 1989).

# Part Four

# Language and Media

# Technical Heterogeneity, Specialization, and Differing Motives: An Examination of the Influences of Technology on Group and Organizational Decision Making

*Roger C. Pace*
*and*
*Steven Hartwell*

A great many scholars have studied the influence of technology on organizational and group decision making, focusing much of their research on communication technology and its impact on the decision-making process.[1] Researchers have examined teleconferencing, computer conferencing, electronic mail, and information systems.[2] The research goals of these studies range from investigating the diffusion of new technologies to the study of how such technologies influence communication styles and decision outcomes.

Technology, however, has many other implications for group and organizational decision making, some of which have been virtually ignored by communication scholars. Advancing technologies produce technical, specialized knowledge. This specialized knowledge creates unique problems for decision-making organizations. Past research into group decision making indicates a substantial influence of technical subject matter on group processes. Baker, et al., Allen and Cohen, Tushman, and Dunne and Klementowski all report important differences between "research and development" groups and other nontechnical groups.[3] Similarly, Allen, Holland, Wolek, Brown and

Utterback, and Pace all discovered unique communication patterns and roles in technical groups.[4]

However, past and present theories on group and organizational decision making do not address these differences. Instead, they assume homogeneity of group and organizational members. These theories fail to account for specialization, plurality of experience, and technical heterogeneity common to many groups and organizations. The most famous work on faulty decision making for example, is Janis's theory of groupthink.[5] But groupthink is a theory about the problems of overly cohesive groups—not about the dangers of too much diversity. Like many group theories, groupthink relies on the premise that members of the group have similar attitudes and predispositions. Few, if any, group theories point out the dangers of overly heterogeneous members.

Similarly, organizational theories fail to acknowledge the problem. Current trends in organizational theory urge participative decision making and adopting a "Japanese" style of leadership. Management theorist Thomas Peters urges: "Involve all personnel at all levels in all functions in virtually everything."[6] Also, he advises: "Organize as much as possible around teams, to achieve enhanced focus, task orientation, innovativeness, and individual commitment."[7] Yet specialization inherently produces autocratic decision making, dependence on authority figures, and a return to vertical hierarchy.

In this paper, we argue that specialization has a profound effect on group and organizational decision making. Specifically, we argue that specialization produces conflicting motives and rationales among decision makers that impede effective communication and adversely alters decisions. First, we briefly discuss the technical causes of specialization. Next, we examine the effects of specialization on communication and decision-making processes. Third, we illustrate the problems associated with differing motives with an extended case study from the American legal system. In the case study, we show how specialized and technical definitions of "causality" impede communication between the American legal system and the scientific community. Finally, we discuss the implications of the case study for other organizations or groups.

## TECHNOLOGICAL CAUSES OF SPECIALIZATION

Advancing technologies, especially the emergence of computers, are changing knowledge and creating a more heterogenous society. Computers are revolutionizing the way we store, process, and retrieve information.[8] Pagels declares: "The computer has indeed created a new class of people who understand and have mastered it. . . .Not only has a new social class been brought into existence by the computer, but the very structure of knowledge is being altered as well."[9]

## Computers and Specialized Knowledge

Computers are reshaping the traditional organization of knowledge in three important ways. First, computers are extending current disciplines and reinforcing existing specialties. This "vertical deepening" of existing disciplines is described by Pagels:

> Using computers, physicists, chemists, and economists can tackle problems that they could not touch before simply because the computational power was not there. The new methods of analysis of complex systems apply not only to the natural sciences— astronomy, physics, chemistry, biology, and the new medicine—but to the social sciences as well—economics, political science, psychological dynamics. Computers, because of their capacity to manage enormous amounts of information, are showing us new aspects of social reality.[10]

Second, computers are creating new disciplines and addressing new problems. Murray Gell-Mann observes, "New subjects, highly interdisciplinary in traditional terms, are emerging and represent in many cases the frontier of research. These interdisciplinary subjects do not link together the whole of one traditional discipline with another; particular subfields are joined together to make a new subject."[11]

Finally, computers are changing the location of knowledge. Many of the early computer enthusiasts argued that computers would unite knowledge and provide central access to a universal storehouse of information. Instead, the proliferation of computers has exceeded the ability to connect them. Rather than a central depository of knowledge, computers have created a patchwork of regional and isolated systems. The lack of universal standards in hardware and software make "connectivity" between these disparate systems difficult. The lack of connection creates problems locating and using available knowledge.

## SPECIALIZATION AND COMMUNICATION

This change of knowledge inherently produces specialization. The amount of stored and available information continues to increase. As a result, one person's ability to master all information, even in a given discipline, is rapidly decreasing. If common life experiences are a necessary base of effective communication, then specialization reduces the chance of accurate or fidelitous information exchange. The problem of communicating between specialties is especially acute in decision-making groups and organizations.[12] Specialization leads to isolated perspectives, differing motives, and new attitudes and beliefs. In extreme cases specialization can lead to the establishment of subcultures, sublanguages, and xenophobic reactions.

## Homogeneity and Effective Communication

Some amount of common experience, culture, or belief between communicators is a prerequisite for effective communication. This shared experience allows communicators to assign mutually understood signs or symbols to referents and then to exchange those symbols with some degree of fidelity. Frentz and Farrell argue: "The concept of communication itself demands a common ontology which is accessible to its participants through form. This ontology includes the total conceptual, aesthetic, and cultural knowledge which a society shares."[13]

Previous research demonstrates that the more homogeneous communicators' attitudes and past experiences are, the more accurate their communication will be. Similarly, the more discrepant the communicators' backgrounds, experiences, and attitudes, the less accurate the communication.[14]

## Heterogeneity

The heterogeneity created by specialization changes decision-making processes and outcomes. Specialization changes many group or organizational members from active participants in the decision-making process to secondhand observers, overly dependent on and mystified by the specialists. Stanley argues:

> Specialization can proceed to a degree that is demonstrably irrational in its effects. Social structural specialization is often replicated on the level of subjective consciousness. Beyond a certain point, specialization can induce in a population general incompetence and helplessness rooted in a sense that it is dangerous to act except in professionalized, role defined, and credentialed settings.[15]

## Differentiation and Conflicting Motives

The heterogeneity created by specialization changes decision-making processes and outcomes. For example, one important decision-making process is differentiation. Differentiation is the process of discovering, identifying, and assessing information related to the decision.[16] In group and organizational decision making, differentiation is also the process of distinguishing and understanding different points of view, delimiting areas of conflict, and identifying coalitions. Walton declares: "The differentiation phase requires not only that a person be able to state his views but also that he be given some indication his views are understood by other principals."[17]

An effective differentiation is a necessary predecessor to consensus formation and decision emergence. The lack of sufficient differentiation is often associated with ineffective decision making.[18]

Specialization produces special interest and commitment which in turn produces coalitions and factions. This often leads to conflicting motives and rationales for decision makers. The divisions and distinctions produced by such differences lead to alternative definitions of the situation and interpretations of fact. Without a basic agreement about the very nature of a problem, groups and organizations are often unable to formulate an effective decision.[19]

## NASA

One case in point is the investigation into NASA's decision making after the ill-fated launch of the space shuttle Challenger. The presidential commission studying the space shuttle disaster discovered astonishing differences between management perspectives and that of engineers throughout the organization. The investigators asked the engineers at the Marshall Space Center to approximate the risk of engine failure in the space shuttle. The engineers were near unanimous in estimating the failure rate at 1 in 200. When managers, however, were asked the same question, they estimated the chances of failure at 1 in 100,000. Astonishingly the managers were not MBA graduates, but all were former engineers with extensive technical backgrounds.

The commission blamed this discrepancy on the different motives of managers and engineers. The primary motive of the engineers was to make the shuttle fly within the parameters of safety. Hence, their estimate was conservative. The primary motive of the managers was to sell the shuttle program to Congress and the American public as a safe, almost routine method of entering space. Hence, their estimate was very liberal. Richard Feynman, Nobel laureate in physics and member of the presidential commission, blamed the differing motives for the loss of the Challenger. "My theory is that the loss of common interest—between the engineers and scientists on the one hand and management on the other—is the cause of the deterioration in cooperation, which as you've seen . . . produced a calamity."[20]

## DIFFERING LEGAL AND SCIENTIFIC UNDERSTANDING OF CAUSATION

Another case in point is the American legal system. Specialization and differing motives often impede communication between the legal community and other groups such as the scientific community. The court system has been reluctant to accept relevant scientific data in many cases. The American trial courts' rejection of empirical information is best explained by their profoundly differing understandings of causation. Although this impediment can be shown to operate in many instances, this paper will present one exemplary case of the impediment, *Lockhart v McCree*,[21] a case dealing with the selection of juries in capital cases.

### The Selection of Jurors in Capital Cases

The prosecution in a capital case has the right to remove any potential juror who has moral or religious scruples against imposition of the death penalty. As prosecutors routinely exercise this right, juries empaneled in capital cases differ in one important respect from juries in all non-capital cases: They consist solely of what are called "death qualified" jurors ("DQs") who have been "qualified" as having no religious or moral scruples against imposition of the death penalty.

All potential jurors who hold such scruples—known after a famous case as "Witherspoon Excludables" ("WEs")—are excluded from participation.[22] The exclusion of WEs in capital cases raises two obvious questions: Does the exclusion of WEs significantly increase the probability that the remaining DQ jurors will find guilt? If so, does the increased probability of finding guilt prejudice the defendant's constitutional right to a fair jury?[23]

## The Court's Invitation to Social Science

When the U.S. Supreme Court first considered these two questions twenty years ago, it concluded that it lacked sufficient empirical data to adjudge whether the exclusion of WEs significantly increased a defendant's risks and, in effect, invited the scientific community to research the question.

In response to the Court's invitation, various researchers conducted a series of studies. These studies found that DQ jurors were more likely to be prosecution prone, less accurate in their results, less representative of women and ethnic minorities, less able to remember the evidence, less rigorous in testing the evidence, more likely to believe that a defendant's silence indicates guilt, more hostile to an insanity defense, more mistrustful of defense attorneys, and less concerned about making erroneous convictions. One study found that the very interrogation process by which WE jurors are removed from the jury increased significantly the probability of a finding of guilt by the remaining jurors. It is clear that juries comprised solely of DQ jurors from which WEs have been systematically removed are significantly more likely to find guilt.[24]

### *The court's rejection of the social science findings*

Despite these uncontroverted findings from social scientists, the U.S. Supreme Court concluded in *Lockhart* that the systematic removal of WEs from the jury did not deny an individual defendant his constitutional right to a fair trial. To establish that he or she has been denied a fair trial, the defendant would have to demonstrate the bias of *individual* jurors because "the Constitution presupposes that a jury selected from a fair cross section of the community is impartial, regardless of the mix of individual viewpoints actually represented on the jury, so long as jurors can conscientiously and properly carry out their sworn duty to apply the law to the facts of the particular case."[25]

What explains the Court's rejection of findings based on the data it had invited the scientific community to provide? Certainly no literal reading of the Constitution mandated the rejection. And, while many courts, including the *Lockhart* Court,[26] have expressed some hostility toward the methodology of social science, this hostility does not explain adequately the Court's decision. The *Lockhart* Court stated candidly that it was prepared to find the bias constitutionally acceptable even though the Court "will assume for purposes of this opinion that the studies are both methodologically valid and adequate to establish that 'death qualification' in fact produces juries somewhat more 'conviction-prone' than 'non-death-qualified' juries."[27]

## Conflicting Models of Causation

*The law model*

The different understandings of causation can be briefly stated. Law-trained people tend to understand causation as deductive and operating in specific instances. The legal model of causality draws from a deductive, Newtonian world perspective. Under the Newtonian model of causation, rules serve to link cause and effect sequentially. The application of a rule to some starting point A leads deductively to some end point B. The model is "Newtonian" in that the laws of gravity that Isaac Newton discovered operate in just this way. The "if, then" Newtonian model of causation permeated science throughout most of the 19th century and continues to permeate legal reasoning to the present.[28]

*The science model*

In contrast, scientifically trained people tend to understand causation as operating not only deductively but also inductively and operative upon a field or over a class of instances. The latter model of causality draws from a technologically more advanced notion of post-Newtonian science and statistical analysis. The discovery of quantum mechanics, particularly, forced science early in the twentieth century to develop this second understanding of causation. The development of statistics permitted scientists to describe the effects of various forces upon a field without the necessity of demonstrating one-to-one Newtonian causation. In quantum mechanics, for example, individual particles simply are not subject to "cause and effect." Only a class of particles, that is, particles within a field, show causal interaction as a statistical probability.[29]

*Probability and causation*

Modern computer technology facilitates the statistical analysis of probability, the inductive analogue to sequential deductive causation. Scientists in many disciplines, including the social sciences, now comfortably think of causation as operating only upon a class and described statistically as a probability, rather than operating in specific instances according to a rule. They are encouraged to think in terms of probabilistic causality by user-friendly statistical software. This modern mode of thinking about causation has created a technology of statistical analysis that is commonly applied to science but not to law.

The *Lockhart* Court clearly rejected as "causative" the inductively based probability of jury bias presented by social scientists. The Court stated most revealingly: "Even accepting [defendant's] position that we should focus on the *jury* rather than the individual *jurors*, it is hard for us to understand the logic of the argument that a given jury is unconstitutionally partial when exactly the same jury results from mere chance."[30] To a scientist trained in the technology of statistical analysis, the element of chance exists in the empirical analysis of virtually all phenomena. To an empirical scientist, the observation that "mere chance" confounds the data does not end the analysis because it is precisely the analysis of chance that leads to the rejection or confirmation of a theory.[31] In contrast, the *Lockhart* Court, once having found the existence of "mere chance," concluded its analysis and summarily rejected the defendant's claim of causality.[32]

The rejection of an inductively based model of causation in *Lockhart* is not a singular aberration of this Supreme Court. The failure of the scientifically trained community to communicate effectively to law-trained decision makers occurs in many areas of law that require statistical and inductive models of causality. Two prominent areas are employment discrimination and environmental pollution.[33]

*Explaining the rejection*

What explains this particular communication breakdown between social science and law? Why does law continue to reject a statistically based inductive model of causation?

One obvious explanation is that the legal system has never created a simple mechanism for introducing probabilistic, inductive information into the trial court. The present mechanism is cumbersome. The current Federal Rules of Evidence provide trial court judges with no clear direction about how to deal with information such as statistical data, or what effect, if any, social science findings accepted in one case should have on later cases.[34]

A second simple explanation lies in the observation that lawyers are not trained in modern scientific inductive reasoning. Although the undergraduate majors of law students are diverse, majors in the empirical sciences are uncommon. Law schools presently provide very little training in social science methods.[35] At present, for example, only one psychologist trained in scientific methodology is on the faculty of a major law school as a tenured, fulltime professor.

A third, perhaps more important, explanation lies in an implicit but powerful underlying premise of the legal system dealing with individual autonomy. The work of science generally and social science particularly is to provide generalized knowledge in the form of principles or rules that describe the world. Social science findings are "positive" in the sense that they describe the world as it "is" rather than as it "ought" to be. In contrast, the work of the law is to do justice; that is, in part, to do what ought to be done in assessing blame in a fair manner for events that have gone wrong.[36] The question in a homicide prosecution, for example, is not what the defendant did, precisely, but how the matter ought to be resolved. While empirical "is" questions are important to a court trial

(Did the defendant fire the gun?), such "is" questions are secondary to and dependent upon the "ought" questions (Should the defendant's illegally obtained confession that he fired the gun be admitted?).

The approach taken in answering these "ought" questions is shaped by the law's implicit understanding of individual autonomy. Because the law views jurors as autonomous agents, personality traits that might only reveal themselves probabilistically among jurors as a class are discounted. In *Lockhart*, for example, the Court discounted the high probability that a jury composed of individuals with personality traits in one domain (condoning of the death penalty) would be significantly correlated with a second trait (being prosecution prone). In contrast, social scientists view individuals as much less autonomous and much more influenced by trait interaction.[37]

The American legal system has traditionally embraced a highly autonomous understanding of human behavior. This understanding is supported by the American ethos of "rugged individualism" and is consistent politically with laissez-faire capitalism.[38] It is consistent with the way in which lawyers have historically structured their profession and their ethics as individual, autonomous professionals. This deeply held ethos supports the Newtonian model of causation exemplified by *Lockhart*.

The present resistance of the legal community to social science findings can thus be seen as several layers thick. The outside, visible layer is the simple lack of procedural devices whereby such data can initially penetrate the legal system. This outer layer is supported by a secondary inner layer of legal education that is mostly ignorant of the basic scientific understanding of inductive and probabilistic causation that might utilize such procedural devices. Finally, at the core, is a belief system that adheres to a highly individualistic and autonomous concept of individual responsibility, a belief system that is hostile to much of what the community of social science might provide.

## CONCLUSIONS

The analysis of technical heterogeneity and the examination of conflicting motives illustrates some tentative conclusions concerning group and organizational decision making. First, specialization inherently produces divisions that impede decision making. Most scholars now recognize that some form of conflict is necessary for effective decision making. In fact, the advantage of group decision making is the diversified perspectives individual members bring to a problem. Conflict allows all points of view to be discussed, provides acceptable ways to release tension, and prevents groupthink.

The success of group or organizational decision making depends on the ability of members to establish common motives of action and similar definitions of the situation. Such agreement makes the integration of opposing points of view possible. But without a common agreement on the basic nature of the situation, integration is impossible. Traditional methods of integration such as consensus seeking, compromise, negotiation, and bargaining are all predicated on the assumption that members have a common interest in the decision. Conflict escalates, however, without the ability to integrate opposing definitions of the situation, further hindering effective communication. As in the case of the American legal system, groups and subgroups frequently ignore each other and make autonomous, ineffective decisions.

The failure to make integrated and cooperative decisions increases pressure toward division. Even if conflicting motives never develop, specialization still produces an overdependence on experts and authority. Members of a group or organization become spectators to the decision instead of active participants. The inability to understand other members or groups hinders the exchange of information and leads to deference of authority and autocratic decisions, thereby negating the advantage of divergent perspectives.

Second, decision-making scholars need to identify the problems associated with heterogeneity and incorporate them in their theories. Janis's concept of groupthink has become so familiar that there is a tendency to attribute all group failures to overly cohesive groups. In reality, many groups are struggling with the inability to understand each other. Similarly, organizational scholars have been preoccupied with the Japanese style of management which functions best in a homogeneous and cohesive culture. But American heterogeneity and plurality make such a culture virtually impossible in most organizations. Scholars should concentrate less on cohesive team building and more on integrated decision making.

Third, the causes of specialization should be examined. While specialization is inevitable, its forces can be diminished with common experience and knowledge. Well-rounded curriculums in universities and job training in industry can reduce the divisions established by specialization. In an increasingly technical world, students should have an increasingly technical education. Managers and group leaders should not settle for autocratic and autonomous decisions but should spend the necessary resources to form truly integrated decisions. Organizational structures should contain checks and balances that structurally enforce the notion of integration.

Finally, the very technologies that produce specialization might be used to reduce division and enhance decision making. Communication technologies are increasing rapidly. New and creative uses of these advances can help bridge long-established distinctions. Emerging technologies are already forcing the merging of established disciplines. Communication technologies frequently lead to new organizational structures and group networks.

The irony of heterogeneity is that while specialization produces division it simultaneously produces dependency. In an increasingly specialized and dependent culture, effective communication is a necessity. Decision makers and academic scholars need to understand the indirect, yet powerful, influence technology has on decision-making and communication processes. Such an understanding will help groups and organizations cope with the forces of technology and enhance their decision-making ability.

## NOTES

1 Mary S. Culman and M. Lynne Markus, "Information Technologies," in *Handbook of Organizational Communication*, ed. Frederic Jablin, Linda Putnam, Karlene H. Roberts, and Lyman Porter (Beverly Hills: Sage, 1987), 420-43; Ronald Rice, "Computer Mediated Communication and Organizational Innovation," *Journal of Communication* 37 (1987): 65-94.

2 Thomas J. Allen and O. Hauptman, "The Influence of Communication Technologies on Organizational Structure: A Conceptual Model for Future Research," *Communication Research* 14 (1987): 567-75; Richard L. Daft and Robert H. Lengel, "Organizational Information Requirements, Media Richness, and Structural Design," *Management Science* 32 (1986): 554-71; Richard L. Daft and Norman B. Macintosh, "A Tentative Exploration into the Amount and Equivocality of Information Processing in Organizational Work Units," *Administrative Science Quarterly* 26 (1981): 207-24; Glen Hiemstra, "Teleconferencing: Concern for Face and Organizational Culture," in *Communication Yearbook* 6, ed. Michael Burgoon (Beverly Hills: Sage, 1982); Starr R. Hiltz, Kenneth Johnson, and Murray Turoff, "Experiments in Group Decision Making: Communication Process and Outcome in Face-to-Face Versus Computerized Conferences," *Human Communication Research* 13 (1986): 225-52.

3 Thomas S. Allen and Stephen I. Cohen, "Information Flow in Two R&D Laboratories," *Administrative Science Quarterly* 14 (1969): 12 19; Norman R. Baker, Jack Siegman, and Albert Rubenstein, "The Effects of Perceived Needs and Means on the Generation of Ideas for Industrial Research and Development Projects," *IEEE Transactions on Engineering Management* EM-14 (1967): 156-63; Edward Dunne and Lawrence Klementowski, "An Investigation of the Use of Network Techniques in Research and Development Management," *IEEE Transactions on Engineering Management* EM-29 (1982): 74-78; Michael L. Tushman, "Technical Communication in R&D Laboratories: The Impact of Project Work Characteristics," *Academy of Management Journal* 21 (1978): 625-45.

4 Thomas J. Allen, "Roles In Technical Communication Networks," in *Communication Among Scientists and Engineers*, ed. Carnot Nelson and Donald Pollock (Lexington, Massachusetts: Health, 1970), 191-208; Winford E. Holland, "The Special Communicator and His Behavior in Research Organizations: A Key to the Management of Informal Technical Information Flow," *IEEE Transactions on Professional Communication* DC-17 (1974): 48-53; James W. Brown and James M. Utterback, "Uncertainty and Technical Communication Patterns," *Management Science* 31 (1985): 301-11; Roger Pace, "Technical Communication, Group Differentiation, and the Decision to Launch the Space Shuttle Challenger," *The Journal of Technical Writing and Communication* 18 (1988): 207-20.

5 Irving Janis, *Groupthink: Psychological Studies of Policy Decision and Fiascoes* 2nd ed. (Boston: Houghton-Mifflin, 1982), 245.

6 Tom Peters, *Thriving on Chaos: Handbook for Management Revolution* (New York: Alfred A. Knopf, 1987), 284.

7 Peters, *Thriving*, 296.

8 Wilson P. Dizard, *The Coming of the Information Age* (New York: Longman, 1985).

9 Heinz R. Pagels, *The Dreams of Reason* (New York: Simon and Schuster, 1988), 36.

10 Pagels, *Dreams*, 41.

11 Pagels, *Dreams*, 36.

12 Peter Schiff, "Speech: Another Facet of Technical Communication," *Engineering Education* 71 (1980): 180-81; Cynthia Selfe, "Decoding and Encoding: A Balanced Approach to Communication Skills," *Engineering Education* 74 (1983): 163-64.

13 Thomas Frentz and Thomas B. Farrell, "Language-Action: A Paradigm For Communication," *Quarterly Journal of Speech* 62 (1976): 333-49.

14 Palteil Lifshitz and Gary M. Shulman, "The Effects of Perceived Similarity/Dissimilarity as Confirmation/Disconfirmation Behaviors: Reciprocity or Compensation?" *Communication Quarterly* 31 (1983): 84-94; Marshall Prisbell and Janis F. Anderson, "The Importance of Perceived

Homophily, Level of Uncertainty, Feeling Good, Safety, and Self-Disclosure in Interpersonal Relationships," *Communication Quarterly* 28 (1980): 22-33; Lawrence R. Wheeless, "The Effects of Attitude, Credibility, and Homophily on Selective Exposure to Information," *Communication Monographs* 41 (1974): 329-47.

15 Manfred Stanley, *The Technological Conscience: Survival and Dignity in an Age of Expertise* (New York: The Free Press, 1978), 253.

16 Joseph P. Folger and Marshall Scott Poole, *Working Through Conflict: A Communication Perspective* (Reading, Massachusetts: Addison-Wesley, 1984); Thomas Scheidel, "Divergent and Convergent Thinking in Group Decision-Making," in *Communication and Group Decision Making*, ed. Randy Y. Hirokawa and Marshall Scott Poole (Beverly Hills: Sage, 1986): 113-30.

17 Richard E. Walton, *Interpersonal Peacemaking: Confrontations and Third Party Consultation* (Reading, Massachusetts: Addison-Wesley, 1969), 105.

18 Jerry B. Harvey, "The Abilene Paradox: The Management of Agreement," *Organizational Dynamics* 3 (1974): 63-80; Janis, *Groupthink*, 245.

19 Pace, "Challenger," 207-20.

20 Richard P. Feynman, "Mr. Feynman Goes to Washington," *Engineering & Science* (1987): 6-22.

21 *Lockhart v McCree*, 475 US 162, 90 L Ed 137, 106 S Ct 1758 (1986).

22 *Witherspoon v. Illinois*, 391 US 510, 20 L Ed 2d 775, 88 S Ct 1770 (1968). The Supreme Court held that the prosecution could remove only those potential jurors who "made unmistakably clear . . . that they would *automatically* vote against the imposition of capital punishment" or that they could not assess the defendant's guilt impartially. (*Lockhart*, at 475 US 162, 184).

23 W*itherspoon v. Illinois, id.*, at 517-8. Justice Stewart, writing for the six-person majority invited further research with this language: "The data adduced by the petitioner (defendant) are too tentative and fragmentary to establish that jurors not opposed to the death penalty tend to favor the prosecution in the determination of guilt. . . . In light of the presently available information, we are not prepared to announce a per se constitutional rule requiring the reversal of every conviction by a jury selected as this one was" (by excluding WEs). Justice Rehnquist's opening line in *Lockhart* acknowledged that invitation: "In this case we address the question left open by our decision nearly 18 years ago in *Witherspoon v. Illinois*." (476 US 162, 165).

24 For example, in a random sample of 2068 adults in a nationwide poll, in which the subjects in face-to-face interviews were asked to find guilt based on written descriptions of the evidence, 65% of DQs found guilt, but only 56% of the WEs (p < .001). In a highly sophisticated study 264 subjects, 104 of whom had been actual jurors, were asked to assume they were actually deciding a capital case. They viewed a two and one-half hour videotaped reenactment of an actual murder trial. On a post-deliberation ballot, only 13.7% of the DQs but 34.5% of the WEs voted "not guilty." (p < .01). Perhaps more significantly, the study examined the effect of the often lengthy juror examination process that precedes the actual trial during which WEs are sorted out. The language employed during this examination implies guilt. A judge may say, "At the end of this trial, I will outline in detail the factors to be weighed in deciding whether to impose the death penalty." Subjects exposed to this examination process were found to be significantly more prone to convict and to believe that the judge and defense attorney thought the defendant guilty. No study reported so far contradicts any of the studies that showed various forms of DQ bias. See "In the Supreme Court of the United States: Lockhart v. McCree. Amicus Curiae Brief for the American Psychological Association," *American Psychologist* 42 (1987): 59.

25 *Lockhart*, 476 US 162, 166. The exclusion of jurors with certain immutable characteristics, such as race or gender, is impermissible under the Sixth Amendment requirement that jurors represent a fair cross section of the community. The *Lockhart* Court held that WEs jurors do not have such characteristics.

26 *Lockhart*, 476 US 162, 168.

27 *Lockhart*, 476 US 162, 173.

28 In the legal literature, "causation" entails questions of fairness as well as questions of causality in the scientific or logical sense. Jurists must consider whether some causal "proof" meets not only standards of scientific causation but also standards of fairness in holding one legally accountable

for a harm done to another. See, for example, Richard W. Wright, "Causation, Responsibility, Risk, Probability, Naked Statistics, and Proof: Pruning the Bramble Bush by Clarifying the Concepts," *Iowa Law Review* 73 (1988): 1001. Most of the questions and concerns raised in that literature are inapposite here because "fairness" does not arise: The defendant in *Lockhart* did NOT protest that the causal argument constructed by the social scientists would hold him responsible for harm done to another. Indeed, he took the opposite view.

29 Troyen A. Brennan, "Causal Chains and Statistical Links: The Role of Scientific Uncertainty in Hazardous-Substance Litigation," *Cornell Law Review* 73 (1988): 469-533.

30 *Lockhart*, 476 US 162, 178.

31 Geoffrey R. Loftus and Elizabeth F. Loftus, *Essence of Statistics* (Monterey, California: Brooks/ Cole, 1982), 7.

32 Justice Thurgood Marshall, in a dissenting opinion joined by Justices Brennan and Stevens, puzzles over the majority's failure to understand the defendant's argument, stating that the defendant "argues simply that the State entrusted the determination of his guilt and the level of his culpability to a tribunal organized to convict." (*Lockhart* 476 US 162, 194).

33 *Watson v. Fort Worth Bank & Trust*, 487 US, 108 S. Ct. 2777 (1987). The U. S. Supreme Court refused to accept that so-called subjective employment assessments (interviews, performance appraisals, experience) can be examined for bias in reliability and validity with sufficient scientific rigor to meet the standards of certain civil rights laws. Reviewed in Donald N. Bersoff, "Should Subjective Employment Devices Be Scrutinized? It's Elementary, My Dear Ms. Watson," *American Psychologist* 43 (1988): 1016-18. Courts experience difficulty in appropriately utilizing statistical inferences in hazardous-substance litigation. Brennan, "Causal Chains," 469-533. "Particularistic evidence" is preferred to statistical evidence in civil litigation. Mary Dant, "Gambling on the Truth: The Use of Purely Statistical Evidence as a Basis for Civil Liability," *Columbia Journal of Law and Social Problems* 22 (1981): 31.

34 Traditionally, facts are introduced into the trial process through witnesses who are found to be competent to testify from their observation about facts that are relevant and material to the specific issues before the court. This traditional process works reasonably well for explicit questions of what are called "adjudicative facts," (whether the defendant had a gun in his hand) and when the witness's powers of observation are important (whether the witness who saw the gun was nearsighted). The process does not work well for what are called "legislative facts," that is, facts which are not explicit (whether DQ jurors are significantly more prosecution oriented than WEs) and when the witness's powers of observation are not important (scores of investigators who conducted studies investigating different aspects of the question). The findings of social science research are most often "legislative facts." See, generally, John Monahan and Laurens Walker, "Social Science Research in Law," *American Psychologist* 43 (1988): 465-72.

35 Symposium, "Social Science in Legal Education," *Journal of Legal Education* 35 (1985): 465-506.

36 Hanna F. Pitkin, *Wittgenstein and Justice* (Berkeley: University of California Press, 1972).

37 W. H. Dray, "Causal Judgment in Attribution and Explanatory Contexts," *Law and Contemporary Problems* 49 (1986): 13; Richard A. Shweder, "Beyond Self-Constructed Knowledge: The Study of Culture and Morality," *Merrill-Palmer Quarterly* 28 (1982): 41.

38 Edward E. Sampson, "The Debate on Individualism: Indigenous Psychologies of the Individual and Their Role in Personal and Societal Functioning," *American Psychologist* 43 (1988): 15.

# Language As and In Technology: Facilitating Topic Organization in a Videotex Focus Group Meeting

*Wayne A. Beach*

Though typically overlooked and thus taken-for-granted as a sophisticated process and product of human innovation, talk (and thus talk-in-interaction) is itself an omnipresent, finely organized, collaborative display of cultural activity.[1] In reference to the analytic exercise of studying conversations directly, Sacks observed over two decades ago that an overriding goal

> ... is to see how finely the details of actual, naturally occurring conversation can be subjected to analysis that will yield the *technology of conversation*. . . . We are trying to *find* this technology out of actual fragments of conversation, so that we can impose as a constraint that the technology actually deals with singular events and singular sequences of events—a reasonably strong constraint on some set of rules.[2] (emphasis added)

Stated somewhat differently, Sacks notes how

> Our aim is to transform, in an almost literal, physical sense, our view of "what happened," from a particular interaction done by particular people, to a matter of *interactions as products of a machinery*. We are trying to find the machinery. In order to do so we have to get access to its products. At this point, it is conversation that provides us such access.[3] (emphasis added)

Upon close inspection, conversation reveals its own technology for getting interactional tasks done—noticeably, in the first instance, by and for participants themselves as they make available to one another their occasioned orientations.

Of the immense variety of social occasions in which conversation is vehicular for achieving understandings, specific gatherings are designed for the explicit purpose of addressing the impact of technological advancements on everyday life. One type of occasion, a Videotex focus group meeting, has been selected for analysis in this chapter for its potential to reveal insights about basic working relationships among communication, culture, and technology. First, Videotex offers an innovative approach to interactive cablevision, one in which current impacts of interfacing televisions and computers in home and work environments are directly assessed. Second, the data to be examined—audio-recordings and transcriptions of the meeting—provide the possibility of understanding how language is relied upon to raise, and resolve, routine problems of an emergent cultural and technological apparatus. In so doing, however, it will be shown how the achieved and thus interactive character of "talking technology" and, conversely, "technological talk" are themselves problematic *as accomplishments*. In this sense, conversation and technology are reflexively coupled: Conversational activities are technological achievements in and through co-participants' methods for getting tasks done, just as descriptions of the impact of specific technologies on everyday life are possible *only* through the language employed to produce such descriptions.

Sacks observed that "whatever humans do can be examined to discover some way they do it, and that way will be stably describable. That is, we may alternatively take it that there is order at all points."[4]  How then, we might ask, does the work of language *as* a technology, *in* a technological occasion, get accomplished? What methods/devices/techniques/practices/strategies are relied upon to get activities—such as "facilitating" a hi-tech focus group meeting—*done* for all practical purposes?

A partial answer to these encompassing questions can be provided by examining the interactional organization of a Videotex focus group meeting. Attention will first be given to a brief overview of Videotex services and discussion of the interactional data to be analyzed in this chapter. Constituent features of "speech exchange systems" will then be described as a way of beginning to understand focus group meetings as occasions displaying characteristics of both casual and institutional discourse. By next examining how the "facilitator" sets up and orients to the business-at-hand, it becomes possible to analyze three problematic instances of "topic organization" as an omnipresent and ongoing achievement—similar to, yet different from, the organization of topics across types of social occasions. Finally, conclusions are drawn about conversation as an intricate technological resource, one in which the workings of communication and culture can be found to be both self-evident and essential in the achievement of ordinary tasks.

## BACKGROUND AND DATA

Within the past several years, pioneering investigations have occurred in the area of *interactive cable television*. One such system has been designed and implemented by Cox Cable Communications, Inc. of Atlanta, Georgia and the Corporation for Public Broadcasting's Office of Science and Technology (the latter having a congressionally assigned responsibility to examine the implementation of new telecommunications technologies, as well as such technologies' potential for educational and public television licensees).[5] Through Cox's INDAX (Interactive Data Exchange) format, the following description has been offered:

> Considered one of the nation's most advanced interactive cable systems, INDAX uses state-of-the-art cable technology to make possible interactive services such as banking, shopping, information retrieval and education. In an interactive system such as INDAX, the cable television viewer can use a keypad (similar in appearance to a remote control unit) to respond to televised material or to request that specific textual information appear on the television screen. A powerful computer located in the cable operator's system responds to viewer requests.[6] (see also Appendix B)

One technique for assessing users' reactions to Videotex services is the *focus group*. These meetings are typically designed to solicit information regarding users' actual "first hand experiences" with Videotex—strengths, weaknesses, frustrations, and suggestions for improvement. Information of this sort can then be cycled back into the technical and planning dimensions of the industry to better refine, market, and advertise this "technological innovation."

The data reported in this chapter were drawn from an occasion in which eight users participated in a two and one-half hour meeting, facilitated by a researcher involved with collecting and analyzing information from the INDAX project. This meeting was audio-recorded with full knowledge and consent by group members, and transcripts were subsequently produced (see Appendix A for Transcription Notation Symbols). Selected segments of these transcripts are provided, for readers' critical inspection, as evidence of the following claims depicting how the facilitator and users co-produce this interactional occasion:

> 1) Facilitator's initial orientations to the meeting are specifically designed to create a sense of order for subsequent talk, and users' hearing, in the course of establishing a format for telling and talking about Videotex experiences;

> 2) What might be taken to be small and apparently insignificant "token" behaviors by the facilitator, such as "okay," "oh great," and "um hmm," are found to be sequentially relevant in the management and organization of "speakership," "recipiency," and "topic" as participants gain access to, and yield, the floor;

> 3) The problematic nature of topic organization is evidenced in circumstances wherein facilitator a) orients to users as having volunteered information prematurely, b) works to mark and thus receipt news while c) also preparing to move from passive recipiency in the preparation, initiation, and carrying out of topic shift;

> 4) The activities noted above are essential in the process of "doing being a facilitator" by moving discussion along and keeping interaction "on track." The manner in which these activities get accomplished reveals, in their sequential organization, a speech exchange system involving both casual and institutional features inherent in the reporting and receipting of "news."

## Focus Group Meetings as Speech Exchange Systems

One of the basic and useful distinctions for examining variations in social conduct involves contrasting talk in "natural/ordinary/casual conversation" with "institutional interaction."[7] Best viewed on a continuum, casual talk displays a wider range of possible and expectedly "appropriate" activities, including recurring displays of affiliation and disaffiliation with speakers' claims. In contrast, institutional talk is constrained by such features as the narrowing of activities—a uniformity of interactional shapes and devices, for example—as specific tasks and roles get noticeably worked-out.[8] Examples of occasions wherein participants orient to institutional constraints include (but are not limited to) classrooms, courtrooms, medical, and news interviews.[9] In each of these settings, turns-at-talk are typically *pre-allocated* in the ways that who speaks, in what order, for how long, and on what topic(s) are more or less pre-specified or constrained. Such candidates might include the explicit purpose for the gathering, orientations to "appropriate" procedure, and the readily apparent use of questions and answers to organize interaction.[10]

These conventionalized forms of talk reflect marked differences with interactions occurring in non-institutionalized settings. Heritage provides a useful summary of several contrasts which are relevant to the subsequent analysis of speech exchange within a focus group meeting.[11] One noticeable dimension of institutional talk is the *reporter-reportee* relationship: The more "formal" the setting, the more pre-established roles of reporter-reportee are expected and interactionally (often rigidly) maintained. Witnesses do not interrogate attorneys, for example, nor do patients diagnose physicians' ill-nesses.[12] Rather, it is assumed that witnesses and patients have some sort of "news" to deliver, as do those being interviewed in mediated and broadcast news events.

A second and related feature of institutional talk involves how those to whom the news is reported *receipt* such news. Heritage and Clayman independently observe how news interviewers display neutrality in receipting news:[13] There is a noticeable lack of "alignment talk" between reporting parties and those relying upon questions to elicit news such as news interviewers and/or attorneys. In these ways, talk is designed for an *overhearing audience* including broadcast audiences, judges, and juries. Those eliciting the news typically *withhold* displays of affiliation and disaffiliation. Specifically, as questions are asked and answers provided, third-turn receipt objects such as assessments ("good"/"how terrible") and various news markers ("oh really"/"I see") are noticeably absent. Also missing are such objects as facilitators and continuers ("um hmm" and "uh huh"), typically provided by recipients in casual talk as displays of passive recipiency, acknowledgement, and/or moving toward gaining the floor and, in turn, assuming "speakership."[14]

The above descriptions of casual and institutional talk provide only a partial characterization of the constituent features of speech exchange systems.

However, this brief summary does offer a lens for assessing the ways in which focus group meetings might be located on the casual-institutional continuum. In the following analysis, an ongoing concern rests with how the talk is adapted to the task-at-hand; that is, how the interactional order is designed to achieve the business of arriving at some sense of understanding and shared orientation to the situation. Concerns rest with the ways in which various turn-taking practices structure participants' opportunities for involvement. We now turn to an analysis of transcribed instances from the focus group meeting, as visual displays of the methods employed by facilitator and users in the course of reporting (and receipting) Videotex experiences. Following this analysis the focus group, as one type of speech exchange system, will be reconsidered in light of the features described.

## ORIENTING TO THE BUSINESS-AT-HAND

Perhaps we should begin with the obvious: Meetings often require some "setting up" prior to "getting on with" the business-at hand. Within the first few minutes of this two and one-half hour meeting, F (the pre-designated "facilitator" of the group) describes to participants what they are here "to *do*:"; namely, "talk about ↑ *In*dax" and "*tell* us about it.":

(1) VT:FGM:I:1,2

```
       F:      Um: what we're here to do is is (.)
               >what we're here today to do:?<
       ==>     is to talk about ↑ Index (1.0) and
               what .hh what we're do:ing? here.
               what this is called this is called
               a focus group.

                       .

                       .
```

```
(2)    F:      Well .hh that (.) format >is kinda
               what we wanta do today? is we wanna
       ==>     just talk to you< .hh about Index
               hh uh? you folks are the experts.

                       .

                       .
```

```
(3)    F:      so: um anyway. what (.) what we're
               gonna do today is just kinduv as:k
       ==>     you to ↑ tell us about it.
```

It is clear that F is attempting to clarify the purpose for meeting, and several important observations might be drawn from (1) - (3). First, while F's talk is produced as a single speaker's narrative, it is nevertheless designed for users' hearing as a means for creating a sense of social order for subsequent talk. In (2), for example, users are referred to as "experts.", the clear presumption being that experts are uniquely qualified to report significant news regarding their practical experiences with Index. Second, the arrows (==>) in each segment draw attention to F's own reference to "talk" as the vehicle for "telling" such experiences. Of course, (1) and (2) are themselves only glossed versions of subsequent, actual activities comprising the group's discussions; they are not designed to extend or elaborate on circumstances that have not yet transpired, including the group discussion itself. And in (3), F appears both to bring an extended turn-at-talk to a close ("so: um anyway.") and to project the relevance of "asking" and "telling" as achievements remaining to be worked-out—in the course of what such a meeting will be shown to be "about."[15]

For these and related reasons, (1) - (3) contribute to what Garfinkel and Sacks refer to as the "accountably sensible character" of an occasion.[16] In and through the ways that F is engaging in activities expected and reserved for the "facilitator" of such a gathering, such as initiating and keeping the meeting going, the encumbrances of such a "role" are exhibited and thereby made available for the group's (and researcher's) inspection. It is the group's *recognition* of F's actions *as* role and task-specific that allows Index users to anticipate what a facilitator, as co-participant, might be aiming or arriving at in setting up a meeting of this sort. And with such recognition comes the possibility of co-producing informative and understandable reportings about various hi-tech experiences. This does not guarantee that reportings will, in all cases, turn out to be relevant and otherwise unproblematic. On the contrary, the subsequent analysis suggests that numerous ongoing troubles emerge involving such phenomena as turn-initiation and completion, topic shift and organization, and in each case these troubles require and receive solutions as the meeting unfolds.

It is the achieved character of such troubles, and their emergent resolution, that can begin to be gleaned from the following extended segment:

(4) VT:FGM:I:5,6

(By F's request, each of the ten participants were asked to introduce themselves. The segment below begins immediately following the last introduction.)

```
        B:   >I'm getting. some< so me
                                [   ]
a==>F:                          Ok ay.
        B:   studies for the (0.2) SD.
        F:   Pardon?=
        B:   =communications.=
        F:   = >You've participated in some?< =
```

```
     B:   Ya I'm getting some now?=
b==>F:    =Oh grea:t ! .h and: > what
               you're doing right now? is: is sort o f on  e.
                                              [     ]
     B:                                        (hmm)
1==>F:    we're .hh we're trying to understa:nd the
          Indax system. (0.5) trying to s:ee what
          potential? it has: for all kinds of uses
          an: (0.2) and >trying to get some feedback
          from you?< .hhh you you folks really do:
          know >m:ore about this system:.< than
          (0.7) >than ↑ almost anybody else in the
          country. there aren't a tho:usand people
          in the country.< .hh   that kno:(w) as much
          about this? (0.4) ↓ potential system as (.)
          you folks do. .hhh >so ↑ that's why we wanna
          talk to you.< .hh an(d) we wanna find out
          everything that we can about. it and then: u:se
          your opinion .hh ta help guide the future
          development of the system.
              (1.0)
     F:   So you're .h  you're sit ting
                            [        ]
     BJ:              I house
     F:   here representing a million people
          $or whatever::$ .hh h
                           [ ]
     BJ:                   I hou s:e
                               [    ]
c==> F:                        >uh huh< =
     BJ:  =foreign students: (.) fr om=
                                 [     ]
c==> F:                          Uh huh?
     BJ:  = >all over the wor:ld.< =
c==>F:    =Uh huh=
     BJ:  =and um: >I use it quite a bit< for:
          the:ir studies (experience in) studying
          English (.) it's the w or:ds
                          [    ]   [    ]
b==>F:                    .hhh     ↑ O::h! interesting=
     BJ:  =the meanings (1.0) uh: they use it
          >quite a bit.<
2==>F:    What else ↓ do the rest of you use it for
              (1.0)
     F:   use it for studying English what (.) you
          know (0.2) or do you no::t use it >cuz
          if you don't use it that's just as important
          to find out.< =
     B:   No I use it myse:lf I use it u:h I like the games
              (1.0)
     Jo:  Um hmm?
        [[   [              ]
```

```
      B:    u::mm (0.2) course my- >I had an eighteen
            year old girl that goes to school she (catches)
            the soap operas (as she sees fit) so we check
            the< .hhh soap operas that she (0.6)
            ju st missed?
                 [          ]
b==>F:        Ah hah? (.) you like the soap opera
            (dia  ries)
                  [   ]
      B:          u:mm=
c==>F:      =uh huh?
      B:    TV guide um:(.) what's goin on in town for
            the foreign students? >(are ready to) go see< =
c==>F:      =Um hmm?=
      B:    =u:mm (.) good restaurants? (.) uh: what's what's
            happening on weekends? (.) that they can go see
            (1.0) a:h >just (a)bout ev'rything< =
c==>F:      =Uh ↑ hu h
                  [ ]
      B:          th at's available=
a/3=>F:     =(o)↑kay .hh and I know you're a real f:an of
            Indax cuz you  wha- what do you use it for
                           [                       ]
      Jo:                   Oh ya I think this-  th at's
            ↓ the thing of the future ((continues))
```

As noted, this segment occurs immediately after group members introduce themselves, an activity that followed F's "setting up" the meeting in (1) - (3). In general terms, (4) above officially marks the initiation of "getting on with" the business of actually talking with users about their experiences. An understanding of how this task gets carried out might begin by reference to the types of activities (arrows 1-3, a-c) in which F, as facilitator, produces *in sequence* as the talk progresses.

Before turning directly to these activities, however, it is important to stress the relevance of "sequence" as consequential to the task-at-hand. Any sense to be made of segment (4) is most obvious when contrasted with examples (1) - (3), for now F's actions noticeably constrain (and are constrained by) the turns-at-talk engaged by other parties. These engagements have much to do with how and when turns begin and are completed through such features as turn transitions, placements, and constructional-units; the ways in which speakers self-initiate and/or are invited to talk through turn distribution and allocation; minimal to extended turn size; and the projectability of each and every utterance-in-sequence, involving the work an utterance might be doing and its possible trajectory. These recurrent features of turn-taking are collaboratively organized en route to getting the meeting accomplished, displaying in their organization participants' orientations to the moment-by-moment contingencies of interaction.

## WORKING THROUGH TOPICS: THREE PROBLEMATIC INSTANCES

Within (4) above, (1==>, 2==>, and 3==>) draw attention to certain behaviors enacted by F that are recognizably facilitator-like. The elaborated turn marked by (1==>), for example, closely resembles F's actions in (1) - (3): Yet another attempt is made to "set up" the discussion by describing to users the value of their knowledge, the desire to "get some feedback from you?" for developing the system, all of which is glossed by "we wanna talk to you" as with (1). Exactly why this turn emerges at just this juncture in the interaction rather than immediately preceding or following group introductions, and how the emergence of this turn marks a trouble-source in this sequential environment, will be addressed shortly.

The turns highlighted by (2==>) and (3==>) indicate what would seem to be basic devices for initiating a topic and keeping a discussion going, namely, asking users questions to elicit information. As with F's actions in (1==>), however, these questions might also be best understood within (not isolated from) the sequential environments in which they are occasioned; that is, by considering what interactional work preceded and thereby paved-the-way for these questions and topic shifts.

However, in accounting for the sequential relevance of F's actions in (1==>), (2==>), and (3==>) it is important not to overlook the comparatively small and (upon first notice) apparently insignificant behaviors marked by (a==>, "okay"; b==>, "oh great", "oh interesting", "ah hah"; and c==>, "um hmm", "uh huh"). These tokens are not randomly or mistakenly placed in the course of interaction, but rather accomplish specific and ongoing work by participants. By examining more closely the three portions of (4) involving (1,==>, 2==>, and 3==>), and the problems implicated in these productions, it is possible to gain an appreciation of the rather intricate technology inherent in a facilitator's attempt to get such a meeting underway.

### Displaying "Not Yet Ready" for Information

Let's once again begin at the beginning:
(5)     VT:FGM:I:5
```
         B:    >I'm getting. some< so me
                               [   ]
   a==>F:                      Ok ay.
         B:    studies for the (0.2) SD.
         F:    Pardon?=
         B:    =communications.=
         F:    =>You've participated in some?< =
         B:    Ya I'm getting some now?=
   b==> F:    =Oh grea:t ! .h and: > what you're doing
                right now? is: is sort
                   o f on e.
                    [    ]
         B:    (hmm)
   1==>F:    we're .hh we're trying to understa:nd the
```

> Indax s̲ystem. (0.5) trying to s̲:ee what
> pot̲en̲tion? it has: for all k̲inds of uses
> an: (0.2) and >trying to get some feedback
> from you?< .hhh you you̲ f̲olks really d̲o:
> know >m̲:ore  about this system:.< than
> (0.7) >than ↑ a̲lmost anybody else in the
> country.  there aren't a tho̲:u̲sand people
> in the country.< .hh that kno:(w) as much
> about this? (0.4) ↓ potential system as (.)
> you folks do. .hhh >so ↑ that's why we wanna
> ta̲lk to you.< .hh an(d) we wanna find out
> e̲verything that we ca̲n about.it and then: u̲:se
> your op̲in̲ion .hh ta help guide the future
> deve̲lop̲ment of the system.

In (a==>), F places an "Okay" in overlap with B's self-selected (and quickly delivered, i.e. > . . . <) attempt to volunteer information and initiate discussion. But why does an "Okay" get placed at just this point?  Upon initial inspection it may appear that F's "Okay" acknowledges and/or affiliates with B's objective—offering a telling about current involvements with "some studies." Yet an examination of what happens next in the sequence, in unison with what occurred prior to B's turn   (introductions) suggests an alternative and even problematic explanation.

Here it is seen that F's "Okay" might be heard as a dual-orientation to this interactional moment: 1) It marks a closing of prior activity (introductions);[17] 2) By so doing, it also displays a transition-readiness to move onto the next activity by once again assuming speakership and hence the role of "facilitator." F does eventually gain the floor as evidenced in (1==>), but this does not occur without interactional work designed for this very possibility.  In particular, F seeks clarification ("P̲ardon?"), perhaps because B's self-initiated turn was unexpected or otherwise not oriented to by F at the precise moment at which J was in transition to next topic/activity.  In not receiving clarification, however, F then issues an "other-initiated repair" in next turn ("You've p̲articipated in some?") by partially repeating a portion of B's earlier utterance.[18] This repair initiation evidences a trouble-source in prior turn(s), and is offered as a means of remedying problems with understanding what B is "up to"—that in fact B is in the process of volunteering information, and appears to be pursuing an opportunity to discuss these experiences.

By turning to (b==>), it is clear that F has now attained the understanding sought through repair.  The task now remains of what to do "next" with such information—to continue or terminate B's volunteered topic?  As a solution to this problem, F's "O̲h grea̲:t !" both marks the news of and positively assesses B's volunteered information.[19]  Yet this assessment token is decidedly not projecting the status of a recipient who is passing the opportunity to take a fuller turn, given that another participants' turn is already in progress.  In fact, "oh" + "assessments" are often placed immediately prior to a topic-shift and function to enforce recipient's coming to a stop (and are thus closure-relevant).[20]  In this specific instance, F simultaneously moves *away* from B's continual volunteering of information and *toward* the elaborated turn in (1==>)—a transition-point at which the closing of prior topic (B's telling) promotes an opening for extending alternative topic(s). In response, B appears to recognize with "hmm," and orients accordingly by not continuing to offer additional information.

One way to summarize the interactional work in this segment is to suggest that F, as facilitator, displayed a "not yet ready" orientation to B's volunteered information. By opting to shift the focus of discussion to what eventually became (1==>), rather than pursue B's experiences in more detail, F effectively deleted the trajectory of B's talk by pursuing an alternative course of action. The "Okay" and "Oh grea:t !" tokens accomplished important work in this segment, and in each case indicate just what F was attending-to at specific points in the talk: Closing down prior and moving to alternative topics.

It should not be surprising that B and F display contrasting agendas in this brief exchange. In (1) - (3) F strongly encouraged participants' involvement, and uncertainty about procedure is expected in light of the fact that the actual discussion—the first order of business—had yet to get underway.  Nor is it surprising that these interactants rather quickly resolved the problematic nature of the talk, for such circumstances are routine in conversational organization. Of course, neither of these observations are intended as a way of discounting B and F's collaborative efforts. On the contrary, they function to complement how B and F were, in the first instance, orienting to the contingencies of these circumstances *at a moment's notice*. The technological details of this work are evident within participants' organization of topic(s).

## On the Fringe of Recipiency and Speakership

That B was not the only participant to volunteer information upon first opportunity, and perhaps "pre-emptively," is apparent in the following and second segment extracted from (4):

```
(6)   VT:FGM:I:5,6
      1==>F:    .
                .
                .
                you folks do. .hhh >so ↑ that's why we wanna
                talk to you.< .hh an(d) we wanna find out
                everything that we can about.it and then: u:se
                your opinion .hh ta help guide the future
                development of the system.
                     (1.0)
      F:    So you're .h  you're sit ting
                             [          ]
      BJ:                     I  house
```

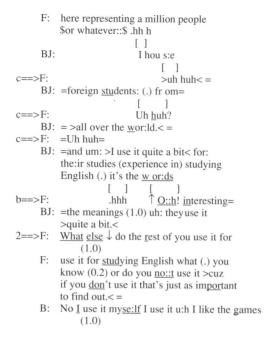

```
    F:   here representing a million people
         $or whatever::$ .hh h
                           [  ]
    BJ:                    I hou s:e
                             [   ]
c==>F:                         >uh huh< =
    BJ:  =foreign students: (.) fr om=
                           [     ]
c==>F:                     Uh huh?
    BJ:  = >all over the wor:ld.< =
c==>F:   =Uh huh=
    BJ:  =and um: >I use it quite a bit< for:
         the:ir studies (experience in) studying
         English (.) it's the w or:ds
                      [   ]    [     ]
b==>F:                .hhh    ↑ O::h! interesting=
    BJ:  =the meanings (1.0) uh: they use it
         >quite a bit.<
2==>F:   What else ↓ do the rest of you use it for
             (1.0)
    F:   use it for studying English what (.) you
         know (0.2) or do you no::t use it >cuz
         if you don't use it that's just as important
         to find out.< =
    B:   No I use it myse:lf I use it u:h I like the games
             (1.0)
```

In overlap with F's formulation ("So you're .h you're sitting here representing a million people $or whatever::$ .hhh") of (1==>), BJ twice prefaces a telling of her experiences. As BJ continues, F immediately and repeatedly (c==>) receipts the telling with three "uh huh's". Once again, the question can be raised: Why are they placed at just this point? One probable answer is that F was simply being attentive to BJ's telling, indicating interest and acknowledgment. As was mentioned earlier, a key dimension to facilitation is receipting reporters' news. However, it may not be sufficient to limit the analysis to F's "being attentive." As Schegloff suggests:

> When 'uh huh's etc. are considered in the aggregate, then, the characterization of the class as signaling attention, interest, or understanding appears equivocal. Although it can be argued that attention and understanding are generically relevant in conversation, no ready account is at hand (when the aggregate of cases is considered) for why these issues need specially to be addressed, why they are addressed with these tokens, why addressed at these particular points (if, indeed, it is at particular points, on this account, that these tokens are placed).[21]

Given the equivocality of "being attentive," then, what alternative characterization (if any) exists for these "uh huh's"?

By initially observing what follows the three "uh huh's," it is apparent that F's (b==>, "O::h! interesting=") occurs immediately prior to a topic shift in (2==>). First, F's inbreath (.hhh) in (b==>) suggests that F was headed for speakership even prior to the oh + assessment, in that inbreaths so placed typically occur prior to turn onset. Second, F's (2==>) is an "unmarked overlap retrieval" that effectively deletes BJ's intermediary turn and, in so doing, displays uncontestable rights to the floor.[22] Strikingly similar to F's "<u>Oh</u> <u>grea:t</u>!" in (5), F's token in (6) also marks the news delivered by BJ, positively assesses the information offered, and works toward a transition to the next topic as F relies upon a question to select the next speaker(s).

With these activities in mind, F's "uh huh's" might now be described as displaying a *preparedness to shift topic*. As Jefferson notes, "uh huh's" routinely evidence a step toward speaker-readiness and thus a "recipient's orientation to some ongoing talk as sufficient and terminable."[23]

Throughout (6), therefore, we have an instance where tokens such as "uh huh" and "oh interesting" simultaneously resolve the problem of attending to the current speaker's talk, while also paving-the-way for moving onto the next topic. In these ways, through the course of BJ's brief telling, F's facilitation amounted to the work of preserving status of recipient, co-participant's role as teller, and yet remaining positioned to quickly assume the next speakership. Though seemingly precarious, working on the "fringe" of recipiency and speakership is par for the course as interactants achieve topic.

## Marking Topic Shifts

The final segment from (4) begins with (2==>) where, as noted previously, F shifts topic by asking a question. Of course, merely asking a question does not guarantee uptake and, following a (1.0) pause, F qualifies a query as a means of eliciting a not-yet-forthcoming response:

```
(7)  VT:FGM:I:6
       2==>F:   What else ↓do the rest of you use it for
                      (1.0)
           F:   use it for studying English what (.) you
                know (0.2) or do you no::t use it >cuz
                if you don't use it that's just as important
                to find out.< =
           B:   No I use it myse:lf I use it u:h I like the games
                      (1.0)
          Jo:   Um hmm?
          [[ [              ]
           B:   u::mm (0.2) course my- >I had an eighteen
                year old girl that goes to school< she misses
                the soap operas >she doesn't get'ta see it so we
                check the< .hhh soap operas that she (0.6)
                ju st missed?
                     [        ]
       b==>F:       Ah hah? (.) you like the soap opera
                (dia ries)
                       [    ]
           B:       u:mm=
```

```
c==>F:    =uh huh?
    B:    TV guide um:(.) what's goin on in town for
          the foreign students? >(are ready to) go see< =
c==>F:    =Um hmm?=
    B:    =u:mm (.) good restaurants? (.) uh: what's what's
          happening on weekends? (.) that they can go see
          (1.0) a:h >just (a)bout ev'rything< =
c==>F:    =Uh ↑ hu h
                   [  ]
    B:           th at's available=
a/3=>F:   =(o)↑kay .hh and I know you're a real f:an of
          Indax cuz you  wha- what do you use it for
                       [                             ]
    Jo:                   Oh  ya  I think this-   th at's
          ↓ the thing of the future (0.8) that thing right
          there (.) with some little impro:vments ((continues))
```

In the next turn, B initiates a response by telling of an eighteen -year-old girl who misses the soap operas. This is followed by Jo's passive recipiency and likely agreement (Um hmm?), and B continues the telling. Though F's overlapping "Ah hah?" in (b==>) appears to be displaying a special understanding or realization, it is unclear whether or not it moves B to closure like "oh great" and "oh interesting" above, just as F's next "uh huh?" following B's "u:mm" may be spurious.[24] In any case, F's next three behaviors in sequence—"Um hmm?," "Uh ↑huh," and "(o)↑kay)"—strongly resemble Jefferson's observations that the movement from "mm hmm ==> uh huh ==> yeah" frequently indicate transitions from "passive recipiency" through "speaker readiness" to "speakership-associated" activities such as "yeah."[25]   In (a/3==>) above, F's "(o)↑kay" functions similarly to "yeah," in that it prefaces what Button and Casey refer to as a "topic initial elicitor"—in this utterance, "wha- what do you use it for"—which frequently occur following closing components in conversations.[26] Prior to (a/3==>), however, notice that F's move to speaker-readiness ("Uh ↑huh") is placed immediately following B's construction of a three part list ("good restaurants?", "weekends?, and a generalized list completer "just (a)bout ev'rything" that is syntactically re-completed with "that's available"). In the recognizeable course of their construction, such listings project completion of an utterance and thus provide "a point at which another can or should start talking."[27]

With this projectability in mind, it now becomes possible to observe how F's "(o)↑kay" in (a/3=>) is placed precisely at the end of B's three part list, with no gap or overlap (thus latched, =), as a pre-closing device for eliciting comments from next speaker.

## Summary: Analyzing Facilitated Interaction

Having worked through several key features of (4), it now becomes possible to describe how the interaction comprising this occasion—as a speech exchange system—might be located on the casual-institutional continuum. This focus group meeting might best be situated *midway* on such a continuum. It displays a variety of features typically associated with each generalized type of speech exchange.

First, it is clear that this discourse is, to some extent, both formalized and task-specific. For example, the meeting would not have occurred if it had not been scheduled, facilitator and users' roles were specified in advance, a sequence of "introductions" occurred, the facilitator set up the meeting for an overhearing "audience" and proceeded to structure the unfolding discussions, and the reporter/reportee relationship was maintained (at least in part) through questions and answers.

Second, and in contrast to more formalized institutional talk including courtrooms and news interviews, the facilitator/reportee did not remain "neutral" by refraining from displays of affiliation. F's frequent continuers and assessments ("uh huh"/ "oh great") displayed considerable alignment with the users' news reports, even though such work was shown to be more than simply affiliative in the ways topics came to be organized. However, just as (4) does represent a sequence involving the first opportunities to offer and receipt news/ tellings, so might F's frequent affiliation and alignment be heard as somewhat "overdone." Two different, though perhaps interdependent, accounts might be offered in this regard. On one hand, in getting such a meeting underway and attempting to set a "proper tone" for reporting and receipting news, it is not difficult to understand F's efforts as moves toward de-formalizing the occasion. Another possible explanation involves framing F's alignment as attempts to minimize possible and emergent troubles with terminating and shifting topics. Put simply: Just because a facilitator is "responsible" for shifting topics and thus moving the discussion along, it does not necessarily imply that users' tellings and experiences be treated as unworthy or in any way inappropriate. Whatever the reasons motivating F's actions, it is also clear that details such as how users self-selected turns-at-talk (at times, preemptively) to volunteer information also differs from more formalized settings.

It remains to be seen how the meeting continues to unfold. Questions may now be raised about what additional methods F employs to elicit news reports, not to mention the kinds of troubles emerging when, for example, users seek clarification of F's topic elicitors and/or are treated by F as though the discussion has gotten "off track" (two instances of which are briefly examined below). These are more encompassing activities comprising such a focus group meeting, extending quite beyond the present analysis. However, it is important to note that these and related activity-types are intricately woven within "topic organization," and in each and every instance require working out by and for the participants themselves.

As noted earlier, F's first and last actions in extended segment (4) involved "okay" as a pre-closing device for topic shift. Numerous additional instances of "okay" have been located throughout these and other transcripts, pointing the direction toward gaining an understanding of "okay" as topic shift-implicative. These data-collection strategies are necessary for locating and substantiating the patterned nature of F's facilitations, as well as how interaction as a "technology" or "machinery" is ordered across occasions. Inherent in the claim that language is "technological" resides the responsibility of searching for recurring instances of a given "phenomenon," locating the occasions of its use, and determining the shape of its organization.[28] Only by so doing can claims for social structure be put forth. While the road toward "universality" is indeed arduous, the goal of advancing each step along the way is to promote a cumulative base for subsequent and more encompassing inquiries.

While it lies beyond the scope of this chapter to offer a comprehensive analysis of "okay" in conversational organization, just as several constituent features of "topic organization" have only been sketched in the analysis of segment (4), it is important to rely upon the analysis provided as a possible map for pursuing such a phenomenon. This pursuit has been extended with the Videotex focus group meeting under investigation, and the instances examined herein begin to evidence how F does, in this occasion and for these tasks, repeatedly rely upon such a device to terminate prior and move on to next topics. The key issue, of course, has more to do with the sequential environments in which marked and unmarked "okays" occur, including any and all sequential consequences and troubles for topic organization, than with simply locating "okay" as a pre-closing device or "discourse particle." [29]

## CONCLUSION

The preceding analysis was offered in order to gain a reflexive understanding of language, technology, and culture by providing empirical (microanalytic) evidence of more global (macroscopic) assumptions regarding technology in everyday life activities.[30] It also displays the intricate and altogether innovative nature of conversation as a technological resource, the organization of which is ultimately rooted in practical circumstances of everyday choice and action. By providing readers with actual instances of interaction, and in so doing inviting critical and shared inspections of the constituent features of a "hi-tech occasion," it becomes possible to formulate relationships among "evidence and claims" in ways having specific, observable consequences. How else might language be described as "technological" if, for example, the facilitation of a focus group meeting was not shown to be an *achievement* comprised of identifiable and recurring features? As it turns out, it is the re-specification of the ordered nature of these features that comprises the analytic exercise of accounting for talk and its manifold possibilities. Without such re-specification, it is likely that everyday conversations—as communication technologies in their own right—would remain taken-for-granted and in these ways glossed as resources for producing and refining technological advancements such as Videotex.

By examining closely F's methods for setting up and getting on with the interaction in segment (4), problems with "topic organization" became apparent. Clearly, working through "topics" is itself a cultural enactment. However, a wide variety of other segments, extracted from the two and one-half hour recording and transcript, can also be examined to reveal unique features of the "culture of technology." Consider the following instance:

```
(8)        VT:FGM:I:21
           Ju:  .

                 .

                 .
                 either vertically horizont(ally) or
                 dia::gonal  ly
                          [    ]
           BJ:             And s:o a:hh=
           Ju:  =and it  won't he:lp.
                     [            ]
           BJ:        you know ↑ I: fe:el the way shc
                 did that'uh we are? ↓ f:ighting the computer:
                 .hh and uh-  but
                          [    ]
           F:             Um hmm
           BJ:  still this um uh. it's a challenge.=
           Jo:  =Or you'll make two  mo::ves on  ya
                                 [                      ]
           BJ:                    >(if you wanna) get ul cers.< =
     ==>   F:   =Uhkay .hh now- ya know those are for the games
                 let's take maybe like the h::ealth games ((continues))
```

Prior to this segment, F was silent for over two minutes. During this time users discussed a variety of experiences with Videotex technology, including the "topic" noted above by BJ: "f:ighting the computer:". This reference begins to indicate ways in which users "anthropomorphize" machines, including computers, by attributing human-like actions and even intentions to how and why the computer "behaves" as it does. For example, consider Jo's "Or you'll make two mo::ves on ya". One avenue for research inquiry, therefore, is to examine how interactants co-produce descriptions, such as BJ and Jo above, as a means of assessing ways in which inanimate "objects" can be oriented to as, essentially, interactional partners.

Yet also notice that within this sequential environment, F does receipt and mark a topic shift with "Uhkay" at the end of this segment, moving the discussion onto "h::ealth games." Need these two activities—"anthropomorphization" and "facilitation/topic shift"—be separated in analysis? Any or all answers seem to depend upon what questions researchers and/or interactants might be asking, and if something like "talking about the computer" may or may not be deemed relevant for a given purpose such as a focus group meeting and/or research

inquiry. In any case, it cannot be overlooked that in this sequential environment the two "activities" emerged interactionally and thus must be treated as displaying interrelationships (at least for the participants involved.)
A final and similar instance appears below:

(9)    VT:FGM:I:[31]

```
       Jo:   .
             .
             .
             >like for instance< the o:ld channel
             F:'uh. fairchild? .hh when it be:at
             you:. they'uh:: o- >one of the games<
             i'wa i'wa (   t) I (think) it was some
             type of ↓ tic tac to:e .hh it would
             sa:y .h YOU: LO::SE TUR:KEY!! ((tough voice))
1==>F:       $Uh okay heh heh$
                   [           ]
       Jo:         an(d)  i-i-i-i-  Go(d) it uh
             ir:i tates
                   [    ]
a==>F:             (And) you feel kinda $ba:  :d h uh$?
                                        [     ]
       Jo:                              $Ya::$
2==>F:       .hh o kay
                   [   ]
       Ju:         >Th at's allright I call my
             computer.< d- dummy?
                   ↑HEH   HEH   uh   heh
             [[ [                  ]          ]
       F:      $Uh h uh  uh heh$              ]
                   [                          ]
       Jo:             >Well see that's a nice th ing about
             it.=
       Ju:   =heh heh ($                        $)
                   [                            ]
       Jo:         say you were uh .hh ah the communica tion
             with it=
3==>F:       =$Okay$ =
       Jo:   you can cuss it ↓ out an it won't answer
             you  back=
                   [    ]
       Ju:         Heh heh  heh heh
                         [      ]
       Jo:               heh heh  heh
                         [    ]
a==>F:                             Um  hmm an' >that's kinda
             nice huh?< .hhh=
       Jo:   =Y a:.
                   [  ]
4==>F:             O kay? .hh um: le- let's talk about (d'er)
             some other issues of concern. We talked a little
             bit about privacy ((continues))
```

Among the rather diverse phenomena that could be examined in this segment, including shared laughter and collaborative descriptions of the computer, it is again apparent that "anthropomorphization" co-occurs with F's repeated attempts to regulate topic. In (a==>) F twice offers affiliative assessments that could be argued as topic-terminal queries. Similarly, each of the "okays" appears to possess "speakership associated" qualities. For example, (1==>, 2==>, and 3==>) get overlapped or latched by the next speaker, and F selects not to produce an "okay + topic initial elicitor." In (4==>), however, once the laughter had "played itself out," F prefaces the topic shift with an "okay" and moves the discussion onto a related set of issues.

In light of F's repeated (and eventually successful) attempts to shift topic, it appears that Jo's and Ju's descriptions were deemed irrelevant and/or somehow "off topic" to the business-at-hand. Clearly, one task of facilitators/reportees is to keep interaction "on track," and this instance provides a clear indication of the methods and persistence involved in "steering" the discussion back to those topics designated as "important" by F. These and related options are legitimately available to those institutionally responsible for an occasion's focus and purpose.

One final note: Upon re-examination of segments (9) and (10), and throughout the entire recordings/transcripts of this focus group meeting, it is clear that Videotex offers technologically sophisticated services to a wide variety of users. As experts, these individuals rely upon their experiences "interacting with the computer" to report news about the system and its operation. As Turkle has noted, "The computer's reactivity and complexity stimulate a certain extravagance of description."[31] Yet, "there is something about people that makes it impossible to capture our intelligence in machines."[32] Through the perspectives and analyses developed in this chapter exemplifying language *as* technology, the conclusion might best be drawn that Videotex interactions both symbolize and embody the "bedrock details" of such an impossibility.

# NOTES

1 See Emanuel A. Schegloff, "Analyzing Single Episodes of Interaction: An Exercise in Conversation Analysis," in a Special Issue on Language and Social Interaction, ed. Douglas W. Maynard, *Social Psychology Quarterly* 50 (1987): 101-114; Emanuel A. Schegloff, "On an Actual Virtual Servo-Mechanism for Guessing Bad News: A Single Case Conjecture," in a Special Issue on Language, Interaction, and Social Problems, ed. Douglas W. Maynard, *Social Problems* 35 (1988): 442-257. With increasing regularity, communication researchers are relying upon conversation analysis (CA) as a viable method for examining the organization of talk-in-interaction. In brief, the basic research procedures of CA include the collection of audio and/or video recordings of interaction, the careful production of transcriptions as textual records of the talk, repeated listenings of recordings in unison with transcripts, and written analyses of findings. See: John Heritage, *Garfinkel and Ethnomethodology* (Cambridge: Polity Press, 1984); Robert

Hopper, Susan Koch, and Jenny Mandelbaum, "Conversation Analysis Methods," in *Contemporary Issues in Language and Discourse Processes*, ed. Donald G. Ellis and William A. Donahue (Hillsdale, New Jersey: L. Erlbaum Associates, 1986), 169-186; *Sequential Organization of Conversational Activities*, ed. Wayne A. Beach, Special Issue of the *Western Journal of Speech Communication* 53 (1989).

2 Harvey Sacks, "On Doing 'Being Ordinary'," in *Structures of Social Action: Studies in Conversation Analysis*, ed. J. Maxwell Atkinson and John Heritage (Cambridge: Cambridge University Press, 1984), 413-414.

3 Harvey Sacks, "Notes on Methodology," in *Structures of Social Action*, 26-27.

4 Sacks, 22.

5 A similar videotext service, Prodigy Services Co., has been designed and implemented by Sears and IBM. As overviewed in the *San Diego Tribune* (9-21-88), the system involves home shopping, information, and entertainment: ". . . San Diegans will be able to let their fingers—tapping happily on the keyboard of a personal computer—do the shopping and banking, make travel arrangements and tap into a variety of sports, news, stock market and lifestyle information." Prodigy's services also include advertising and distribution networks via the Prodigy Interactive Personal Service. These and related options are available through "a low-cost modem the size of a cigarette pack that is simply plugged into the back of the user's PC and has no settings to be adjusted." As a result of spending "more than $30 million on advertising over the next 15 months in order to stage a media blitz unparalleled in the videotext industry," the goal is to appeal to a market broader than "gadget-happy computer buffs. . . . We're looking to attract folks beyond the computer hobbyist."

    AT&T's contributions to the videotex industry include the Sceptre terminal and keypad, as utilized by Gateway Southern California (see Appendix B).

6 "Project Report," University of Nebraska Educational Television Network and Division of Continuing Studies (1982), 1. This project report provided a useful overview of current activities in interactive cablevision. However, the actual audio-recordings and opportunity for micro-analysis were provided through a grant from the Communication Research Center, San Diego State University, directed by John Witherspoon.

7 See J. Maxwell Atkinson, "Understanding Formality: Notes on the Categorization and Production of 'Formal' Interaction," *British Journal of Sociology* 33 (1982): 86-117.

8 See Paul Drew and John Heritage, eds, *Talk at Work: Social Interaction in Institutional Settings* (Cambridge: Cambridge University Press, forthcoming).

9 In respective order, see: Alex McHoul, "The Organization of Turns at Formal Talk in the Classroom," *Language in Society* 7 (1978): 182-213; Hugh Mehan, *Learning Lessons: Social Organization in the Classroom* (Cambridge: Harvard University Press, 1979); J. Maxwell Atkinson and Paul Drew, *Order in Court: The Organisation of Verbal Interaction in Judicial Settings* (London: Macmillan, 1979); Douglas W. Maynard, *Inside Plea Bargaining* (New York: Plenum Press, 1984); Richard M. Frankel, "Talking in Interviews: A Dispreference for Patient-Initiated Questions in Physician-Patient Encounters," in *Interactional Competence*, ed. George Psathas (New York: Ablex, forthcoming), 162-231; Christian Heath, "Talk and Recipiency: Sequential Organization in Speech and Body Movement," in *Structures of Social Action*, 247-265; John Heritage, "Analyzing News Interviews: Aspects for the Production of Talk for an Overhearing Audience," in *Handbook of Discourse Analysis, Volume 3: Discourse and Dialogue*, ed. Teun van Dijk (London: Academic Press, 1985), 95-117; David Greatbach, "A Turn-Taking System for British News Interviews," *Language in Society* 17 (1988): 401-430; Steven E. Clayman, "Displaying Neutrality in Television News Interviews," in Special Issue of *Social Problems*, 474-492.

10 See Harvey Sacks, Emanuel A. Schegloff, and Gail Jefferson, "A Simplest Systematics for the Organization of Turn-Taking for Conversation," *Language* 4 (1974): 696-735.

11 Heritage, "Analyzing News Interviews," 112-116.

12 Of course, both witnesses and patients can and do rely upon specific interactional techniques for dealing with constraints enacted through the pre-allocated nature of the setting. In courts, for

example, defendants being cross examined routinely orient to attorneys' questions as accusatory by constructing excuses (justifications, rationalizations) into their answers (see Atkinson and Drew, *Order in Court*, Chapters 2,3), and/or by generating alternative and competing descriptions of past scenic details (see Paul Drew, "Analyzing the Use of Language in Courtroom Interaction," in *Handbook of Discourse Analysis*, 133-147). And patients have been shown to employ various body movements such as postural shifts, kicks, and gestures in order to display "recipiency," and in so doing eliciting involvement from co-interactant (see Heath, "Talk and Recipiency").

13 Heritage, "Analyzing News Interviews," 96-101; Clayman, "Displaying Neutrality."

14 See Gail Jefferson, "'Caveat Speaker': A Preliminary Exploration of Shift Implicative Recipiency in the Articulation of Topic," End of Grant Report—Mimeo (London: Social Science Research Council, 1981); Emanuel A. Schegloff, "Discourse as an Interactional Achievement: Some Uses of 'uh huh' and Other Things that Come Between Sentences," in *Analyzing Discourse: Text and Talk—Georgetown University Roundtable on Languages and Linguistics* (Washington D.C.: Georgetown University Press, 1982), 71-93.

15 See Jenny Mandelbaum, "Interpersonal Activities in Conversational Storytelling," in *Sequential Organization of Conversational Activities*, 114-126.

16 Harold Garfinkel and Harvey Sacks, "On Formal Structures of Practical Actions," in *Theoretical Sociology*, ed. J.C. McKinney and E.A. Tiryakian (New York: Appleton-Century-Crofts, 1970), 338-366.

17 For a discussion of "okay" as a pre-closing device in telephone conversations, see Emanuel A. Schegloff and Harvey Sacks, "Opening up Closings," *Semiotica* 7 (1973): 289-327.

18 See Emanuel A. Schegloff, Gail Jefferson, and Harvey Sacks, "The Preference for Self-Correction in the Organization of Repair in Conversation," *Language* 53 (1977): 361-382. See also Schegloff's discussion of remedying problematic hearings and understandings in "Discourse," 87-89.

19 For a more detailed analysis of "oh" in conversation, see John Heritage, "A Change-of-State Token and Aspects of its Sequential Placement," in *Structures of Social Action*, 299-345.

20 Jefferson, "Caveat Speaker," 62-66.

21 Schegloff, "Discourse," 79.

22 See Gail Jefferson and Emanuel A. Schegloff, "Sketch: Some Orderly Aspects of Overlap in Conversation," Mimeograph in author's possession (1975).

23 Jefferson, "Caveat Speaker," 26.

24 See Jefferson, "Caveat Speaker," 30-31, for a discussion of how certain tokens come off "as utterly spurious and may be produced to be seen as utterly spurious . . . given our still lingering sense of the triviality and transiency of acknowledgment tokens in general."

25 Jefferson, "Caveat Speaker," 32.

26 Graham Button and Neil Casey, "Generating Topic: The Use of Topic Initial Elicitors," in *Structures of Social Action*, 169-177. See also Douglas W. Maynard, "Placement of Topic Changes in Conversation," *Semiotica* 30 (1980): 263-290.

27 Gail Jefferson, "List Construction as a Task and Interactional Resource," in *Interactional Competence*, 13.

28 See Wayne A. Beach, "Orienting to the Phenomenon," in *Communication Yearbook 13*, ed. James A. Anderson (Beverly Hills: Sage Publications, in press).

29 See Deborah Schiffrin, *Discourse Markers* (Cambridge: Cambridge University Press, 1987), 102, 138.

30 See, for example, Emanuel A. Schegloff, "Between Macro and Micro: Contexts and Other Connections," in *The Micro-Macro Link*, ed. James Alexander, B. Giesen, R. Munch, and N. Smelser (Berkeley: University of California Press, forthcoming).

31 See Sherry Turkle, *The Second Self: Computers and the Human Spirit* (New York: Simon and Schuster, 1984), 14.

32 Turkle, 19.

## APPENDIX A

*(From *Sequential Organization of Conversational Activities,* ed. Wayne A. Beach, *Western Journal of Speech Communication* 53 (1989): 89-90.)*

### Transcription Conventions

The transcription system employed for data segments is an adaptation of Gail Jefferson's work [see J. M. Atkinson and J. Heritage (Eds.), *Structures of Social Action: Studies in Conversation Analysis,* London: Cambridge University Press, 1984, pp. ix-xvi]. Symbols are employed to provide vocalic and prosodic details (e.g., pauses, word stretch and emphasis, intonation, aspiration, etc.) so as to preserve the integrity of recorded interaction. The orthography is designed to capture how words sound, but not at the expense of making the transcript unreadable. Abbreviated information, provided prior to transcribed segments, index location and original source from which data were drawn.

| *Symbol* | *Name* | *Function* |
|---|---|---|
| 1. [    ] | Brackets | Indicate beginnings and endings of overlapping utterances. |
| 2. = | Equal signs | Latching of contiguous utterances, with no interval or overlap. |
| 3. (1.2) | Timed Pause | Intervals occurring within and between same or different speaker's utterance, in tenths of a second. |
| 4. (.) | Micropause | Brief pause of less than (0.2) |
| 5. ::: | Colon(s) | Prior sound, syllable, or word is prolonged or stretched. More colons indicate longer prolongation. |
| 6. . | Period | Falling vocal pitch or intonation. Punctuation marks do *not* reflect grammatical status (e.g., end of sentence or question). |
| 7. ? | Question Mark | Rising vocal pitch or intonation. |
| 8. , | Comma | A continuing intonation, with slight upward or downward contour. |
| 9. ↑↓ | Arrows | Marked rising and falling shifts in intonation. |
| 10. °° | Degree Signs | A passage of talk noticeably softer than surrounding utterances. |
| 11. ! | Exclamation | Animated speech tone. |
| 12. - | Hyphen | Halting, abrupt cut off of sound, syllable, or word. |
| 13. *cold* or ___ | Italics Underline | Vocalic stress or emphasis. |
| 14. OKAY | CAPS | Extreme loudness compared with surrounding talk. |
| 15. > < < > | Greater than/ less than Signs | Portions of an utterance delivered at a noticeably quicker (> <) or slower (< >) pace. |
| 16. hhh ·hhh ye(hh)s | H's | Audible outbreaths, possibly laughter. The more h's, the longer the aspiration. Aspirations with superscripted period indicate audible inbreaths. H's within parentheses mark within-speech aspirations, possibly laughter. |
| 17. ((noise)) | Scenic details | Transcriber's comments (e.g., gestures, non-speech sounds). |
| 18. (    ) | Parentheses | Transcriber is in doubt as to word, syllable, or sound. Empty parentheses indicate indecipherable passage. |

| 19. | pt | Lip Smack | Often preceding an inbreath. |
| 20. | hah<br>heh<br>hoh | Laugh syllable | Relative closed or open position of laughter. |
| 21. | $ | Smile voice. | Laughing talk between markers. |
| 22. | [[ | Simultaneous<br>utterances | Double lefthand brackets undicate utterances linked<br>together, begun simultaneously. |

## APPENDIX B

As evident from the 1984 Gateway Southern California brochure advertisements below, the "interactive" in Videotex cable services consists of "communicating with your TV" through the Sceptre terminal and keypad (a trademark of AT&T company). Described as "two-way communication [that] opens new channels of communication," it is useful to contrast these technological orientations and achievements with conversational activities comprising a focus group meeting. While each is a collaborative production, the technology of conversation is a unique exchange system in which speakers and hearers co-produce and thereby constrain turns-at-talk—and the activities accomplished in and through the sequencing of such turns. Just as the approaches to communication are markedly different, so are the purposes of the technologies in everyday life, and these differences allow for yet another perspective on the possibility of a reflexive view of "language as and in technology."

## Introducing Gateway.<sup>SM</sup>
## It opens new channels of
## communication.

Welcome to Gateway. The exciting new two-way communication service from a Times Mirror company.

It's not cable. It's not a personal computer or a video game.

Gateway is a brand new technology called videotex. And it helps you with the things you do every day.

Like paying your charge accounts. Buying tickets to a Rams game or a concert. Catching the latest news. Finding tomorrow's bargains at the neighborhood mall. Helping your kids with their homework.

You already have the two main things you need for Gateway, your TV and your phone. All you need now is a Sceptre™ videotex terminal and keypad, made by AT&T.

There are more than a hundred services on Gateway now, and more are added every day.

Every service is designed to help you save time, and manage it better. Help you improve the ways you save, invest and spend your money. And help make your life a lot easier and a lot more enjoyable.

Come see a live Gateway demonstration today, and try it for yourself. Then become a subscriber. And open new channels of communication.

# Telephone Speaking and the Rediscovery of Conversation

*Robert Hopper*

Humans "discovered" speech communication a very long time ago—in the sense that bees discovered flying a long time ago. According to a self-conscious academic perspective, however, bees have not yet re-discovered flying in the sense that humans have rediscovered language and conversation. The "linguistic turn" in twentieth-century thought may begin such rediscovery.[1] The present chapter implicates telephone speaking in this rediscovery process. We have rediscovered the human conversation in the past century—the very period of time during which we have routinely used telephones to converse with one another. This chapter draws circumstantial connections between telephone speaking and the appearance in scholarship and popular culture of such notions as language, communication, medium, and conversation.

Scholars have advanced claims about the social impacts of electronic mass media, including television, movies, music, and the press.[2] We have perhaps overlooked the social consequences of the primary electronic medium for interpersonal communication. Meanwhile telephone experience massages our consciousness as surely as does any electronic medium. Telephony, in fact, requires more active moment-to-moment participation than any other electronic medium.

The current claim, analogous to historians' claims about clocks,[3] is that we have rediscovered the human conversation through our telephone conversations. To experience telephone conversation is to heighten awareness of speech communication. Telephone speakers conceptualize communication as a two-party exchange of spoken signals that travel from party A to party B. During the past century, scholars, journalists, and others have reconceptualized speech communication along these lines.

The present essay begins with descriptions of four ideas about communication that have spread since telephone conversation has become commonplace.

Second, I describe details of a comic sketch, published in 1897, that prefigure contemporary conversation analysis. Third, I detail evidence of telephone experience in the lectures of Ferdinand de Saussure—lectures that have been credited with opening the "linguistic turn." Each segment of the chapter provides evidence that telephone experience leads to the rediscovery of conversation.

## Four Lessons from Telephone Conversation

Four propositions shared in many contemporary notions of language, communication, and conversation may be stated as "lessons" that telephone experiences make irresistible to intellection:

1. Communication occurs when messages travel through channels.
2. Speech sounds provide central components of communicative interaction.
3. Speech communication is dyadic.
4. Speech events display beginnings, endings, and interruptions.

### Lesson One: Communication is Message Travel

Since the advent of telephones we have grown used to diagrams picturing the travel of messages along "channels." These channels are frequently pictured as electric wires. Shannon's (1948) Mathematical Model of Communication, published in the Bell System Technical Journal (Figure 1), describes communication as the passing of a signal through a conduit that connects two speakers.[4]

FIGURE 1     **Mathematical Model of Communication**

**Shannon, 1948**

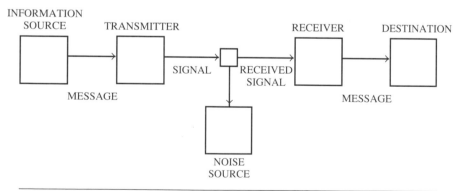

Only what passes through this middle-place deserves consideration as communicative. Has this conduit metaphor become available to us precisely due to experience with telephones?[5]

There must have been diagrams of communication-across-wire-channels adjacent to the development of telegraphy and also celebrations of the telegraph's speed. But with the coming of the telephone, the experiences of vocal immediacy across distances lent to telephone interaction an element of miracle. This miracle was described by the *Electrical Review* in 1889:

> That Hello! . . . went down through the desk, down through the floor . . . then out into an underground conduit . . . then up the Hudson . . . crossed rivers and mountains . . . went through lonely forests . . . heard the music of the sea, . . . and it caught a glimpse of Bunker Hill monument before it plumped into the city of baked beans and reached its destination in the ear drum of a man seated in a high building there. . . .
>
> In about one millionth of the time it takes to say Jack Robinson, it was there. . . . It was as if by a miracle the speaker had suddenly stretched his neck from New York to Boston and spoken gently into the listener's ear.[6]

This writer emphasizes the word "hello" as the passenger travelling this lengthy route. Phone experience emphasizes the verbal-vocal immediacy of hearing the familiar other's voice. This passage shows pleasure at the miracle and also celebrates the gentleness of the spoken word's arrival at a distant ear. Early telephone users expressed amazement that they could recognize a familiar voice over the phone. Telephone interaction increases consciousness of voice's role in speech communication.

## Lesson Two: Speech Sounds are Central to Communication

Telephone conversation splits speech from the rest of action. If, as McLuhan suggests, print gives us an eye for an ear, then the telephone resists the eye and returns to the ear. The phone world is a sightless world, a world of sound only. Telephone speakers may gesture and move while talking, but this action is ordinarily unavailable to a telephone partner.

One surprise in telephone communication is how much is left in sounds when you split off the rest of experience.[7] The telephone teaches us about the centrality of sounds to speech communication. To study the medium that splits sound off from the other senses is to study what is most basic about speech communication. Sacks quips:

> There's a view going around to which some of you may subscribe that pretty much all of what's interesting in conversation doesn't happen by the use of words. . . . [But] there are some delicate, interesting things that happen, distinctly with words.[8]

Speech is, in fact, the central ordering system for interpersonal communication.[9] This claim's consequences for communication theory grows as one contrasts oral with visual signs. We may interpret numerous visual signs at once—a crowd of people or a row of trees appears to vision in an instant.[10] You can see somebody speak, smile, and wave an arm at the same time. But sound is linear; it strings signs out across time.[11] Splitting sound from the rest of communication emphasizes the linearity of signs, the sequential unfolding of speech communication. We argue below that Ferdinand de Saussure, in founding the contemporary study of language, stressed (1) the primacy of spoken sound over writing, and (2) the linearity of signs, or their sequential arrangement. Experiences of telephone speaking instantiate these insights.

## Lesson Three: Speech Communication is Dyadic

The telephone is a medium built for two. The phone call cuts a dyad out of the speech community and wires it into an exclusive, private, two-person mini-community.

In face-to-face speaking any number of co-present participants may enter into (or observe) an encounter. The telephone bystander does not have the same access to a conversational encounter as does a face-to-face bystander. A Mark Twain story published in 1880 decries the frustrations of overhearing just one end of a telephone conversation.[12]

The telephone isolates two people and wires them together in secret. You can overhear a face-to-face conversation by accident, but if you overhear a phone call you are eavesdropping. The rapid spread of the telephone, much of it employing multi-party lines, spawned renewed fascination with eavesdropping.

*The World's Work* reported in 1905 that a farm wife who had telephone service just a few months was asked how her family liked it and replied, "Well, we liked it a lot at first, and do yet, only spring work is coming on so heavy that we don't hardly have time to listen now."[13]

The telephone conversation, for better and worse, carves the dyad from the rest of the speech community and creates an event intended for just two pairs of ears. There are no innocent bystanders in telephone conversation. Bystanders are either excluded or eavesdropping.

The experience of telephone conversation suggests the centrality of two-party encounters, which seem simpler to describe than multi-party encounters.[14] Consequently, the dyad forms the basis of twentieth-century models proposed by Saussure, Shannon, Wendell Johnson, Wilbur Schramm, David Berlo, and others. In sum, since the invention of the telephone, the dyadic conversation has joined the oration and the sentence as theorists' central exemplars of human speech communication.

## Lesson Four: Telephone Encounters Begin, End, and Interrupt

Telephone call beginnings and endings are more distinct than those in face-to-face encounters. If you drive with colleagues to College Station, or do the dishes with your daughter you can drift in and out of conversation. In those environments you speak as you do other actions. The theorist who attempts analytic separation of specifically communicative action from the rest of life faces formidable conceptual difficulties. Much action has semiotic properties: fiddling with hair, falling asleep, getting up to answer the door—all these may draw interpretation without becoming primarily communicative acts. Actually conversation participants win few advantages for attending to which details of action are specifically communicative.[15]

But the telephone provides experience in pure speech communication. Telephone participants are doing virtually nothing but communicating with a specific partner. Before or after the call they do not interact with that partner. This contrast "splits" communicative interaction temporally, perceptually, and conceptually from other actions. Hence, telephone experience promotes notice of what is specifically communicative about communicative activity.

A telephone call starts and ends at definite moments, whereas co-present talk is braided into other action. For instance if you encounter a friend on the street, your gaze shows recognition before speech begins.[16] In telephone talk, however, there are definite moments in which sounds (and especially speech) perform opening and closing for encounters. These definite beginnings display contrast between communicative activity and other action, and suggest that communication may be conceptualized in its own right.

Early telephone users developed awareness of the beginnings, endings, and interruptive powers of telephone communication. A journalistic anecdote that appeared in 1897 told of a would-be male suitor who had spent an hour in futile attempts to tell a young lady of his passion for her:

> "Excuse me for a moment, Mr. Featherly," she said, "I think I hear a ring at the phone." And, in her queenly way, she swept into an adjoining room.
> Presently she returned and his mad passion found a voice.
> "I am sorry, Mr. Featherly," she said, "to cause you pain, but I am already engaged. Mr. Sampson, learning that you were here, has urged his suit through the telephone."[17]

This tale slurs a woman's fickleness, but it also shows that the telephone may co-opt or interrupt an ongoing conversation, and even affect its course. This modest tale tells a universal experience. Who has never been interrupted by the telephone? When these interruptions occur, cannot they intrude into and disrupt the most intimate kinds of encounters? Will not the party who answers the phone likely say "excuse me" when withdrawing from present activity into a telephone dyad?

The experience of telephone speaking provides background knowledge essential to this story. However much the tale relies upon, say, gender stereotypes for its impact, it relies on the experience of telephone users in its formal details.

The coming of the telephone gives its users a kind of special knowledge, perhaps analogous to things a user of word processing knows. The vulnerability to interruption by phone ring becomes a taken-for-granted aspect of experience with this medium. It's a bit like the word-processing user's vulnerability to accidental file erasure—that is, you can scarcely use the medium without encountering this experience, and others' recountings of it. A present-day joke in which a character "loses" a file, for instance, may be appreciated more fully by a "user" than by a non-user. Further, the user-mentality created by routine experiences with a medium allows these vulnerabilities to be exploited in fiction or criticism.[18]

## FLETCH AND SID: TELEPHONE TALK IN FICTION

Telephone experience enhances its users' appreciation for the details of spoken speech. Consider the following dialogue that appeared in the publication "Tit-Bits" in 1897:

Halloa Fletch! Do you hear me?

Yes.

This is Sid. Thought I'd call you up.

Glad to hear from you, Sid. How are you?

First-rate. How's things?

Oh, nothing especially. Hadn't anything

to do, you know, and thought I'd call you up.

(Pause)

Yes. (Another pause.) Everything going on about as usual in the old town?

. . . [ten lines deleted here] . . .

Weather's fine where you are, is it?

Splendid.

(Pause)

Well, I must be going now. Awfully glad to have had a chance to talk to you, old fellow.

Glad you called me up.

Good-bye!

Good-bye![19]

This sketch jibes people who would call each other with nothing to say. Such encounters become pure forms that we may celebrate for their triviality. This scene's humor springs from accurate observation. Several particulars of this sketch foreshadow transcription practices of contemporary conversation analysts. Each new paragraph in the sketch shows speaker change; and many pauses occupy a separate line, so as not to be interpreted within the turn of either speaker.[20] The author of the piece also foreshadows Jefferson's transcription system by using parentheses around the word "pause."

Observe how these speakers move toward closing the encounter.

Weather's fine where you are, is it?

Splendid.

(Pause)

Well, I must be going now . . .

When a topic dries up as "weather" (pardon the pun) does here, a pause provides a sort of passing function. The next speaker begins with "Well" and offers an excuse for ending the call. This excuse is taken up by the other, and leads to an exchange of goodbyes. In these particulars, the Fletch and Sid sketch resembles transcriptions with which Schegloff and Sacks represent "possible pre-closings" in contemporary telephone talk.[21]

Perhaps more remarkable, this sketch captures many details of Schegloff's canonically routine telephone opening.[22] Here is Schegloff's illustration of that form alongside the first lines of the antique sketch about two bumpkins.

| Schegloff, 1986 | Fletch & Sid, 1897 |
| --- | --- |
| 1 R Hello | |
| 2 C Hello Ida? Halloa | F! Do you hear me? |
| 3 R Yeah | Yes. |
| 4 C Hi,=This is Carla | This is S. Thought I'd call you up |
| 5 R Hi Carla. | Glad to hear from you, S. |
| 6 C How are you. | How are you |
| 7 R Okay:. | First-rate |
| 8 C Good.= | |
| 9 R =How about you. | How's things |
| 10 C Fine. | Oh,   nothing especially. . . . |

Schegloff characterizes a canonical telephone opening as four interconnected adjacency pairs: summons-answer; identification-recognition; greeting-greeting; and exchange of "how are you" inquiries and answers. These features in Fletch and Sid's dialogue show precise fit with Schegloff's descriptions.[23]

In sum, this sketch illustrates a keen ear for conversation's details. The experience of talking on the phone is a workshop for such details. At the time of this sketch's publication, American society was undergoing a craze of telephone adoption. This created a community of readers who could appreciate this writer's humor in the context of their own experience. These readers could appreciate the incisive description that undergirds the sketch. The telephone, in sum, makes its users into self-conscious consumers of speech. This medium's massage constitutes the rediscovery of conversation's details.

Scholars are subject to the same cultural curricula as their fellows. It is not too surprising, therefore, that scholarship conceptualizing speech communication is affected by widespread experience in telephone conversation.

## SAUSSURE: THE TELEPHONE'S ECHO IN SCHOLARSHIP

The "linguistic turn" in contemporary thought may be rooted in the experience of the telephone. Ferdinand de Saussure, whose lectures[24] open modern linguistics, semiotics, structuralism, and communication theory, posited a model (Figure 2) that in many respects specifies the components later to appear in Shannon's diagram (Figure 1).

---

FIGURE 2

### The Speaking Circuit
### Saussure, 1908-1910

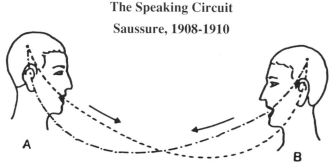

. . . il faut se placer devant l'acte individuel qui permet de reconstituer le circuit de la parole (p. 27).

---

Like the Shannon model, Saussure's representation shows two parties speaking to one another. Saussure depicts speech travelling along dotted lines. Shannon's lines explicitly represent telephone wires. In Saussure's model the lines of message-flow, "look suspiciously like telephone wires."[25] Saussure, like Shannon, explicitly distinguishes between sources and encoding, a distinction suggested by the encoding capacities of telephones.

How strong is the circumstantial link between Saussure's approach to language and speech, and his experiences of telephone use? Ferdinand de Saussure lived the middle third of his life in Paris during the very years that saw that city become festooned in telephone wires.

In Saussure's *Cours de Linguistique Générale* (delivered in 1908-1910, first published in 1915), Figure 2 is labeled with the electronic metaphor: "speech circuit."

> . . . *il faut se placer devant l'acte individuel qui permet de reconstituer le circuit de la parole.*[26]

Baskin translates this passage as: "We must examine the individual act from which the speaking-circuit can be reconstructed." Harris writes this translation: "We must consider the individual act of speech and trace what takes place in the speech circuit." Both translators maintain the circuit trope only as a dead metaphor, ignoring its probable novelty for the Master of Geneva. Further, these twentieth-century translators abandon the possible phenomenology in Saussure's phrasing: "*Il faut se placer devant.*" Our translation attempts to keep Saussure's metaphor alive and provisional: "We must place ourselves in front of (in the presence of) the individual act that allows us to construe speech communication as a circuit."

What is it that leads Saussure to compare speaking to a dyadic circuit? Given the chronology of his life, it seems likely that telephone speaking fed this metaphor. One begins to speak on the telephone only when an electronic circuit is completed. What travels on that circuit is the spoken voice.

In sum, Saussure's diagram of the "circuit de la parole" prefigures the subsequent information model (Figure 1) in constituting a dyad in which one speaker's phonations travel through an electronic conduit toward the recipient's ear.

Perhaps telephonic experience also led Saussure to ground his notions of linguistics and semiotics in spoken utterances. Saussure criticizes thinkers who emphasize writing over speech, and notes that the only purpose of writing is to represent speech.

> *l'objet linguistique n'est pas défini par la combinaison du mot écrit et du mot parlé; ce dernier constitue à lui seul cet objet* (p. 45).
>
> The object of study in linguistics is not a combination of the written word and the spoken word. *The spoken word alone constitutes that object.*[27]

Linguistics before Saussure had been dominated by considerations of writing and print. The advances of nineteenth-century linguistics arose from examination of written texts. Written messages could survive over time and be repeatedly re-examined. Saussure's *Cours,* however, turns its back on writing, and argues for a science of speech based on the facts of speaking.

As Saussure noted the centrality of the spoken word to semiotics, he also described spoken signs as "linear." The chapter in the *Cours* that is much quoted for its insistence that the sign is arbitrary actually lists two definitive principles of signs. The second of these is linearity.

> The linguistic signal, being auditory in nature, has a temporal aspect. . . . This principle is obvious, but it seems never to be stated, doubtless because it is considered too elementary. However . . . its importance equals that of the first law. The whole mechanism of linguistic structure depends upon it. (See p. 121) Unlike visual signals (e. g. ships' flags) which can exploit more than one dimension simultaneously, auditory signals have available to them only the linearity of time. The elements of such signals are presented one after another: they form a chain.[28]

The experience of telephonic communication, the experience of interaction in sound only, suggests this linearity. When interaction is truncated into sound, then the sequential structures of interaction are laid bare. Speech speaks in sequence. This is illustrated in the *Cours* by Saussure's discussion of syntagma:

> In its place in a syntagma, any unit acquires its value simply in opposition to what precedes, or to what follows, or to both.[29]

Here, the conversation analysts' insistence on the importance of sequential occurrence follows Saussure's insight.[30] But where does this insight come from? Why is it an insight of the present century? Does telephonic experience teach us to apprehend the linearity of signs?

It is not intended that these notes discredit Saussure's insights: au contraire. Saussure's genius grew during a generation in transformation. The transformation came about because a speech teacher invented a new electronic device for speech communication.[31] This machine's use grew frequent and widespread, and it carried pure speech—speech split off from the rest of action, and therefore available for inspection. Telephone experience reframes the possibilities of interaction. This reframing puts speaking soundly at the center of what makes interaction work.

Speech has long been important to humans, but we could not hear speech speak[32] until the telephone blew its cover. Written communications were, until Saussure's day, the main reminders that speech leaves traces of its passage.[33] The telephone brings experiential attention to speaking.

## CONCLUSIONS

The telephone was invented in the U.S.A. late in the nineteenth century and its use spread in this country faster than anywhere else. Perhaps uncoincidentally, the contemporary disciplines of speech communication also sprang up in the U.S.A. early in the twentieth century, and established a secure pedagogical tradition predominately in this country. It seems that contemporary America's fascination with "communication" has been facilitated by routine and repeated telephone speaking.[34]

Let us summarize the characteristics of phone conversation that undergird notions such as communication, language, and conversation. Telephone speech sends messages travelling through a conduit. It splits sound from the rest of the senses. It splits the dyad from the rest of society. It splits communication from other activity. Telephone conversation is pure speech communication.

### Limits And Benefits of the Analysis

There are limits to the present analysis: a few sentences from one thinker, and several anecdotes about early telephone speakers. But the conclusions about the telephone's impact on orientations toward communicative interaction go beyond these instances. The four lessons of telephone speaking have now become stable mythology about conversation. There is good news and bad news about this.

The good news is that the telephone helps us to find the centrality of speech to all our disciplinary concerns. The difficulties concern how far we may go with the conduit metaphor and with the focus on the dyad. When we base theory in machine metaphors we may expect mixed outcomes.

We pass by the conduit metaphor quickly: Is everything about messages "in" a channel? Does the channel notion apply with much force in face-to-face interaction? Does the conduit notion bias our attention toward single messages and away from interaction?

There are also costs for focusing attention on the dyad. Researchers must compare and contrast two-party speaking and multi-party speaking. With three or more participants in conversation, the achievement of each turn is different than when there are just two speakers. In a two-party conversation, if the other is not continuing to speak, it is up to you to speak next; but in a multi-party conversation there are others eligible to take the floor.[35] Further, the existence of third parties leads to various possibilities of alignment, celebration, or teasing. Glenn demonstrates that the achievement of shared laughter is different in dyads than in multi-party conversation.[36]

These limitations are offset by numerous advantages of telephone talk as a focus for communication scholarship. Our telephonic experiences help us pay attention to the importance of audition to speech communication, and to the centrality of turn taking in interaction. Hence, descriptions of telephone conversation should play a vital role in communication theories. Additionally, there is one methodological benefit to seeking the structures of semiotics in the sounds of telephone speaking. Because phone talk is limited to speech, an audiotape recording of a telephone conversation displays quite precisely those signals available to the conversation's participants.

> Materials drawn from telephone conversation . . . [allow] the possibility of successfully studying conversation and its sequential organization without examining gesture, facial expression and the like. Telephone conversation is naturally studied in this manner, and shows few differences from conversations in other settings and media.[37]

Audiotape recordings of telephone calls capture a high percentage of textual detail with low intrusion into a setting. This represents a methodic alliance of convenience between the technological capacities of telephone and tape recorder.

## The Telephone and the Tape Recorder

A distinct minority of current scholarly literature on the telephone gives the reader any sense of listening to the sounds of speech.[38] Rather, contemporary accounts focus on functions of telephone communication and its correlations with factors of economics and demography. If a Martian, one thousand years from now were to learn about telephones only through such works, how good an understanding could be reached? How much better might the Martian understand our experience if a three minute tape recording of a phone call survived to be replayed?

There is a principled relationship between the lessons of the telephone and the tape recorder's emergence as a convenient device that makes possible the detailed study of telephone speaking from a participant's-ear hearing.

Contemporary conversation analysts revolutionize communication scholarship from the inside out because certain facts emerge through an alliance between the tape recorder and the telephone.

A machine to audio-record human speech was invented at about the same time as the telephone. It was patented by Thomas Edison as the "phonograph." This device did not become a common household object nearly as quickly as the telephone, but Saussure seems to have been aware of its possibilities for preserving speech samples. Saussure argues that to displace written records linguists would have to study recorded samples ("*échantillons phonographiques*") which were, he noted, already being collected in Vienna and Paris.[39]

The importance of sound recording to studies of language is celebrated by Bernard Shaw's professorial anti-hero Henry Higgins. Higgins and his colleague Pickering are discussing language varieties and how to study them when Higgins' servant announces that a cockney flower girl wishes to speak to him. Higgins exclaims:

> This is rather a bit of luck. I'll shew you how I make records. We'll set her talking; and I'll take it down first in Bell's Visible Speech; then in broad Romic; and then we'll get her on the phonograph so that you can turn her on as often as you like with the written transcript before you.[40]

Two basic principles of contemporary conversation analysis are illustrated in Higgins' speech: First, a transcription of speech can often provide clarity to analysis, and special transcription systems (of which Bell's visible speech is the major forerunner) can be helpful in this regard. Second, analytic leverage might be gained by playing a recording multiple times, especially if the scientist examines the transcription during the replayings.

The machine Higgins uses for his repeat listening exercises, the phonograph, is the direct ancestor of today's cassette tape recorder. The audio recording preserves, as it happens, precisely that same sounded dimension of speech communication that is carried over the "speaking circuit" provided by the telephone. For this reason, the application of tape-recording technology to telephonic communication is particularly fortunate.

As it turned out, sound recording was much slower than the telephone to achieve widespread utilization. Reasonably high-fidelity portable tape recorders were still somewhat rare in the 1960s. This writer borrowed a fifty pound reel-to-reel in 1964 to study oral interpretation; and lugged similar machines to class to play speech samples in the early seventies. Presumably, Sacks, Jefferson and others used such cumbersome machines to record telephone speaking in the same era. The contemporary study of recorded speech began when Harvey Sacks gave Gail Jefferson a recording of a group therapy session and asked her to write down just what she heard.[41] Jefferson has made this task into a life's work. Much of Jefferson's transcription corpus turns out to be telephone conversation.

Communication media piggyback with each other. Perhaps the greatest benefit of the alliance of the telephone and the tape recorder in current conversation analysis is that it stimulates concerted description of the most venerable communicative technologies—those of human speech.

## NOTES

1 Richard Rorty, ed. *The Linguistic Turn: Recent Essays on Philosophical Method* (Chicago: University of Chicago Press, 1967), 4.

2 See essays by Gronbeck and by Carey in the present volume.

3 Arguments that European clock-making technology gave birth to contemporary notions of time appear in Daniel Boorstin, *The Discoverers* (New York: Random House, 1983), 71, 311, 403; David Landes, *Revolution in Time* (Cambridge: Harvard University Press, 1983), 79ff; and Hugh Kenner, *The Mechanical Muse* (London: Oxford, 1987), chapter 2. The arguments include: First, that there had to be reliable mechanical clocks before the universe could be described as built by a perfect clockmaker; Second, that the universe-as-clock metaphor was taken up by Descartes and Kepler during eras in which clocks became increasingly inexpensive and portable; Third, further developments in clock technology made possible such contemporary artifacts as the rush-hour and the set-length workday. In sum, clock technology leads to reconceptualizations of time and the rhythms of living.

4 Claude E. Shannon, "A Mathematical Theory of Communication" *The Bell System Technical Journal* 27 (1948): 379-423.

5 M. J. Reddy, "The Conduit Metaphor—A Case of Frame Conflict in our Language about Language"; in *Metaphor and Thought*, ed. A. Ortony (Cambridge: Cambridge University Press, 1979); George Lakoff and Mark Johnson, *Metaphors We Live By* (Chicago: University of Chicago Press, 1980).

6 Carolyn Marvin, *When Old Technologies Were New: Thinking About Electric Communication in the Late Nineteenth Century* (New York: Oxford University Press, 1987), 196, quotes from "From an admirer of the telephone," *Electrical Review*, Nov 23, 1889, 6.

7 Albert Mehrabian, *Nonverbal Communication* (Chicago: Aldine-Atherton, 1972), provides a limited, yet oft-overgeneralized claim that most of communication is nonverbal. This claim is based on questionnaire variance accounted for by the verbal (transcripts), vocal (masked tonal patterns), and visual aspects of a video signal. Derek R. Rutter, *Communicating by Telephone* (Oxford: Pergamon Press, 1987), reports a line of research that attempts to show that telephone conversation is "cueless," but most of his predictions fail. Despite these (failed) claims, speech remains the essential basis of ordinary speech communication.

8 Harvey Sacks, Unpublished lecture, Fall, 1971, lecture 3, page 1. Sacks' lectures have been transcribed by Gail Jefferson.

9 Frank E. X. Dance, "The Centrality of the Spoken Word," *Central States Speech Journal* 23, (1972): 197-201; John Modaff and Robert Hopper, "Why Speech is Basic," *Communication Education* 33 (1983): 37-42.

10 Aron Gurwitsch, *The Field of Consciousness* (Pittsburgh: Duquesne University Press, 1964). Gurwitsch is summarizing passages of Husserl's *Philosophy of Arithmetic* that have not yet been published in English.

11 There are certain sounds, such as the sounds of the instruments of an orchestra or singers making harmonies, that are simultaneous combinations, but these strike most ears as singular phenomena: the orchestra plays, the barbershop quartet sings. In most speech communication, like most telephone conversation, one party speaks at a time (See Harvey Sacks, Emanuel Schegloff & Gail Jefferson, "A Simplest Systematics for the Organization of Turn-taking for Conversation," *Language* 50 (1974): 696-701.)

12 Mark Twain, "A Telephonic Conversation," in *The $30,000 Bequest and Other Stories* (New York: Harper, 1905). The story was first published in *Atlantic* Magazine in 1880. The implication is that Twain himself was a very early adopter of the telephone, which became commercially available in 1878. Could there be a relationship between Twain's telephone experience and his pioneering work in developing the "colloquial style" in American Literature—a style based on concrete particulars of dialogue? See Richard Bridgman, *The Colloquial Style in America* (New York: Oxford University Press, 1966), pp 5-12. Bridgman quotes Hemingway, Mencken, and Faulkner as writing that this colloquial style originates from one book, *Huckleberry Finn*, published by Twain in 1884. Was Huck's conversational style the first gift of the telephone to literary realism?

13 John Brooks, *Telephone: The First Hundred Years* (New York: Harper & Row, 1975), 116-117.

14 The limit to two parties limits the tasks of description. One may presume that telephone utterances are for the ears of the partner, and conceptualize speaker-listener relationships without describing alignment shifts among speakers and overhearers.

15 In fact, there may be pervasive constraints to precisely ignore the message status of actions. See the discussion of "let it pass" in Harold Garfinkel, *Studies in Ethnomethodology* (Englewood Cliffs: Prentice-Hall, 1967), chapter 1. (This may help us to understand why much communication scholarship describes so very little about messages.)

16 Erving Goffman, *Behavior in Public Places* (New York: Free Press, 1963), chapter 2.

17 Marvin, 1988, p. 72 reprints this story from the *Electrical Review*, March 3, 1888, p. 11, and includes a note that it originally appeared in the *New York Sun*.

18 One contemporary example appears in the film, "Stand by Me." At the end of the film, the narrator-author turns off his computer without saving the file first. Is this "play," showing the self-destruction of the story; or a failure of verisimilitude?

19 Marvin, 30, quotes this instance as "At the Telephone," *Tit-Bits* (London), Oct. 9, 1897, 28.

20 The very earliest telephones used the same speaker for reception and sending. The user had to alternate them between mouth and ear at projected transition-relevance places.

21 Emanuel A. Schegloff, and Harvey Sacks, "Opening up Closings," *Semiotica* 4 (1973): 289-327.

22 Emanuel A. Schegloff, "The Routine as Achievement," *Human Studies* 9 (1986): 115.

23 Schegloff's 1986 model also motivates noticing that the probable first utterance of the call, the answerer's "hello," is not shown in the fictional instance. In this one particular, science supplements art.

24 Given the centrality of "speaking" to contemporary theory, it seems no accident that such figures as Sacks, Saussure, and Austin advanced their path-breaking insights in lectures, not writings.

25 Roy Harris, *Reading Saussure* (LaSalle, IL: Open Court, 1987), 216.

26 Ferdinand de Saussure, *Cours de Linguistique Générale*. Publie par Charles Bally et Albert Sechehaye (Paris: Payot, 1915); Edition critique prepare par Tullio de Mauro (Paris: Payot, 1985; 1972), 27. Translations: *Course in General Linguistics*, trans. Wade Baskin (1959). New York: McGraw-Hill, 11; and *Course in General Linguistics*, trans. Roy Harris (1986), LaSalle, IL: Open Court Publishers, 11.

27 Harris translation, 24-25, emphases added.

28 Harris translation, 69-70. The passage appears on page 103 in French editions, most of which retain uniform pagination.

29 Harris translation, 121.

30 Of course, Saussure in no way expands his insight to dialogic structuring, say, utterance pairs by different speakers. But the insight he developed may be applied to the notion of "adjacency pair" as described in Schegloff and Sacks, "Opening up Closings." (See note 20).

31 Alexander Graham Bell, who is credited with inventing the telephone, was neither an electrician nor a theorist, but a third-generation speech teacher. His father, elocution professor Alexander Melville Bell, devised "Bell's visible speech," a precursor of contemporary transcription systems used in phonetics and discourse analysis. Graham Bell's financial backing to invent telephones came from his students' grateful parents. Bell's invention of the telephone achieved notice at the 1876 exposition because the Emperor of Brazil walked by his display and recognized him as a

teacher of the deaf. Bell's elocution training also proved useful as he stimulated early interest in the phone in a dramatic series of lecture-demonstrations. Finally, his victories in patent squabbles owed his effectiveness as a courtroom orator.

32 Martin Heidegger's sentence *"Die Sprache spricht,"* loses some of its oral resonance when translated as "Language Speaks." [I owe this insight to Kerry Riley.] See Martin Heidegger, *Poetry, Language, Thought.* Translated by Albert Hofstadter (New York: Harper & Row, 1971), p. 199. Heidegger's insights on the nature of speech and language also occur in the post-telephone age.

33 Saussure's attempt to found a science in speaking fails to free itself from writing. See Jacques Derrida, *Of Grammatology.* Trans. G. Spivak. (Baltimore: Johns Hopkins University Press, 1976), 35-70. But Saussure's insistence on the centrality of speech continues to provide directions even for such critics. V. N. Volosinov, *Marxism and the Philosophy of Language,* trans. Ladislav Matejka and R. R. Titunik (New York: Seminar Press, 1973; first published in Russian, 1930), 45-99, takes Saussure to task for his failure to incorporate dialogue into his theory of semiotic opposition. Again, the open texture of Saussure's observations provide grist for moving his concepts forward. And, the direction of that movement toward dialogue semiotics may also owe telephone experience.

34 Tamar Katriel and Gerry Philipsen, "What We Need is Communication: 'Communication' as a Cultural Category in Some American Speech," *Communication Monographs* 48 (1981): 301-317.

35 Harvey Sacks, Emanuel A. Schegloff, and Gail Jefferson, "A Simplest Systematics for the Organization of Turn-taking for Conversation," *Language* 50 (1974): 712-713.

36 Phillip Glenn, "Initiating Shared Laughter in Multi-party Conversations," *Western Journal of Speech Communication* 53 (1989): 127-149.

37 Emanuel A. Schegloff, "Identification and Recognition in Telephone Conversation Openings"; in *Everyday Language: Studies in Ethnomethodology,* George Psathas, ed. (New York: Irvington-Wiley, 1979), 24.

38 For example, in the retrospective assessment of the telephone by Pool and his colleagues there seems not to be one printed quote from a telephone conversation. (Ithiel de Sola Pool, "Retrospective Technology Assessment of the Telephone, Volume 1." Washington D. C.: National Science Foundation Project Report #ERP75-08807, 1977). In Hudson's account of telecommunications in rural development, there is little notice taken of how speech over the phone sounded to its new users. See Heather H. Hudson, *When Telephones Reach the Village* (Norwood, N.J.: Ablex Publishers, 1984).

39 Saussure, de Mauro edition, 44.

40 George Bernard Shaw, *Pygmalion.* (New York: Vintage, 1916), 16.

41 Gail Jefferson, Lecture to Conversation Seminar, University of Texas, May, 1988.

# Part Five

# Media and Consciousness

# The Technological Shadow in
# *The Manchurian Candidate*

*Thomas S. Frentz*
*and*
*Janice Hocker Rushing*

Practically since its conception, the Hollywood film has occupied a paradoxical position in America's affair with technology. On the one hand, the medium itself is an apotheosis of technological progress, thrilling audiences with special effects that must constantly be superseded if boredom is to be forestalled. Film breeds utopian narratives of the triumphs of technique, for the big screen loves the grandeur of scale and speed, the mystery of the unrealized but not unimagined. On the other hand, movies have long cradled dystopian visions of the machine as a menacing plague, a sinister agency that propagates itself and threatens to strip humanity of its soul.[1] In this role, film critiques the very world of which it is a part.

Recently we examined the Promethean myth, known in modern times as "the Frankenstein complex," as it has evolved in contemporary cinema.[2] We argued that this myth is part of what C. G. Jung calls the cultural *shadow*—the dark fear we all share, but repress, that technology is taking on a will of its own. This shadow myth reveals an entelechial impulse toward perfection, or the gradual division of agency from agent, until an end point is reached at which technology is completely autonomous, the master-slave relationship between person and machine is reversed, and humankind is eliminated by obsolescence or force. Prometheanism is a predominantly *masculine* myth that mirrors the present patriarchy's preoccupation with scientific expertise and the domination of nature. Enfolded within the myth, however, is an opposing archetypal feminine principle that is typically associated with nature, and with movement toward identification and inclusion rather than separation and fragmentation. The myth not only prophesies the technological apocalypse, but implies that the tragedy may be averted if feminine values are integrated with masculine values into consciousness, and if the shadow is faced and accepted as a projection of ourselves.

The Frankenstein myth is constantly evolving. In our earlier work, we analyzed three films that exemplify important phases in the unfolding story. *Rocky IV* (1985) portrays the recreational technology of pleasure and the *re*-creational technology of physical conditioning as making over the human body in the image of the machine. In *Blade Runner* (1982), genetic engineering manipulates the DNA molecule to manufacture synthetic persons who are dependent on the engineers only for longevity and reproduction. At this point, if *Blade Runner* is representative, the human can still reverse the deleterious effects of progress by incorporating the feminine principle, that is, by ceasing to dominate nature and by recognizing the need to withdraw the projection of the shadow. Finally, *The Terminator* (1984) depicts the horrific consequences for humanity, should the shadow myth ever reach its completion: in a post-nuclear war begun by computers, the machines have learned how to reproduce themselves, and are in a battle to eliminate the human beings. Questions of gender politics are beside the point, *The Terminator* implies, when things have gone this far.

Quite obviously, the shadow myth of Frankenstein is multifaceted and ongoing, and three films do not exhaust its content. In this essay, we will examine a relatively early phase of the story as expressed in *The Manchurian Candidate* (1962). Developmentally, this episode in the myth documents the dehumanization of the *mind*—both individual and cultural—through technological conditioning. The film presents a portrait of the psyche as losing its identity with the eruption of the technological shadow. This identity may be reclaimed by attending to dreams, and by contacting the feminine principle that technology has repressed. But such reclamation is perilous, for if the ego is weakened by dependence on the *Great Mother* archetype, as the film seems to suggest, then the psyche is ripe for victimization by technology. As a rhetorical statement, *The Manchurian Candidate* suggests that the psyche must revitalize itself by integrating the adult form of the feminine principle—the *anima*. While it is clear what path the ego must choose in relation to the shadow, the identity of the shadow is misrepresented in the film to the extent that its causal agent is projected onto irrelevant enemies. Thus, the advocacy is not one that will ultimately heal the psyche. After a brief exposition of the plot, we analyze *The Manchurian Candidate* for its portrayal of dehumanization through Pavlovian brainwashing and political image making, for its representation of potentials for healing through dreams and the integration of feminine aspects into the psyche, and for its projection of the shadow onto inappropriate enemies.

## THE STORY

*The Manchurian Candidate* has been called "neither satire nor suspense thriller nor science fiction fantasy nor identity-puzzle nor allegory but something of all five."[3] Indeed, its somewhat bizarre plot fuses all these aspects. A small American patrol led by Captain Bennett Marco is captured by Chinese Communists near the end of the Korean War. Soon after the war's end, the patrol's sergeant, Raymond Shaw, is publically welcomed home in Washington as a Medal of Honor winner. Even though he recommended Raymond Shaw for the honor, Ben Marco distrusts the circumstances surrounding the incident. Along with another member of the patrol, Al Melvin, Ben has nightmares that eventually reveal to him that he and his entire patrol had been brainwashed by a Manchurian psychiatrist skilled in Pavlovian conditioning. Although the conditioning process has obliterated his memory of it, Raymond killed two fellow soldiers in Manchuria, as well as his newspaper columnist employer, Holburn Gaines, in New York, all as tests of the effectiveness of his programming. As Ben unravels the enigma, we learn that Raymond has been conditioned as an assassin—part of a Communist plot to take over the American government.

The Communists' American operator is none other than Raymond's own mother, who manipulates his stepfather, the McCarthyish Senator Johnny Iselin, into the Republican nomination for vice president. In her mad ascendancy to power, she programs Raymond to kill Johnny's rival, the liberal Senator Thomas Jordan; Raymond also kills Jordan's daughter, Jocie, whom he has just married, because she witnessed the crime. The conspiracy comes to a head at the Republican National Convention, where Raymond is supposed to shoot the presidential nominee so as to propel Johnny Iselin (and thus the Communists and Mrs. Iselin) into power. Unbeknownst to his mother, however, Raymond has, with the help of Ben and cooperating U.S. Army psychiatrists, overcome his conditioning. Instead of murdering the presidential nominee, Raymond shoots Johnny Iselin, then his mother, and finally himself.

## TECHNOLOGICAL DEHUMANIZATION

As the repressed fear that technology will destroy the human being, the shadow materializes in *The Manchurian Candidate* in several characters—both those who perform the conditioning and those who are the victims of that conditioning. Viewed in the context of the larger myth, the film creates two Dr. Frankenstein characters—the Communists (primarily Yen Lo, the North Korean psychiatrist) and Raymond's mother—who use advanced technological agencies to turn human beings into tools for the furthering of their own power. The victims of their machinations—the Frankenstein-made monsters—are Ben Marco, Raymond Shaw, Johnny Iselin, and, by implication, the American culture at large. We examine first the conditioning itself, and then its consequences. The effects of the dehumanization are manifest not only in the individual victims, but in the scenic instability surrounding them.

## The Conditioning

Although the actual process is not shown, Raymond explains to Ben retrospectively that the soldiers were brainwashed over a three-day period with a combination of drugs and light-induced descent into the unconscious. As the Communists' crowning achievement, Raymond has been trained to play a card game whenever he hears the words, "Raymond, why don't you pass the time by playing a little solitaire?" He follows strictly any order given him immediately after turning up the Queen of Diamonds. Once the command has been obeyed, he promptly forgets everything that has happened since the first cue, thus suffering no guilt or fear of capture.

Several reviewers noted the melodramatic implausibility of the brain-washing sequences.[4] But the question, "Could this really happen?" is less interesting than the symbolism of brainwashing as the ultimate extension of the classical conditioning theory that had thoroughly permeated the fields of academic and applied psychology by the time the film was released. Not only could perfectly decent soldiers' behavior be controlled, even to the extent of causing them to commit acts they would normally deem morally repugnant, but so could their unconscious minds be "re-tooled" through the rigorous and ruthless application of external stimuli.

Quite clearly, Raymond's captors regard him not as a person, but as a valuable piece of equipment. The Chinese psychiatrist, Yen Lo, continually refers to Raymond as "the mechanism." In Richard Condon's book, upon which the film was based, Yen Lo tells the Communist bureaucrats assembled for the demonstration of Raymond's conditioning that he "operates" Raymond through the "remote control" device of the playing cards.[5] Yen boasts (as if Raymond were a robot) that "the most admirable, the most far-reaching characteristic of this extraordinary technology of mine is the manner in which it provides for the refueling of the conditioning, and this factor will operate wherever the subject may be—two feet or five thousand miles away from Yen Lo—and utterly independently of my voice or any assumed reality of my personal control."[6] He asks his Soviet accomplice in America (Zilkov), "Do you realize, Comrade, the implications of the *weapon* that has been placed at your disposal?" He reassures Zilkov that, because Raymond has been relieved of "those uniquely American symptoms, guilt and fear," he is "entirely police-proof. His brain has not only been 'washed,' as they say, but 'dry cleaned.'" Even Ben Marco, who is more sympathetic about the theft of Raymond's mind, accepts the fact of Raymond's mechanization. When Ben attempts to extinguish the Communists' conditioned responses by inducing Raymond to play a game of solitaire with a deck of fifty-two Queens of Diamonds, Ben pronounces, "You don't work any more. That's an order. . . . Remember, Raymond, the wires have been pulled."

The makeover of Johnny Iselin from infantile imbecile to public relations phenomenon is the handiwork of his wife, every bit as gifted in mediated image construction as is Yen Lo in psychological reconstruction. She would undoubtedly still draw top dollar today as a political campaign consultant. The shrewdness with which she maneuvers Johnny into the media limelight is rather astonishing for 1962—only two years after The Great Debates and six years before Joe McGinniss' *The Selling of the President 1968*.[7] We see her operate early in the film when Raymond steps off the plane in Washington to the spectacle of a cheering crowd and a covey of reporters. She hustles Johnny up to Raymond so the cameras can get a shot of them underneath the huge and misleading banner reading "Johnny Iselin's Boy." (Johnny is, of course, Raymond's stepfather.) Later, Mrs. Iselin attends a press conference; when she gives him the cue, Senator Iselin interrupts the proceedings by charging the Secretary of Defense with harboring known Communists in the Defense Department. At this point he has to give his full name, for the Secretary does not know who he is. But Mrs. Iselin is certain they soon will. She switches her approving gaze from Johnny to his image on the television screen in front of her; she is much more interested in the pseudo-event than in her husband.[8]

Mrs. Iselin also understands the importance of sly innuendo in accusing one's opponents. Poor Johnny is frustrated because she keeps changing the number of Communists that he is supposed to claim are in the Defense Department. The boys in the cloakroom are making fun of him, and he's looking like an idiot.

> *Mrs. Iselin:* Well, you're going to look like an even bigger idiot if you don't get in there and do *exactly* what you're told.
>
> *Johnny:* Babe . . . .
>
> *Mrs. Iselin:* Who are they writing about all over this country, and what are they saying? Are they saying, '*Are* there any Communists in the Defense Department?' Of course not. They're saying, 'How *many* Communists are there in the Defense Department?' So just *stop* talking like an expert all of a sudden and get out there and say what you're supposed to say!
>
> *Johnny:* Oh, come on Babe, I . . .
>
> *Mrs. Iselin:* I'm sorry, Hon. Would it really make it easier for you if we settled on just one number?
>
> *Johnny:* Yeah. Just one real simple number, that'd be easy for me to remember.

She inadvertently glances at Johnny smothering his steak with Heinz ketchup. The next scene finds Johnny in the Senate chamber proclaiming, "There are exactly fifty-seven card-carrying members of the Communist Party in the Defense Department at this time!"[9] Although the theme is not developed, she apparently intuits the peculiarly American blend of commercial pop culture and political sloganeering that would eventuate a couple of decades later in ubiquitous sound bites such as "Where's the beef?" and "Go ahead . . . make my day."

The plot makes Raymond Shaw and Johnny Iselin antagonists; Johnny can't seem to remember who Raymond is, and Raymond keeps protesting to reporters that the Senator is *not* his father. In terms of the Frankenstein myth, however, the two are consubstantial—at least when Raymond's mechanical alter-ego is activated. They are both "rats" (or perhaps monkeys, as Johnny's middle name of "Yerkes" would indicate) to the conditioning schemes of the dual practitioners who stake out their respective mazes. Like lab animals, neither has any ontological status beyond that of tool. The connection between the two characters is drawn by Mrs. Iselin herself when Raymond accuses her of organizing "this disgusting three-ring circus" at the airport in order to get more votes for Johnny by linking him with the Medal of Honor recipient. She replies deceptively, "Raymond, I'm your mother, how can you talk to me this way? You know I want nothing for myself. You know that my entire life is devoted to helping you and to helping Johnny. My boys. My two little boys." The tie is indicated more subtly through the metaphor of *cards*. Raymond has been conditioned to *play* cards, and Johnny to *shout* about the "card-carrying Communists"—both on cues from their "programmers." The ordinary playing deck Raymond uses is descended from the Tarot, an ancient game for the divination of the unseen and the prediction of the future. As such, the deck of cards has retained its mysterious association with *fate*, as in the saying, "he's playing with a stacked deck," or "he got dealt a rotten hand." Indeed, both Johnny and Raymond are players in a game of fate that is bigger than themselves, and, as is typical of humans in the myth of technology on its way to autonomy, they cannot see what that fate might be.[10]

### Consequences of Conditioning

When a person is retooled into a machine, the common effect within the Frankenstein myth is a loss of human identity.[11] In *The Manchurian Candidate*, Ben and Raymond provide close-up views of the consequences of brainwashing. Although he was brainwashed along with Raymond and the others in his squad, Ben's mind has not been as radically mechanized as Raymond's; his function for the Communists is merely to perpetrate the fraud of Raymond's heroism. Nevertheless, he is clearly disturbed when he returns to America. His confusion takes the form of a breakdown in his *persona*—Jung's term for the roles a person plays, the masks he or she wears in order to present an appropriate face to society.[12] In this case, Ben's persona has been manufactured by his captors. Whenever anyone mentions Raymond, Ben robotically recites (as does fellow soldier Al Melvin): "Raymond Shaw is the kindest, warmest, bravest, most wonderful human being I've ever known in my life." Ben senses the sham even as he repeats it, however, and his dreams, along with a letter sent by Melvin to Raymond, confirm his hunch. When his army superiors chalk his disturbance up to "a delayed reaction to eighteen months of continuous combat in Korea," his psyche, aided by insomnia, disintegrates into a bundle of nervous ticks. By the time he leaves New York on a train for an ordered vacation, he cannot even light his own cigarette, and is overcome by nausea. When he visits Raymond's apartment but finds instead the traitorous Chunjin (who, under the guise of an interpreter, led the American patrol to the Communist forces in Korea, and now monitors Raymond while disguised as his valet), Ben savagely attacks him, then attempts to fight off the police who come to stop the ruckus.

Although Raymond is much less conscious than Ben, he too seems vaguely aware of the ambiguity of his affairs. He has been conditioned to believe he saved the lives of his troops while killing large numbers of the enemy, but he refuses to wear his Medal of Honor. As he reads Al Melvin's letter describing the reality behind their conditioning, the lamp prominently foregrounded in the frame is circumscribed with mask-shaped designs, as if to announce the ruse of Raymond's false identity scenically. But Raymond, seeming to desire some clarity, turns on all the lights in the room. The letter says Raymond is the best friend Al had in the army, Raymond later tells Ben. But Raymond knows this is not true. "You know how much the guys in the outfit hated me," he says, "well, not as much as I hated them, of course."

In fact, Raymond's cold and mechanical personality has isolated him from all normal human ties. Ben demands that Raymond show him the letter, but Raymond, as if to disdain any overtures of friendship, reveals that he never saves letters. When his American operators fabricate a car accident in order to get him into a hospital so they can check up on their mechanism, Comrade Zilkov explains to Yen Lo that the elaborate setting was constructed to look realistic in case the patient has any visitors. But he can't imagine anyone coming to visit Raymond. As "The Twelve Days of Christmas" plays in the background while Ben accompanies Raymond's intoxicated stupor, Raymond scoffs, "Twelve days of Christmas. One day of Christmas is loathesome enough for me!" And the fact that Raymond plays *solitaire* whenever his conditioned reflexes are reactivated is the most poignant image of the fragmenting and isolating effects of mechanization on the individual.

The film presents this erosion of identity, not only in an individual, but also in a cultural sense. From a mythological perspective, the entire cinematic atmosphere is permeated with the surprising and upsetting characteristics of the *trickster* archetype—an archaic form of the collective shadow. The trickster is a malicious, mocking, amusing, and sometimes daemonic shape shifter that is forever fracturing identity by disrupting the orderly world of conscious life.[13] The trickster continually manifests itself in the film through character anomalies and scenic incongruities.

This mercurial mood is established near the beginning in the brainwashing demonstration scene, where the editing creates the illusion that Yen Lo and the other Communists are changing back and forth between themselves and the benign members of the Ladies Garden Club discoursing on the growing habits of hydrangeas; the alternations establish the permeable line between appearance and reality. In Ben's dream the ladies are white, and in the dream of Al, who is black, they change to black. The quaint, old-fashioned comfort of the ladies' hotel contrasts starkly with the stiff, efficient room of the Manchurians, with its larger-than-life posters of Stalin and Mao Tse-tung. This capricious feeling is continued with the multiple appearances of Chunjin, the Korean who betrays the American army unit, takes notes at the brainwashing demonstration session, and then appears in New York asking Raymond for a job.[14] More alarmingly, it is difficult to tell who the villains are in this story. As one reviewer points out, the consulting psychiatrist who assists Marco bears an uncanny facial resemblance to Yen Lo.[15] Certainly one of the two most heinous characters in the film is Yen Lo, but he also sports the only real sense of humor; he is a Chinese who enjoys mocking American commercial slogans. "Tastes good . . . like a cigarette should," he laughs when he explains that the brainwashed soldiers are smoking cigarettes made of yak dung. "If kill we must for a better New York," he mirthfully tells the humorless Zilkov, then let Raymond exterminate his respected employer as a test. He is a Communist whose wife left him a shopping list for Macy's. And of course, Johnny Iselin is the Communist-baiting "patriot" who is really a dupe for his Communist-aiding wife.

This sense of the trickster at play is heightened throughout in the odd visual compositions of the *mise en scène*. The opening scenes before the credits are of a Korean brothel with a picture of General MacArthur on the wall. Raymond enters to roust his troops from their debauchery, and he is shot from a low angle so that the flag hanging below MacArthur sprouts from Raymond's head like a horn. This motif is repeated later in the dining room of Senator Jordan, whose shoulders fledge the huge wings of an eagle on the wall behind him as he tells Raymond he once sued his mother for defamation of character. Johnny looks at a picture of Abraham Lincoln as he answers protests about his accusations concerning Communists in the Defense Department. And as he starts to whine to his wife about settling on an exact number, he is superimposed as a ghostly apparition on top of the large picture of Lincoln hanging in their living room, implying that politics-by-pseudo-event brings out the shadow behind even the most revered "face" of our democratic system.

J. H. Fenwick has noted that anachronisms in setting highlight the incomprehensibility of scientific technology and its over-sophistication in comparison with everyday life. For instance, Raymond's house and office are dignified brownstones, but they are filled with "the jumble of contemporary embellishments—lamps, typewriters, dictaphones, all clashing with their background."[16] The military headquarters from which Marco works and the police station in New York are Edwardian in decor, but "the mess of electrical wiring and low slung lights, make it look as though those who worked there had just camped down."[17] And the room where Raymond is hospitalized is comfortable, with old-fashioned diagonal lines of chimney and sloping ceiling, but "this cosiness is utterly destroyed by the harsh verticals and horizontals of the tubular steel scaffolding round Shaw's bed."[18]

The mass confusion in identities—both scenic and personal—is summed up metaphorically in the pivotal masquerade party Mrs. Iselin throws for Jocie in an attempt to derail her father, arch-rival Senator Jordan, from blocking

Johnny's nomination for the vice presidency. In the first scene, Johnny scoops into a "cake" in the shape of an American flag, but made of Polish caviar. Mrs. Iselin is dressed as the harmless Little Bo-Peep (her sheep, of course, is Johnny), Johnny as Lincoln, Raymond as a gaucho ("Gaucho Marx," he later jokes incongruously), an American reporter as an Arab, and innocent Jocie as the Queen of Diamonds (a symbol the audience has by now been "conditioned" to associate with Raymond's evil mother). The persona often appears in dreams in images of clothes, uniforms, and masks.[19] Here, in one surrealistic and clownish jumble, the various constructed personae are revealed; they also begin to disintegrate, for it is at the party that all the "best-laid plans" of the conspirators begin to go awry, spelling their ultimate doom.

## POTENTIALS FOR HEALING

The otherwise bleak progress of events in the Frankenstein myth is impeded in some stories, as we mentioned earlier, when the characters dehumanized by technology contact some representative of the feminine principle. This more hopeful turn is generally correlated with motifs of *waking up* from sleep, or of *seeing* the effects of mechanization.[20] In *The Manchurian Candidate*, it is, paradoxically, sleep that awakens Ben Marco, Al Melvin, and eventually even Raymond Shaw, for paying attention to their dreams—"seeing" the inner mind—reveals to these victims of conditioning the falsity of their re-manufactured psyches.[21] In addition, both Raymond and Ben are awakened by their female "love interests," who perform the psychic function of the projected *anima,* the feminine aspect of the male psyche. The *anima* is the "soul image" of the man's unconscious, which is compensatory to the persona, and can both haunt him (if she is not faced) or heal him (if she is accepted).[22] In marked contrast, Johnny, the victim of mass media conditioning, is never awakened at all; he has no dreams, and remains under the control of his domineering wife, who is even more the Great Mother to him than she is to Raymond.

### The Dreams

At the beginning of his first nightmare, Ben sees only the deception he has been conditioned to see—the members of the Ladies Garden Club droning on about plants. But then the dream begins to unravel the truth, as the vision shifts to Yen Lo explaining that the soldiers think they are waiting out a storm in the lobby of a New Jersey hotel. In the next cut, the ladies are speaking the words of Yen Lo, and then the action shifts back and forth between the insidious Yen Lo and the harmless women. Thus, whereas the scene is initially confusing to the viewing audience, it reveals to Ben that "the most wonderful human being I've ever known" strangled his fellow soldier, Ed Mavole, to death. Ben then wakes up, both literally and figuratively, although the knowledge initiates his nervous disintegration. Al Melvin's dream picks up where Ben's left off, and similarly breaks through the deception of the ladies to divulge that Raymond also shot the patrol's "mascot," Bobby Lembeck. This disclosure wakes Al, too, and although he recites the lie he has been taught about Raymond, he decides, at his wife's urging, to write Raymond about the dream. But Raymond himself is still too much under the influence of his captors to believe Al's story, as suggested when he plays a game of solitaire on top of Al's letter.

Dreams, however, ultimately do penetrate Raymond's conditioned unconscious. After an "activated" Raymond kills Jocie and her father, he calls Ben from a hotel room across from Madison Square Garden and says, "I . . . I think maybe I'm going crazy. I'm having terrible dreams like you used to have." Ben arrives, and through use of the forced card deck, induces Raymond to remember killing Mavole, Lembeck, Gaines, and Jordan, before he mercifully orders him not to remember anything else. Raymond asks tearfully but truthfully, "They can make me do anything, Ben, can't they? Anything?" At that point, Raymond recognizes the extent of his enslavement.

## The Feminine Archetypes

The potential of each male victim to be healed by the female principle is best understood in relation to the archetypal roles played by the dominant women in their lives. The first bearer of the soul-image for a man is always the mother, says Jung; as the healthy male matures, the image shifts to the woman who arouses the man's feelings, whether in a positive or negative sense.[23] In myth, the mother takes on the archetypal aspects of the Great Mother. She may appear as the Devouring or Terrible Great Mother, who refuses to let her children grow up, or as the Nurturing or Good Great Mother, who cares for and protects her young from harm. Over time, a man's inner image of the feminine normally detaches itself from the mother archetype and emerges as the *anima*. "The anima," writes Erich Neumann, "is the vehicle par excellence of the transformative character. It is the mover, the instigator of change, whose fascination drives, lures, and encourages the male to all the adventures of the soul and spirit, of action and creation in the inner and outer world."[24] As a transformer of the personality, the *anima* is dangerous and often leads to peril, but if she does destroy the ego, it is because her detachment from the Great Mother is incomplete; that is, the maternal is preponderant over the *anima* in the man's unconscious.[25]

"My mother, Ben, is a terrible woman," Raymond offers as the understatement of the millennium. Indeed, there is abundant evidence in the film associating Mrs. Iselin with the Terrible Great Mother. When she forces Raymond to break off his first romance with Jocie, she wears a Chinese kimono emblazoned with dragons, a typical Devouring Great Mother motif in fairy tales. As novelist Condon puts it, "It [the house coat] had a deep black Elizabethan collar that stood up straight behind and around her shining blonde head, in the mode of wicked witches."[26] She is linked throughout, furthermore, with the Queen of Diamonds. When Ben and the police psychiatrist start to unlock the riddles of the solitaire game, Ben suddenly remembers that Yen Lo had said in his brainwashing demonstration, "The Queen of Diamonds is reminiscent in many ways of Raymond's dearly loved and hated Mother . . . and is the second key to clear the mechanism for any other assignment." This connection is made visually when Mrs. Iselin gives Raymond the assignment to kill the presidential candidate; Raymond is (she wrongly presumes) under the influence of the Queen of Diamonds card—this time the much larger costume Jocie discarded from the masquerade party. Mrs. Iselin is shot from a low angle so that she appears to be rising from the card, and the resemblance between the blonde queen and the blonde mother is unmistakable. All face cards in Tarot are traditionally related to the person playing the game, especially with the player's *mask*, or another individual upon whom the player has projected part of his or her personality.[27] The Queen of Diamonds, in particular, is descended from the Queen of Pentacles in the Tarot deck; the pentacle is the star connected with various Great Goddesses, including Ishtar, Isis, and Morgan of the Celts. Eventually it became associated in medieval times with the Witch's Cross or the Witch's Foot.[28]

Against this powerful and evil figure, Johnny stands no chance. At the masquerade party, she tells him, "Run along, the grownups have to talk," as she queries Senator Jordan on his intent to block her husband's nomination. As Condon has it, Johnny is impotent with her (although not with other women, whom she is more than willing to supply). He has no mature *anima* to compensate for his persona; indeed, his persona is *constructed* by the Terrible Great Mother, and she controls his every move. When the Devouring Great Mother keeps her offspring's ego weak and dependent, the film implies, she prepares him as psychic "bait" for technology, and thus is complicitous in the process of dehumanization.

Raymond, too, is largely controlled by his mother, even before his mechanization, and in spite of the fact that he feebly resists her power throughout the film. He was chosen as "the mechanism" partly because of his general weakness of character. When Yen Lo asks in the brainwashing demonstration who Raymond dislikes the least among his men, Raymond chooses Captain Marco. With his usual biting humor, Yen Lo implores the crowd, "Notice how he is drawn always to authority?" But Raymond has a better opportunity to transform his personality than does Johnny, even given that his crimes are worse, because he meets Jocie. He chances upon her the summer before he goes into the army when he is bitten by a snake near his lake home, and she appears, as he later tells Ben, "unexpectedly with a razor blade in her hand." The snake is commonly linked to the Great Mother; its symbolism is ambiguous, relating to its obvious danger, but also to the numinous process of growth whose purpose is inaccessible to consciousness.[29] As Jocie expertly extracts the venom from Raymond's leg, she symbolically rids him of his mother's poison. At nineteen, Jocie is an innocent and alluring *anima* figure for the mother-dominated Raymond.

Jocie immediately has what Condon calls the "concentrically transforming" effect on Raymond that is characteristic of the *anima*, for the *anima* compensates for what the persona lacks.[30] The person upon whom a man projects his *anima* is always the object of intense love or hate, and her influence "is immediate and absolutely compelling."[31] As Raymond later tells Ben while showing him Jocie's picture, "Some people are loveable, and other people are not loveable. I am *not* loveable. Oh, but I was very loveable with Jocie. Ben, you cannot believe how loveable I was, in a way." In other words, Jocie helps him to contact that part of himself that is opposite from his sterile persona.

Unfortunately, Raymond cannot fully differentiate his *anima* figure from his mother. The image of one seems to evoke the image of the other. As Senator Jordan explains to Raymond that he had to sue Mrs. Iselin for defamation of character, Raymond abruptly asks Jordan for permission to marry Jocie. During his otherwise happy summer with Jocie, Condon informs us, "Whatever they did together he held himself rigid, awaiting the scream of his mother's rage, and it cost him thirty pounds of flesh because he could not keep food down as he battled to hold the thoughts of his mother and Jocie apart."[32] Thus, when his mother orders him to give up Jocie (the *anima* is a grave threat to the Devouring Mother), the older woman initially wins the battle over his soul. Even after he has been fully mechanized, however, Jocie reasserts her influence when she materializes at the masquerade party dressed as the red queen. Entranced and confused, Raymond impulsively marries her that night, and again becomes loveable and fun-loving. But Jocie proves an unequal match for Mother's next move. Mrs. Iselin regains control of "the mechanism," and the Raymond/Romeo and Jocie/Juliet story of children of feuding parents ends in tragedy.[33]

In the battle for Raymond's soul, it is unclear whether Jocie or Mother have the ultimate upper hand. As noted above, when Mrs. Iselin activates Raymond and instructs him to kill Jocie's father, Jocie perishes for witnessing the crime. Although Jocie is not ultimately able to save Raymond, her sacrifice does rehumanize him; it is after his mechanized self kills her that he begins having the nightmares that soften his hardened persona enough for Ben to deprogram him.[34] He recovers feelings—the guilt and fear that Yen Lo said had been forever conditioned out of him. With the reclamation of these emotions, he also gains a moral conscience and an individual will. Thus, even in death, Jocie continues to influence Raymond by leading him toward recognition of his murderous alter ego.

This struggle between the Great Mother and the *anima* over Raymond's soul is portrayed in the scene of Mrs. Iselin's last words to Raymond. As she cradles Raymond's face in her hands and laments that the Communists took his soul away from him (she did not know they would pick him as the assassin), the deceased Jocie's Queen of Diamonds costume looms in the background of every frame, visually coming between Raymond and Mother. Even as Mother concludes her remarks by kissing her son as an implied prelude to incest, their lips meet over the backdrop of Jocie's costume. Shortly thereafter, Raymond departs for his last assignment.

As Raymond climbs the steps of Madison Square Garden to reach the cubicle where he will commit his final act, officials in the arena test the microphones and lights to see if they are working. They are, but Raymond-the-mechanism is not. When Ben and his army superior arrive at the stadium, the contrast between the confused pandemonium on the floor and the calm in Raymond's cell are established with sound and editing; short takes of frenzied fans, determined cameramen, and a very nervous Johnny Iselin are interspersed with single shots of Raymond waiting—quiet, composed, and single-minded. He sights his rifle on the target intended for him (the presidential candidate), but on the spoken cue his mother had given him, he moves it to the real villain and her pawn. He acts, as he has always done, in solitaire, but this time of his own volition. For the first time since he took it off upon returning to America, he places the Medal of Honor around his neck. To Ben, who has seen the light in Raymond's lookout and by now has reached it, Raymond reveals his heroic motivation: "You couldn't have stopped them; the army couldn't have stopped them, so I had to. That's why I didn't call. Oh God, Ben. . . ." But guilt has caught up with him, and the last one he stops is himself.

When Frankenstein's monster awakens and experiences the predicament of his unhuman existence, he grows more noble in character than Dr. Frankenstein; he turns on his maker, however, when he is unable to persuade him to make a place for himself in society. Both motifs are widely repeated in early phases of the Frankenstein myth, as when the replicant Roy Batty of *Blade Runner*, a terrifying but poignant fallen Adam, kills Eldon Tyrell, the genetic engineer who refuses to prolong Roy's lifespan. Although the theme of created turning against creator provokes fear and horror, it is a "positive" sign that the myth-making culture sees how serious the technological threat has become. Things have gotten so bad, the myth is saying, that humanity may be replaced by its machines. In *The Manchurian Candidate*, Raymond's awakening is actually our own, and symbolizes a recognition of the consequences of technological dehumanization.

Raymond's awakening occurs too late for him to save himself or those he loves. But Ben fares better than Raymond, for he falls in love with Rosie, who is, like Jocie, charming and caring, but also older and wiser. Some reviewers saw Rosie as a superfluous character, inserted in the film primarily as a romantic interest for Frank Sinatra (Ben Marco).[35] It is true that she does not tangibly advance the plot, but her importance is her inner impact on Ben. Appearing out of nowhere as Ben's psyche deteriorates on the train (just as her younger look-alike counterpart Jocie appeared mysteriously to Raymond), she is ethereal and otherworldly, as is typical of the *anima,* who is often felt as an autonomous entity, leading "an entirely independent existence, perhaps in a world of invisible things."[36] A strange conversation ensues. Ben asks abruptly, "Are you Arabic?" She tells him her name is "Eugenie Rose," but friends call her Rosie. "Are you Arabic? Let me put it another way. Are you married?" she queries just as nonsensically. But their interchange becomes more comprehensible if she is seen as Ben's positive *anima* projection, and a steadying influence on his crumbling persona. The *anima* has a collective, in addition to a personal quality, and often appears as a historical figure.[37] According to Condon, Ben fantasizes that Rosie is a German woman he had seen in a magazine twenty-three years ago, or an angel, or an Arabic queen he has dreamed of, but who died in 1395. She is also wearing a silver belt buckle shaped as Quetzalcoatl, the feathered serpent god of the Aztecs, who was husband or son to the Great Goddess.[38] Her odd name seems to be derived from Saint Eugenia, a Goddess whose shrines were attributed with the healing miracles of its "eugenic" springs, and from the rose, which is, of course, an ancient and enduring symbol of the Goddess, and particularly of the female soul-image.[39]

Rosie inexplicably knows Ben intimately from the beginning: "I told the authorities I knew all about you," she tells Ben, and she seems intent on healing him. Realizing his mental state, she "conditions" him on the train to remember her address and telephone number, breaks off an engagement with her fiancé, rescues Ben from jail and tends to his wounds after he attacks Chunjin, and serves as a catalyst for his unraveling of Raymond's (and thus his own) dilemma. She is still thinly connected with the Great Mother in Ben's mind, for he proposes to her in a non sequitur, much as Raymond did to Jocie, right in the midst of talking about the effects of the Queen of Diamonds on Raymond. But her maternal aspects are the opposite side of Mrs. Iselin, well-tempered with the more mature *anima,* and she successfully nurtures Ben back to health.

## PROJECTING THE SHADOW

Raymond Shaw is the hero of this tale. As if in an ancient Greek tragedy, he is tangled up in Fate, and unknowingly commits sins for which he must thereafter pay. Ben makes this clear in his eulogy, which he composes for this Medal of Honor winner as the thunder of Raymond's rifle blends into the thunder of the gloomy storm enveloping Ben and Rosie's apartment: "Made to commit acts too unspeakable to be cited here, by an enemy that had captured his mind and soul. He freed himself at last and in the end, heroically and unhesitatingly gave his life to save his country. Raymond Shaw . . . hell . . . hell!" Given the plot of this story, Raymond did what he had to do, and, as a tragic hero, he evokes our pity, as well as our fear.

It is easy to overlook, then, the fact that Raymond's heinous creators are not the typical "mad scientists" in our midst, but the external projections of Communism and maternal forces. Interpreted mythically, this means that the *causes*, the loci of technological creation, are not Promethean figures, but the archetypes of the Enemy and of the Terrible Great Mother.[40] Thus, while the film acknowledges the threats of a soulless psychology and of a *polis* manufactured by technological images, it is unwilling to face the real source of the shadow. While Communism may indeed participate in the technological dehumanization that is here deplored, it remains in the film as effective a red herring against recognizing the log in our own eyes as were the accusations made by the famous Senator from Wisconsin upon whom Johnny Iselin is modeled. And even more questionable is the projection of mechanization upon the Great Mother. It is true, as we noted above, that when the possessive, devouring aspect of the Mother archetype keeps the male ego weak and dependent, it can, perhaps unwittingly, aid technology in its dehumanizing course. But the mechanization itself—be it psychological conditioning, the proliferating influence of mediated images, or any other form of technique—is anathema to the feminine principle in any form. The Great Mother may be Terrible, but she is certainly not technological. To the extent that the film projects the capacity for mechanization upon the maternal aspect of the feminine archetype, it, too, is complicitous in hiding from us the root cause of the shadow.

Since everything unconscious is projected, it could be argued that projecting the shadow into dramatized characters is a first step in facing it. If it is not personified, it cannot be seen. But the ultimate goal of a maturing individual or culture is to recognize the projection as a part of oneself. The shadow, when realized, is the source of renewal, although it must not be acted out indiscriminately.[41] If we get stuck in the projection, the crimes cannot be owned and the innocent may pay. "The effect of projection," writes Jung, "is to isolate the subject from his environment, since instead of a real relation to it there is now an illusory one. . . . [Projection] is an unconscious factor which spins the illusions that veil his world. And what is being spun is a cocoon, which in the end will completely envelop him."[42] The culture in this case ends up, ironically, as isolated from its technological reality as friendless Raymond, playing solitaire with himself.

What is more, the shadow cannot be eliminated; it can only be suffered through and allowed to transform hubris into an attitude of humility, discipline, and responsibility.[43] Before applauding Raymond for killing its source, we might do well to recall an offhand comment Ben makes to Raymond when the latter, drunk and complaining about his mother, says that Ben certainly doesn't want to hear about this. Ben replies, "Of course I do—it's rather like hearing Orestes gripe about Clytemnestra." Raymond does commit Orestes' crime, and the much-offended Furies will no doubt hound him in his grave.[44]

## CONCLUSION

Although *The Manchurian Candidate* obscures the real source of the modern technological shadow by projecting it onto the dual forces of maternalism and Communism, time has proven it to be uncannily prophetic in its depiction of the peculiarly American form of technological dehumanization. The film is apparently wiser than its critics who denounced the blend of brainwashing with political intrigue as absurd. While the fantastic excesses of Pavlovian conditioning have not eventuated in as smooth a weapon as Raymond Shaw, does not our modern political election process rely on similar techniques, applied to the mass public rather than to a single unfortunate individual? Just as Yen Lo operated Raymond and Mrs. Iselin manufactured Johnny by feeding them their familiar cues, sophisticated media "handlers" now evoke predictable responses from the voting public by cueing their candidates to spout familiar slogans which stand, in shorthand form, for certain "button-pushing" emotional issues—the most recent being family, drugs, taxes, the Pledge of Allegiance, and flag burning. The phenomenon of Johnny Iselin no longer seems that far-fetched; almost anyone can get elected, it seems, given the right campaign aides who understand the vulnerabilities of our mass-mediated psyche, and can persuade their identity-drained tools to "play their cards right" and avoid mistakes.

The film also portrays accurately, at least to some degree, the interaction of technological dehumanization with the feminine archetypes. Whereas few real mothers would wish their children to become Raymond Shaws or Johnny Iselins, it is difficult for most of us to avoid the extreme dependency on our favorite technologies—indeed, on the very technologized environment in which we live—that results in arrested maturation, a sort of perpetual state of childishness that interferes with the development of an adult identity. As many social critics muse (but few know how to avoid), too much stay-at-home "couch potato" spectatorship robs us of responsible participation in the *polis*, pulverizes rhetorical argument into visual and audial particles, and may even threaten the foundations of democratic government.[45] Thus, while maternalism is not itself responsible for the dehumanizing effects of technology, the "innocence" of an extended childhood plays its part whenever we substitute dependence on technology for dependence on the Mother. However, the rising concern for the natural environment over the last two or three decades, and particularly the acknowledged link between the values of nature and the feminine, provide some hope that a conscious integration of mature feminine values may yet restore an awareness of our Frankenstein-monster predicament, and a will to withdraw our shadow projections.

There is even some evidence that the American public at large is seeing the technological shadow that it has itself created—at least that part of it related to the political scene. We write this only a short time after the 1988 presidential campaign, which provoked an almost universal outcry, among professional commentators, political watchdogs, and everyday citizens alike, against such "dirty tactics" as sound bites, name-calling, emotionalism, and rigorous avoidance of issues. To us, the troublesome irony is that we protest a process that we ourselves perpetuate. After all, unlike Raymond as he climbed the stairs for his final act, this process "works." Moreover, the most outraged critics are often members of the media machine, who know they are being manipulated by political image makers, who know that they in turn help to condition the electorate, but who cannot escape the mechanical maze they construct. A pressing question, then, with regard to the issues of *The Manchurian Candidate*, as well as of other episodes of the Frankenstein myth, is whether seeing is a prelude to repossessing technological agency as a part of the human agent or whether we approach the frightening point of no return.

# Notes

1 "Utopian" and "dystopian" are categories used by scholars of science fiction such as Patricia Warrick, *The Cybernetic Imagination in Science Fiction* (Cambridge: MIT Press, 1980), 130-202. However, we have found that the categories extend to genres other than science fiction; see Janice Hocker Rushing and Thomas S. Frentz, "The Frankenstein Myth in Contemporary Cinema," *Critical Studies in Mass Communication* 6 (1989): 61-80.

2 Rushing and Frentz, "The Frankenstein Myth"; Isaac Asimov, "The Myth of the Machine," in *Asimov on Science Fiction* (Garden City: Doubleday, 1981), 153-163.

3 Peter John Dyer, "The Manchurian Candidate," *Sight & Sound* 32 (1962-1963): 36.

4 J. H. Fenwick, "Black King Takes Two," *Sight & Sound* 33, (1964): 115; Brendan Gill, "The Current Cinema: Bad Men and Good," *New Yorker*, 3 November 1962, 115-116; Arthur Knight, "The Fu Manchurian Candidate," *Saturday Review*, 27 October 1962, 65.

5 Richard Condon, *The Manchurian Candidate* (New York: Signet/New American Library, 1959), 58. In certain cases, the book departs significantly from the plot of the film. Because our interpretation focuses on the film, we cite relevant passages from the book only when they are consistent with the plot of the film.

6 Condon, *The Manchurian Candidate*, 57.

7 Joe McGinniss, *The Selling of the President 1968* (New York: Pocket Books, 1969).

8 We would like to thank our colleague, Professor Frank Scheide, for calling our attention to the complex images in this scene. See also Daniel J. Boorstin, *The Image or What Happened to the American Dream* (New York: Atheneum, 1962).

9 Dyer, "*The Manchurian Candidate*," 36.

10 Rushing and Frentz, "The Frankenstein Myth," 67.

11 Although he does not relate the film to the Frankenstein myth, Dyer, "*The Manchurian Candidate*," finds the originality of the film to lie "in the changes it rings on its central theme of identity," 36.

12 C. G. Jung, *Psychological Types*, a revision by R. F. C. Hull of the trans. by H. G. Baynes, Vol. 6 of *The Collected Works of C. G. Jung* (Princeton: Princeton University Press/Bollingen Series XX, 1976), 465-467.

13 C. G. Jung, *Four Archetypes: Mother/Rebirth/Spirit/Trickster*, trans. R. F. C. Hull, from Vol. 9, Part I of *The Collected Works of C. G. Jung* (Princeton: Princeton University Press/Bollingen Series XX, 1970), 133-152.

14 Dyer, "*The Manchurian Candidate*," 36.

15 Fenwick, "Black King Takes Two," 116.

16 Fenwick, "Black King Takes Two," 116.

17 Fenwick, "Black King Takes Two," 116.

18 Fenwick, "Black King Takes Two," 116.

19 Edward C. Whitmont, *Symbolic Quest: Basic Concepts of Analytical Psychology* (Princeton: Princeton University Press, 1978), 156.

20 Rushing and Frentz, "The Frankenstein Myth," 67, 69-70, 72-77.

21 The Freudian aspects of the film have been widely noted, for it is generally assumed that Raymond and his mother have an incestuous, Oedipal relationship, and the book confirms this, 326. See, for example, Andrew Sarris, "Film Fantasies, Left and Right," *Film Culture* (1964): 28-34. A Freudian interpretation cannot explain the import of the dreams, however, for Freud believes dreams hide the truth from the dreamer, whereas it is quite clear that they reveal it in this case.

22 C. G. Jung, *Aspects of the Feminine*, trans. R. F. C. Hull, from Vols. 6, 7, 9i, 9ii, 10, 17 of *The Collected Works of C. G. Jung* (Princeton: Princeton University Press/Bollingen Series XX, 1982), 78-81.

23 Jung, *Aspects of the Feminine*, 86.

24 Erich Neumann, *The Great Mother: An Analysis of the Archetype*, trans. Ralph Manheim (Princeton: Princeton University Press/Bollingen Series XLVII, 1972), 33.

25 Neumann, *The Great Mother*, 33.

26 Condon, *The Manchurian Candidate*, 113.

27 Mary K. Greer, *Tarot for Your Self* (North Hollywood: Newcastle Publishing Co., Inc., 1984), 75-76.

28 Barbara G. Walker, *The Woman's Encyclopedia of Myths and Secrets* (San Francisco: Harper and Row, 1983), 782.

29 Neumann, *The Great Mother*, 328. The snake plays a symbolic role in the "fall from grace" that leads to consciousness in the Garden of Eden myth, as the Kundalini serpent of Tantric Yoga, and in the healing staff of Hermes and Asclepius; see also C. G. Jung, *The Development of Personality: Papers on Child Psychology, Education, and Related Subjects*, trans. R. F. C. Hull, Vol. 17 of *The Collected Works of C. G. Jung* (Princeton: Princeton University Press/Bollingen Series XX, 1981), 125.

30 Condon, *The Manchurian Candidate*, 111.

31 Jung, *Psychological Types*, 471.

32 Condon, *The Manchurian Candidate*, 113.

33 In the book, the newspapers report the marriage as a Romeo and Juliet story; Condon, *The Manchurian Candidate*, 305.

34 This action is parallel to the rehumanization of Rick Deckard after he kills the female replicants in *Blade Runner*; see Rushing and Frentz, "The Frankenstein Myth," 75.

35 Dyer, *"The Manchurian Candidate,"* 37.

36 Jung, *Aspects of the Feminine*, 80.

37 Jung, *Aspects of the Feminine*, 81.

38 Condon, *The Manchurian Candidate*, 184-189.

39 Walker, *The Woman's Encyclopedia*, 286, 866-869; Jung, *Psychological Types*, 223.

40 Whitmont, *The Symbolic Quest*, 168.

41 Whitmont, *The Symbolic Quest*, 164, 167.

42 Jung, *The Portable Jung*, ed. Joseph Campbell, trans. R. F. C. Hull (New York: Viking, 1971), 146-147.

43 Whitmont, *The Symbolic Quest*, 167-169.

44 Janice Hocker Rushing, "Evolution of 'The New Frontier' in *Alien* and *Aliens:* Patriarchal Co-optation of the Feminine Archetype," *Quarterly Journal of Speech* 75 (1989): 1-24.

45 Bill Moyers, "The Prime-Time President," PBS, 1988.

# The Ultimate Technology: Frederick Wiseman's *Missile*

*Thomas W. Benson and Carolyn Anderson*

Technology is imagined by culture and in turn constructs culture. The technology of nuclear weapons defines the conditions of modern existence, provoking us to invent a culture that can exist alongside our moment-to-moment ability to destroy all humanity and all its cultures. The role of the cultural critic in an apocalyptic situation may seem to be tertiary, at best—removed from direct participation in the technology itself and from the immediately pragmatic decision making that governs its control and use; still, even a tertiary role may be useful in articulating the ways in which our terministic screens, our cultural practices, have brought us to this brink, and how they might perhaps be re-articulated to help us avoid the plunge.

We shall take as the central text for our analysis Frederick Wiseman's 1988 documentary *Missile*, about the training of ICBM launch crews at Vandenberg Air Force Base in California.[1] Wiseman's continuing fascination, in a long series of distinguished films, with the culture of institutions, and, in *Missile*, with the culture of technology, suggests that his film may be an especially rich text for analysis.

*Missile* presents itself to us as an invitation to speculation and interpretation. We wish to inquire in this essay into the rhetoric of *Missile*. Rather than presupposing that *Missile* is a conventionally persuasive documentary, we shall argue, rather, that the film is a cultural discourse, and take it as our purpose to inquire into the rhetoric of that discourse and its reading as situated cultural actions. *Missile* appears to ask how a culture forms itself around the fact of the nuclear missile.

It is easy to fall into high theory or a high moral tone when discussing the ultimate. So we may as well say at the outset that we do not think we or Wiseman offer the answer. On the other hand, the subject is important, and it seems to us that Wiseman has important and original ways of interpreting the situation we are all in together. So we would like to see if we can make some sense of his view, and do so from our perspective as observers of communication, which means trying to understand his film as a communicative action, and as implying the actions of its subjects and its viewers in a cultural and potentially rhetorical situation. And we will, with apologies, do this in a way that is rather self-conscious, by examining some troublesome and puzzling aspects of our own theories and methods along the way. We do not think of ourselves as theorists or even as methodologists, but as interpreters, historians, and critics of culturally and rhetorically situated communication. Even so, this work seems to require self-conscious reflection about theory and method—usually unstated in a finished piece of interpretation, just as the doubts and puzzlements are often left unstated.

We have found it difficult in writing this essay to express the appropriate tone of moral seriousness and attention to first principles without lapsing into a tone of ponderous certitude. That problem itself has told us something about the achievement of Wiseman's film, because it was a problem he had to overcome in his own way.

This, in turn, has led to a recognition that it has been impossible to avoid our own ideological position. Our profession and our preference is to doubt and question—and to advocate curiosity and doubt. Doubt about nuclear weapons can seem to lead only to a certainty that they should not be used—or, to state it as a doubt, that there could not be a situation in which it is possible to achieve certainty that they should be used. It is not our purpose to argue that view here, however. Rather, it seems to us, this position provides a way of understanding what Wiseman is up to and up against in this film, and makes it quite clear that not everyone is going to see the film in the same way, since we all live together in a culture that hinges its fate on convincing the Soviet Union that certain actions of its own will, with absolute certainty, bring down upon it the finality of nuclear annihilation—even at the cost of our own annihilation.

A clue to the film's central enigma arises from a conflict that Wiseman builds into the form of the film. On the one hand, the film presents itself as a bland and uncomplicated observation of the day-to-day activities of a group of young Air Force officers who are undergoing training to become launch officers for nuclear missiles. The understated normality of the film can be taken at one level as an insistence by Wiseman on the descriptive potential of documentary, its obligation to see what is in front of the camera straight on, to seem to deny any interpretation that goes beyond simple attention—itself a radical and paradoxical claim, of course. The descriptive normality of the film is also an implicit demand upon our own actions as spectators—a demand that we attend to the empirical, material reality as a first condition of viewing, rooted in the place we find ourselves and in the moment in time that is the film's present tense. But against the empiricism, immediacy, and normality of the film, Wiseman places its contrary: a demand that we reflect upon the film as a construction—a text—and that we see it as resonating with a variety of other texts and contexts.

There are a number of ways in which this second mode of reading is offered to us, but none, perhaps, is more striking than a peculiarly assertive gesture that Wiseman makes at the very end of the film. In the last extended scene in *Missile*, a general is speaking to the student officers who are now presumably graduating from the course. Wiseman quite unmistakably cuts off the general before he is through speaking, just as the general is celebrating the values that motivate and justify his own and his audience's willingness to push the button on command. He says, "We are a people who are concerned about our God . . ." Wiseman's interruption of the general in mid-speech is a clear violation of any assumption we may have made that the film is merely descriptive. The utterance of the word "God," and Wiseman's evident interruption of the general and termination of the film at this point (after a short montage of transitional shots), appeals for interpretation as seeming to assert, at the interpretive level, that there is something peculiar here, and further as asserting that the only way we are going to be able to make sense of Wiseman's gesture is to use it as a way of re-interpeting the film we have just seen by reflecting on it as a whole, and in a context of other texts. The text, of course, that Wiseman's placement and interruption of the invocation of God alludes to is in a way the definition of textuality itself for western culture, the foundation of hermeneutics, the assertion of the word itself in the Christian gospels: "In the beginning was the word, and the word was with God, and the word was God."[2] Wiseman ends by reminding us of the terms of our beginning, both forcing us backwards, again, through the film and forcing us to begin that backwards reading by acknowledging as its terms for doing so that the new beginning of our reading, the word, is not only an assertion of a beginning but the rehearsal of an ending, both the film's and, if the button is pushed, our own.

Wiseman's formal conceit, then, is to have created *Missile* as a mirror-structure, a deliberately non-preachifying sermon whose "text," in the sense of a scriptural passage that in a sermon would come at the beginning, comes instead at the end, so that we first encounter the film as an enigmatic description and then, on encountering its end, find ourselves reflected back into the text, to begin again. It is difficult to avoid the impression that Wiseman's *Missile*, therefore, is deliberately deconstructing itself, and deliberately asserting a doubleness of structure, in which the two texts, one forwards and one backwards, exist in a simultaneous, intertextual relation to one another. It is not, then, that the second, begun-again text, the text that we construct when we recall and reflect upon *Missile*, replaces the first, empirical and time-bound text, but that the two coexist, informing one another.

To have claimed, as we feel forced to by Wiseman's text, that it deliber-ately asserts not only its textuality but also its reflexive intertextuality with itself, may seem to have given us, at the outset, quite enough to untangle in a critical analysis. But it gets more, rather than less, complicated—and not, we think, simply as an exercise in academic deconstruction, but as a requirement for understanding how Wiseman's text presents itself as a rhetoric, how he engages us as participating spectators, and how he draws us into his own peculiar contemplation of the apocalypse.

The term *intertextuality* is used by Julia Kristeva to describe what she calls "the three dimensions of textual space" in the novel.

> These three dimensions of textual space are writing subject, addressee, and exterior texts. The word's status is thus defined horizontally (the word in the text belongs to both writing subject and addressee) as well as *vertically* (the word in the text is oriented toward an anterior or synchronic literary corpus).

Kristeva goes on to make it clear that the writing subject and the addressee are both, themselves, discourse, and that the texts implicated vertically comprise not only literary texts but the whole of culture.

> The addressee, however, is included within a book's discursive universe only as discourse itself. He thus fuses with this other discourse, this other book, in relation to which the writer has written his own text. Hence horizontal axis (subject-addressee) and vertical axis (text-context) coincide, bringing to light an important fact: each word (text) is an intersection of word (texts) where at least one other word (text) can be read.[3]

Figure 1. Intertextuality in *Missile*: Generic

|  | Strategic Air Command | *Films* Dr. Strangelove | Missile |
|---|---|---|---|
| *Locus of Discourse* | | | |
| The film | | | |
| Characters | | | |
| Technology | | | |

*Missile* is read as part of a genre of nuclear war films, represented here with two films that are closely linked structurally and ideologically—*Strategic Air Command* and *Dr. Strangelove.*

Reading across, the film is situated in the genre; reading up, the spectator attempts to understand three simultaneously occuring locations of discursive action: the technology itself; the world of the speaking/acting characters in the film; and the film itself as an interpretive action.

Any audience making sense of *Missile* is likely to do so by actively reading it as part of a multidimensional matrix of contexts.[4] Kristeva writes of Mikhail Bakhtin's challenging of the laws that govern texts and separate them from the world of action: "Bakhtin," she writes, "situates the text within history and society, which are then seen as texts read by the writer, and into which he inserts himself by rewriting them."[5] The text of *Missile*, we have suggested, is already a multiple text, reflexive and ironic. In addition, as indicated in figure 1, the film embodies at least three levels or loci of discourse: the technology itself (which is partly a discourse), the actions and words of the characters, and the film form itself.[6] This multiform film is read, as well, intertextually, in conversation with other films, both Wiseman's (see figure 2) and, for example, fiction films about nuclear war (see figure 1). These multiple forms are in turn in conversation with the contexts that the viewer brings to the film—cultural, ethical, political, and so on (see figure 3). Our reading of the film will try to suggest the interpretive actions that the spectator is invited to engage in, using these multiple contexts as a resource, both in performing a response to the film and, in so doing, becoming

Figure 2. Intertextuality in *Missile*: Auteur

|  | *Films* |  |
|---|---|---|
| *Locus of Discourse* | Wiseman's other works:<br>*Titicut Follies*<br>*High School*<br>*Basic Training*<br>*Sinai Field Mission*<br>*Manoeuvre*<br>etc. | *Missile* |
| The film |  |  |
| Characters |  |  |
| Technology |  |  |

*Missile* is read as part of Wiseman's entire body of work, taking his films singly and together.

its interlocutor, in something like the sense that Wayne Booth suggests in his *The Company We Keep: An Ethics of Fiction*. Booth argues that reading is an ethical act.

> My main effort is to find ways of talking about the ethical quality of the experience of narrative in itself. What kind of company are we keeping as we read or listen? What kind of company have we kept? [And, Booth seems to suggest, what kind of company are we, that is, who are we as we assume the values of a text even if only tentatively?].[7]

Booth's book seems likely to put discussion of ethics and audiences in a sense familiar to rhetorical scholars back on the critical agenda, and although he is careful to qualify his agenda by speaking primarily about the ethical dimensions of texts and their reading, we would like to suggest that, when we are viewing Wiseman's *Missile*, we are in the immediate presence of more than one sort of text: in addition to this film, and the films of Wiseman and others on which

Figure 3. *Missile*: Intertextual screens

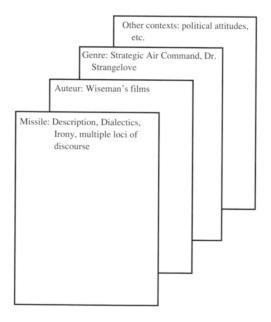

In the first screen, *Missile* is read as a dialectical text that is both descriptive and ironic, and that embodies its own discourse and/as a reading of the discursive actions of its characters and of technology. This dialectical text is read intertextually as part of the other screens: Wiseman's other work, other films and texts, and the contexts in which *Missile* is taken by the reader to be situated.

its reading relies, the film treats, as a text, the actions and words of its characters and the material circumstances in which they find themselves—in this case, the technology of nuclear war. In the view propounded by Booth, Kristeva, and Bakhtin, and, for rhetorical scholars, by Kenneth Burke, it is not just that we tell stories; we are stories.[8]

The notion of intertextuality has some linked advantages and disadvantages for rhetorical critics; it can be a way of attending to the as yet not very well articulated problem that texts offer themselves for interpretation but that readers arrive at their own, often contrary or idiosyncratic readings. Intertextuality offers a way of getting at this issue of varieties of readings, but, once begun, it also undermines any interpretation by asserting the instability of the text. Intertextuality is a dialectical, conversational process, rather than a simply comparative, confirmative one.

What does intertextuality do to the rhetoric of watching films? Let us glance at the method of reading developed by William Rothman in his *Hitchcock—The Murderous Gaze*. The intertextual grid evoked by Rothman in his reading of each of five Hitchcock films is the unfolding body of Hitchcock's work, regarded as informing the reading of any particular film.[9] In addition, although Rothman is scrupulous in following each of his five films sequentially, from beginning to end, and tracking not just the form of the film but the experience implied for the viewer, he is, at the same time, reading the film as an unfolding narrative and reading it with the privilege of knowing how it has already revealed itself to a viewer who has seen it through to the end. For example, Rothman opens his reading of *The Thirty-Nine Steps* with the first images of the film:

> *The Thirty-Nine Steps* opens with a panning movement along an electric sign. Against a black background, flashing letters spell out the words "MUSIC HALL," announcing a setting (one to which the film returns at its climax) and a metaphor (the mood and tone of the film and its organization as a series of "numbers" designed to entertain, continually draw inspiration from the music hall).
>
> The next few shots echo the introduction of *The Lodger*. They follow a mysterious figure—we have no view of his face—into a theater and end with him ensconced in his seat, back to the camera, ready for the show to begin. . . . But when a few moments later the camera frames him frontally, we discover, not Ivor Novello's anguished mask but the intelligent, humane face of Robert Donat, with its characteristic look of dispassionate amusement. . . . The break with *The Lodger* is declared.[10]

With these sentences Rothman opens his analysis of *The Thirty-Nine Steps* by asserting the immediate experience of seeing what is on the screen as a lighted sign spells out the words "MUSIC HALL," but then he adds a parenthesis that looks to the end of the film by revealing that the film will return to this setting at its climax. He tells us that the sign announces both a setting and a metaphor—the metaphor of the film itself as a series of music-hall turns. In this single, opening sentence, Rothman has positioned his criticism in a peculiar way. He clearly takes it to be important that the film reveals itself to us in sequence;

equally important, he comments on the sign as realistically motivated—that is, it is a picture of something that is present to the characters in the world of the fiction. But he then asserts a symbolic, metaphorical reading of the sign, a reading that could not be made without a knowledge of the film as a whole and that is not realistically motivated but formally motivated—it is a symbol that is not available to the characters in the film and is available to us, as viewers, only if we reflect on the film as a whole. The sign can be a metaphor at the outset of the film only if we are not beginners, and only if we are willing to include in our conversation about the film both its narrative unfolding from moment to moment in time, and its symbolic implications as a whole and in context. With this move, Rothman addresses the text of *The Thirty-Nine Steps* as being in an intertextual relation with itself; it is simultaneously what Peter Rabinowitz calls the text before and the text after reading, the text as a story and the text as a formal design and a philosophical reflection. Rothman then immediately asserts a further grid of intertextuality by invoking our (potential) awareness of Hitchcock as the auteur and implied author not only of this film but of a body of films forming a larger text.

Rothman does not simply compare *The Thirty-Nine Steps* with *The Lodger*; rather, *The Lodger* becomes part of the text of *The Thirty-Nine Steps* at the same time that Rothman re-positions the spectator back in the film and back in this moment in time: "From this moment on," he says, "we accept *The Thirty-Nine Steps's* protagonist, Richard Hannay, as exempt from the lodger's anguish."[11] This sentence may be read simply as a critic's construction, that is, as merely asserting that what the innocent first-time viewer is seeing and feeling is different from what the viewer would see and feel in watching *The Lodger*, with no particular presumption as to whether he or she had seen it. But it could also be read, and Rothman seems to want us to read it, as asserting at least the legitimacy of an intertextualized reading, a reading richly appreciative of the experience of *The Thirty-Nine Steps* as realism, and, at the same time, as reflectively appreciative of the larger forms of the whole film and of the body of Hitchcock's work, of Hitchcock as his own creation and of ourselves, his viewers, as both exposed and recreated by Hitchcock.

What Rothman alleges about the intertextuality of the work of a single auteur, Stanley Cavell seems to allege about the interaction of works in a genre. In Cavell's *Pursuits of Happiness: The Hollywood Comedy of Remarriage*, it may be that the least interesting passages are those devoted to arguing that the comedy of remarriage does, indeed, constitute a genre.[12] Categorizing has always been the beginning, and too often the end, of genre-alyzing.[13] But Cavell does not stop there. Instead, having established the comedy of remarriage as a body of texts sharing themes and forms, he addresses them as constituting a serious philosophical conversation, into which he enters, respecting each film as an aesthetic experience in its own right, but also arguing that the experience of each of the films is part of the experience of the others, as well as being a transformation of Shakespearean romance and the history of philosophy.

Cavell and Rothman show us, in their detailed but also highly speculative readings, how intertextuality works at the level of the single film, of the film as part of the ongoing work of a single auteur, of the film as part of one or more genres, and of the film as intertextually implicated in the dialogue of the culture as a whole. And they show us how all these interacting levels of textuality free the film for reflection and discussion, as well as for a variety of interpretations.

We are going to argue that Wiseman's *Missile* presents itself as a self-conscious text, dialectically transforming itself, Wiseman's other films, the genre of nuclear films, and the culture of the bomb. Admittedly, such a critical position must be offered with some hedges. *Missile* is a documentary film, not a theatrical fiction feature, and so we have no wish to argue that it fits neatly into a simple category.[14] Rather, we shall argue that it presents itself to be read in relation to the genre—but not that it is entirely reducible to a single category. And, although we shall argue that *Missile* is self-conscious and reflexive, seeming to demand close study and reflection, the circumstances of its broadcast and distribution inhibit access to it as a text. Wiseman's *Missile* was shown on public television—in some markets, only once. Unlike a large body of comparably interesting fiction films, *Missile* is not available from video shops, nor is it rerun on television. Wiseman, who himself complained in the *New York Times* about the difficulty he had getting the film on the air, closely guards the distribution of his films, which are available for rental and lease only through his own company, Zipporah Films. And so he is put in the position of inhibiting, for reasons of business, the access to his films that his own urgency to get them broadcast and the complex design of the films seem to demand. We do not offer this as a complaint against Wiseman—he's got to make a living—but as an observation about the somewhat contradictory circumstances in which this peculiar filmic rhetoric is enacted.[15]

The contrary combination of its form and its circumstances of access make *Missile* a generically peculiar text. On the one hand, *Missile* appears to share formal qualities both with postmodern literary artifacts and with Hollywood films—rendering it a text to be viewed and reflected over at length. On the other hand, its ostensible subject matter and its actual circumstances of viewing give it something of the presence of a pragmatic public speech, existing in and for a particular moment in time, coming into being to do something rather than to be interpreted as a formal text. In *Missile*, each of these positions undermines the other in such a way as to reflect directly upon the film's meaning and effect.[16]

It seems to us that Wiseman's *Missile* not only reads itself, as we have alleged, but that it reads other films, and offers itself to be read in conversation with them. Two films, for example, seem to stand in clearly dialectical relation to each other and to *Missile*—Anthony Mann's 1955 *Strategic Air Command* and Stanley Kubrick's 1963 *Dr. Strangelove, Or, How I Learned to Stop Worrying and Love the Bomb.*[17]

*Strategic Air Command*, starring Jimmy Stewart and June Allyson, is an ode to the Strategic Air Command in the context of a domestic melodrama. *Dr. Strangelove* answers *Strategic Air Command* by turning it into comedy, intrigue, and disaster; in significant ways, *Dr. Strangelove* is an interpretation of *Strategic Air Command*, almost a remake in other terms, in something like the way that Martin Scorsese's *King of Comedy* could be understood as a comic remake, a transformative re-reading, of his earlier, darker *Taxi Driver*.[18] One might expect, knowing Wiseman's work, his taste for the oppressive, peculiar, and surrealistic in American institutions, that if he undertook to make a film on a nuclear missile launching facility, he might produce a remake of *Dr. Strangelove* that drew on his own earlier films, such as *High School* or *Welfare*, to emphasize bureaucratic bullying, bickering, boredom, and incompetence. Instead, Wiseman's joke is to make a film that is not only less zany and less explicitly critical than *Dr. Strangelove*, but that is less dramatic than the rather plodding *Strategic Air Command* and all the more chillingly comic because of its surface normality.

*Strategic Air Command* puts James Stewart into the familiar role of the reluctant hero. A professional baseball player, he is called back to active duty in the Air Force to become a pilot in the new program of strategic nuclear deterrence. *Strategic Air Command* is cold war with the *Good Housekeeping* seal of approval. Stewart and his new bride, spunky June Allyson, resent the way the Air Force tears him away from career and family. He goes, but reluctantly, at first, until he rediscovers the pleasures of flying and of patriotism, while trying to prevent Allyson's romantic and domestic longings from distracting and embarrassing him. He has overcome his reluctance, but on his own terms, and when an arm injury grounds him, he resigns from the Air Force rather than accept a desk job. Stewart's reluctance to be called up from the reserves, complicated by the job of having to manage June Allyson, is a tricky role that perhaps only Stewart could have pulled off; he is perhaps the only heroic American film actor who can whine without seeming spineless.[19]

Robert Ray, in *A Certain Tendency of the Hollywood Cinema, 1930-1980*, provides a compelling analysis of the ideology of the reluctant hero in American film. According to Ray, the reluctant hero story is, in

> its most typical incarnation, . . . the story of a private man attempting to keep from being drawn into action on any but his own terms. In this story, the reluctant hero's ultimate willingness to help the community satisfied the official values. But by portraying this aid as demanding only a temporary involvement, the story preserved the values of individualism as well.[20]

For Ray, what is crucial about the myth of the reluctant hero is that it acts as an ideological device to suggest that we can avoid having to choose between individualism and the needs of the community.

How do Stanley Kubrick's characters compare to the reluctant—but steady and competent—James Stewart? In George C. Scott's General Buck Turgidson, reluctance is replaced by overeagerness; in Sterling Hayden's General Jack D. Ripper, individualism is transformed into madness.

If we look at *Missile* for clues as to the issue of reluctance, especially in the light of Ray's depiction of the hero's motivation as an ideological account of choice, we see a striking pattern. Wiseman depicts the officers of *Missile* as sober professionals working their way through a rational career path. The issue for them is not whether they are overeager or reluctant so much as whether they are competent, whether they can make the grade in the system—and, if they have difficulty, the system seems determined to do everything it can to help them pass the tests. James Stewart's reluctance seems to be a reasonable problem in the context of the Hollywood paradigm, because we have seen it so often. But in the context of *Missile*, it is clear that any sign of reluctance would be enough to disqualify an officer from further participation in the system. In the opening scenes of the film, the officer-trainees participate in an orientation session and what is called a "professional responsibility seminar." In the first scene, Colonel Jim Ryan introduces himself and tells the officers that the training program for launch officers of the land-based Minuteman ICBM is a tough one, leading to a "challenging career" (scene 1).[21] The next day, Ryan introduces what is called a "professional responsibility seminar":

What we're going to do today is talk about the awesome responsibilities that you're going to have as crew members. And we're going to have a little seminar and hopefully we'll get good dialogue going here on, you know, the moral responsibilities. And then at the end of this day, we're going to ask you to sign a piece of paper that says that you have thought of all the moral implications about inserting launch keys, and that you have no reservations. That if the President of the United States deems that our way of life is threatened, and that it's about to be over, that you have no hesitation, once that you have authenticated the message and know that it's the president talking, that you insert those launch keys and launch your missiles. And you know full well the consequences of launching those missiles, which are equipped with nuclear warheads, and the great devastation that that will bring. And we want you to think about that. We don't want you to capriciously go through this program and be robots in inserting launch keys. We want you to fully comprehend the awesomeness of this responsibility. (scene 2)

In the intertextual context of *Strategic Air Command* and *Dr. Strangelove*, does the marked absence of the reluctant hero story in *Missile* indicate that these officers have found a way to restore choice to the roles that they play? On the surface, yes. Each of the officers, we are told, is to accept the responsibility to launch only after having first reflected on what that choice means. But in the context of the film, that choice is made irreversibly as the price of continuing with this career. At the end of the day, they will sign a piece of paper, in effect an informed consent form, that says they have already made the choice to launch. Once that paper is signed, there is no more choosing; there is only the response to the orders of the president. However reasonable and necessary this system of informed consent may be as an operational matter, it has clear, perhaps primarily metaphorical, consequences for the depiction of the culture of the bomb.

The peculiar world of *Missile* is one in which there is a seminar with no disagreement, a career of performance but no action. The great action of which all these officers are a part is the action of launching nuclear missiles, but the film and the training program make it clear that although these men and women are willing participants, they have also willingly put themselves in the position of waiting for the word to come to them from elsewhere. As one watches the film, it is hard not to feel that we are all in the same position as these officers; we give a sort of consent to the launching of missiles, and yet we are not in any realistic sense part of a decision to launch. It is up to us to reflect whether, unlike the officers in the film, it is even possible for us to discuss the bomb meaningfully.

Deterrence has become a commonplace that is seemingly beyond disagreement in mainstream American rhetoric, just as it is for the officers in *Missile*, and this despite the expensive and unresolved debate about the Strategic Defense Initiative as an alternative to deterrence. What is the rationale for deterrence as a practice in *Missile*? In this opening seminar, it appears in all its contrariness: we will launch the missiles, on the order of the president, if it is determined that our way of life is about to end. Of course, deterrence has already failed if that point is reached; it is hoped that the threat of a response will act as a sufficient deterrent. To be credible, the threat must be supported by an evident ability and willingness to launch our missiles in response to a nuclear attack.

The logic of this deterrence, although it is stated by the officers in the film as part of a historical continuity, is a sharp break with earlier strategic concepts in popular thought. The missiles are so destructive that they would not be useful as weapons if they were not a code, just as we believe they would not be a useful code if they were not weapons. But the nuclear situation turns war as an instrument of policy into a new domain of paradox. War, if it is not conducted to send a message, is simply murder. But if one country destroys another to revenge a nuclear attack, is that not simply murder, too? Consider Van Johnson, standing on the deck of a carrier steaming toward Japan in *Thirty Seconds Over Tokyo* (1944). Johnson discusses with his fellow B-25 pilot Robert Mitchum whether they feel any hesitation about bombing Japanese cities, with inevitable loss of civilian lives. He says, "It's a case of drop a bomb on them or pretty soon they'll be dropping one on Ellen [his wife]." Johnson states a sort of deterrence theory that aims to prevent damage to one's own country by an enemy in a war situation. The theory of nuclear deterrence is that we would use strategic nuclear weapons only if we were attacked by strategic nuclear weapons. The bombings of Hiroshima and Nagasaki, of course, were subject to no such constraints. The only time that nuclear bombs have been employed in war, they were not part of the doctrine of nuclear deterrence, but an extension of the World War II concept of strategic bombing.

The theory of nuclear deterrence as it is now stated is that it is designed to prevent strategic nuclear weapons from being used by either side. That concept of strategic deterrence was seen, from before the production of the first nuclear bomb, as the only way to control it—and, significantly, as being required not simply as a matter of policy, but as a necessary outcome of the existence of the technology. In his history of the bomb, Richard Rhodes writes that as early as 1940, long before a sustained chain reaction had been achieved, allied scientists were predicting that "nuclear weapons would be weapons of mass destruction against which the only apparent defense would be the deterrent effect of mutual possession."[22] The technology itself, then, is seen as dictating policy, not just constraining choice but eliminating it.

*Missile* echoes another aspect of choice in the nuclear age, whose technology dictates a system of control. From the beginnings of nuclear weapons development in America, as Richard Rhodes tells the story, the decision to develop and deploy the bomb was the president's alone, as commander in chief.

> Thus at the outset of the U.S. atomic energy program scientists were summarily denied a voice in deciding the political and military uses of the weapons they were proposing to build. . . . A scientist could choose to help or not to help build nuclear weapons. That was his only choice. The surrender of any further authority in the matter was the price of admission to what would grow to be a separate state with separate sovereignty linked to the public state through the person and by the sole authority of the President.[23]

Wiseman's *Missile* looks closely at the effect of the system of nuclear control on culture at the microcosmic, institutional level, and at the microcosm as a metaphor for the culture as a whole.

Consider, for example, the language used by the officers. The launch procedures are clearly conditional, to be employed *if* the president authorizes a launch, but the language of launching is almost always in the simple future tense. In scene 2, Colonel Ryan is discussing the moral responsibility of nuclear missiles, which he contrasts to the holocaust. Then he says, without intentional irony,

> Okay now, what makes this world that we're in very, very complex is that as you go down in the bowels of the earth every night, you may not get an opportunity as such to evaluate the situation. Although, you're going to be getting intelligence reports; you're not going to be working completely in a vacuum. You're not, you're not going to get that klaxon at night that says okay, our way of life is threatened—probably—insert your keys and launch your weapons. You're going to be getting the intelligence buildup and so forth that goes along with that. . . . You have to, as you go through our program, decide in your own mind that, that our way of life is such, and that our command and control system on our weapons system is such, that the President of the United States is not going to ask you to insert those launch keys until, you know, there is just no other option—it's the final solution.

The use of the phrase "final solution" to describe a missile launch may be appropriate, but it seems at least historically tone deaf in a rhetoric that seeks to inculcate a notion of moral responsibility among officers.

In scene 5, an officer again describes the unit's mission in the simple present tense:

> INSTRUCTOR: What is the mission of the Minuteman weapons system?
> STUDENT: The mission, on page one dash three, states: the mission of the Minuteman hardened and dispersed weapons system is to deliver thermonuclear warheads against strategic targets from hardened underground launchers in the continental United States.
> INSTRUCTOR: When the time comes to launch your missiles against targets as designated by higher authority. . . . Once we get the word to launch it should not take that long to launch your missiles, as opposed to Titans. Titan, in a word, is a volatile system. If you drop a wrench you could launch a missile, something like that. No, I'm just kidding.

In scene 11, an instructor says, "That's the only time that a deliberate release is authorized. When we have received emergency war orders that are authorizing the release of nuclear weapons from a competent authority, then that's when we'll be able to do it." In all of these scenes, the assurance of safety and of presidential control is offset by the assumption that sooner or later the missiles are going to be launched. It is only a question of when. Of course, it could be argued, and perhaps should be argued, that this use of the future tense is only a figure of speech, perhaps designed to reinforce the continuing readiness of our deterrent forces to act if necessary. Perhaps their morale requires such a support, simply as a way to stay used to the idea that they might one day get the order to turn their keys. But it is hard not to reflect that the officers controlling our nuclear weapons systems, from those with their hands on the keys up through the chain of command to those who advise the president, speak of nuclear war as a certainty. In making deterrence credible, we have created a military culture that speaks of its failure as inevitable.

As if to prompt our worries, Wiseman inserts an ambiguous pair of scenes just after the professional responsibility seminar of scene 2, in which Colonel Ryan links deterrence and the final solution. In scenes 3 and 4, the students are instructed in the use of Smith and Wesson Combat Masterpiece revolvers—scene 3 in the classroom, scene 4 on the firing range. The first words we hear, after Colonel Ryan's reassurances in scene 2, are from the shooting instructor, who says, "Never point a weapon at anything unless you intend to shoot it." In the context of a speculation about whether these missiles are ever likely to be fired, the balanced assurances on the one hand that deterrence will work and the repeated assumptions on the other hand that nuclear war is only a matter of time, it is hard not to hear the shooting instructor's words about these handguns as a metaphor for the nuclear missiles—which are pointed and ready to shoot. Is their training contradictory, or are these officers being trained to point the missiles, and must therefore be understood as intending to shoot them? How are we to interpret this handgun training? Is the shooting instructor's remark equivalent to the famous line of Walter Brennan as Ike Clanton in John Ford's *My Darling Clementine*: "When you pull a gun, kill a man"? The two lines are different, but perhaps too subtly so: Ike Clanton's line is a prescription, the shooting instructor's a proscription, but the difference may be lost on those who speak and those who hear the lines. Again, is the training simply for morale purposes? Who are these officers going to shoot, once they are locked up in their underground launch capsules? And why? Those pistols have been seen before in Hollywood films. In *A Gathering of Eagles*, Rock Hudson, as an Air Force officer, shows a group of visiting dignitaries a pistol that stands, apparently, as a guarantee of safety for SAC officers—perhaps to ward off intruders. But in the more recent *WarGames*, a prologue to the film shows one officer threatening to shoot his team partner who is refusing to turn his key on the command to launch. Because Wiseman provides us with no explanation, we are left to rely on the evidence of the film itself and on our experience of war movies.

One of the most peculiar features of *Missile* is Wiseman's self-imposed practice of staying within the confines of this single institution, in marked contrast to classical narrative technique. The film begins by entering Vandenberg Air Force Base and all its scenes are enacted there, mostly in classrooms, offices, barracks, corridors, and launch control capsules. Even when we see a backyard barbecue and a softball game, the fact that we are with a small group of familiar people sustains the impression of a tightly constricted spatial perspective. In *Strategic Air Command,* we find ourselves in a variety of locations spread over a wide geographical area: a ball park, two homes, in various airplanes, at or over Alaska, Greenland, and Japan, and in a variety of locations on a military base. The impression created is one of mobility over vast spaces and of the distance this puts between James Stewart and June Allyson, at the same time that it testifies to the global reach of the *Strategic Air Command* and to the ubiquity and omniscience of the narrating camera. In the typical Hollywood fashion, we are able to intercut between two locations to indicate parallel and simultaneous lines of action, a practice that Wiseman denies himself.[24]

In *Dr. Strangelove*, we also visit a number of locations in the nuclear chain of command—Burpelson Air Force Base, from which General Jack D. Ripper (Sterling Hayden) initiates the unauthorized attack on the Soviet Union; the B-52 bomber piloted by Major T. J. "King" Kong (Slim Pickens); General Buck Turgidson's bedroom; and the war room at the Pentagon, from which the President (Peter Sellers) tries, with the aid of his advisors, to abort the attack. Kubrick cuts freely among the three major scenes of action—the base, the bomber, and the war room—often using his power as the narrator to observe simultaneous lines of action when the people involved are unable to communicate with one another—this is especially the case with the conference in the war room, which is unable to contact the men on Major Kong's B-52, though we can move freely between the two locations.[25]

The effects of Wiseman's strategy of narrative constriction and confinement are subtle and pervasive. The film creates an abiding sense that we are observing a culture no longer in charge of its own most important actions and therefore as having relinquished its ability to take action. Everything hinges on something that will happen elsewhere and on a word that will be given from elsewhere. At some point, the president will make a decision, the klaxon will sound, and the word will be passed through the system—a word that this culture is entirely structured to wait for and act upon. The job of the participants is to be prepared, both morally and technically, to do the bidding of that word. Having made their choice, they are not asked to choose again, are not asked to debate, are not asked for their advice.

More than in any other of his documentaries Wiseman visibly makes absence his subject, thereby establishing the philosophical grounds of the film and reflecting once again on the nature of his medium, for, if film is the most concrete form of documentary, it is the one that most clearly establishes the rule:

you cannot take a picture of the subject of a documentary. The subject is always absent.

In choosing to locate his film in a single institution, Wiseman avoids the sort of evasion that can result from stitching together a variety of locations and times, and covering them with a voiceover narration, resulting in an implied sense that the omniscience of our point of view stands for the hope of orderly hierarchy and ultimate control. It is a common device of the nuclear war film to stage part of its action in a large room in which the status of the world is displayed on gigantic TV screens, so that even in the crisis there is at least a sense of being able to see beyond the present setting.[26] Wiseman's formal attitude forces us to look at the cultural circumstances of a particular time and place. And yet his attitude differs markedly from the radical filmmakers of the 1930s who thought that to show us problems and victims would prompt us to adopt solutions. Leo Hurwitz, for example, commenting on the goals of the left documentary of the Film and Photo League, said that "the world had to be shown what its eyes were turned away from."[27] In contrast to the confidence that 1930s filmmakers showed in the rhetorical effect of documentary concreteness, Wiseman uses the resources of documentary against itself. He imposes on himself and on us the rule of simply looking straight at what is here and now; what he finds, here and now, is evidence of absence. Wiseman himself has commented that *Missile* is a film about "the absent guest," leaving it unsaid whether, in the tradition of the genre, the guest is death, God, the president, the Russians, the filmmaker, or the audience.[28] Elsewhere, Wiseman has said that when he sought funding for *Missile* from the Program Fund of the Corporation for Public Broadcasting, one panel member who voted against him remarked, "Who cares about the training of these men and women? They're not the decision-makers."[29] In the context of the film as it was eventually completed by Wiseman, the reluctant panelist appears exactly to have missed the point. The point is not to offer reassurances that decision makers are somewhere in charge or even to imply that they are inept, but to contemplate the material reality of the situation the rest of us are in, far from the locus of decision making, wherever that may be. The method creates the peculiar sensation that perhaps, no matter what single location Wiseman had chosen to look at in the chain of command, he would have found this strange absence.

In the context that the films of nuclear fear have created, we have learned to feel no particular confidence that there is, somewhere, a center of decision making that resolves all doubts—though the instructors in *Missile* certainly talk as if there were such a place. Even when we turn to the testimony of those most likely to be part of the decision process, we find absence. Richard P. Feynman, a Nobel laureate in physics, a participant in the Manhattan Project that created the bomb, was appointed in 1986 to be a member of the presidential commission investigating the *Challenger* accident. Feynman, an experienced inquirer and a respected scientist, gave an account of his work on the commission as a long process in which he was unable to find out who was making decisions about a report to which he was expected to sign his name. "I got the feeling we were being railroaded: things were being decided," he says, "somehow a little out of our control."[30]

In *Missile*, Wiseman employs a formal device that deepens the film's curious sense of absence. Before and after all but two of the thirty-five narrative episodes in the film, which we have called "scenes," are sequences of transitional shots, which we have called "inter-scenes," since they go far beyond a merely transitional function. The inter-scenes combine visual and sound material that is simultaneously descriptive and metaphoric. The thirty-three inter-scenes include from one to eight shots, usually photographed at a considerable distance, of street traffic, airplanes flying overhead, the exteriors of buildings, the gates of the base, and so forth. These images provide a limited and ambiguous sense of spatial relationships among various base locations, a sense of context and place, and thus allude to the expository but ironic function served by the cross-cutting and war-room-overview techniques of *Strategic Air Command* and *Dr. Strangelove*. Sometimes that function is especially pointed, as, for example, when base signs are featured. Frequently an inter-scene ends by showing the exterior of a building followed by a scene that supposedly takes place in the interior of that same building.

Yet these inter-scenes are often far longer than they need to be to serve as either visual breaks between scenes or as spatial orientation. They slow the film down, especially in the middle section of *Missile*, when many of the scenes are less than two minutes long and some of the inter-scenes last more than thirty seconds. Although the images in the inter-scenes often include motion (for instance, tracking shots of airplanes or automobiles), it is a strangely purposeless motion—from nowhere and to nowhere that can be determined. The images of roads and traffic create a pattern of connectedness that is somehow simultaneously isolated. We cannot see the people who are piloting the planes or driving the trucks and cars and buses. Rarely is a human figure pictured in an inter-scene shot. The guards at the gate to the base are an exception, and the sight of a man jogging or riding a bicycle is an anomaly in this barren environment. In some of the inter-scenes photographed indoors and, therefore, taken from a closer distance, the people standing in the corridors outside the classrooms are backlit and thus pictured as anonymous silhouettes, shadows, their voices overlapping and indistinct.

The inter-scenes emphasize the paradox of routine danger and the haunting sense of absence that pervades the film. The flashing red lights over the doors of the underground training facility seem not so much a danger signal as simply part of the expected visual environment. One inter-scene includes three shots of a burning building. The fire seems unattended and, in one shot, a jogger moves through the foreground between camera and fire. That inter-scene follows a launching practice that ends with the instructor saying, "Just get the task done and leave it at that. Get a successful launch and take care of any other problems that happen afterwards."

There are many indications throughout *Missile* that despite the apocalyptic mission of its inhabitants, the base is a microcosm of average American society. The inter-scene images of barracks, a movie theater, classrooms, and a church emphasize this impression, as does the final speech of the film when a general speaks proudly of the base record of community involvement and claims "we're devoted fathers and mothers." The last image of *Missile* is of a school bus moving away from the camera.

Wiseman's close observation of reality goes beyond description to achieve something akin to the sort of detached engagement sometimes attributed to absurdist drama. Wiseman has often spoken of his admiration for the work of Eugene Ionesco and has referred to his own work as surrealist, saying that he finds surrealism in the normal. David Grossvogel, in a passage on Ionesco, writes that "As defined by Andre Breton, surrealism is the refusal of an individual existence to submit to the posited limitations of existence. This refusal supposes a program of both active destruction and original creation. . . . It represents the modern poetic craving, the desire by the poet to express the unexpressed and the inexpressible, the desire by man to grow beyond himself. This beyond is to be sought within the everyday realm, in the objects, the circumstances, the familiar patterns that are to be 'defamiliarized.'"[31]

Wiseman's films, despite their disavowals of simple description and their echoes of existential "engaged detachment," retain a striking similarity to the aspirations of postmodern ethnography. The anthropologist Paul Rabinow, following Foucault and Rorty, writes that "we need to anthropologize the West," a project that would result not simply in description, but in a sort of detached self-knowledge, something akin to defamiliarization, or, in Kenneth Burke's phrase, "perspective by incongruity." To "anthropologize the West," writes Rabinow, would be to "show how exotic its constitution of reality has been; emphasize those domains most taken for granted as universal (this includes epistemology and economics); make them seem as historically peculiar as possible; show how their claims to truth are linked to social practices and have hence become effective forces in the social world."[32]

But despite strong intertextual echoes, Wiseman is neither an absurdist dramatist nor an ethnographer. His documentaries operate by a method of detached, self-effacing, and deniable subversion, refusing to argue, and instead offering a defamiliarizing gaze that is constantly surprised by the familiar. Hence, in examining, through Wiseman, the culture of the bomb, we all risk what may be, from the culture's point of view, a potentially fatal error—the subversive error of seeing it as, after all, merely a culture, rather than as a kind of well-adjusted pragmatic truth or a universal necessity.

What of the ethics of the audience? That the public is disenfranchised by the nuclear situation now seems obvious. But to argue, as we think it fair to do, that the audience of this film may be placed in a subversive situation is, quite possibly, a version of self-congratulation. The subversive, existential, anthropologizing doubt that we have described may actually, if it is itself anthropologized, be seen as existing in a condition of bad faith, for, if there are going to be missile launch officers we presumably do not want them to be subverted as we have, through this reflection on Wiseman's *Missile*, found ourselves subverted. Wiseman shows us that we place ourselves in a condition of absence.

For these people of the bomb, who stand, perhaps, as a metaphor for the condition of our culture, motivation is constituted by an absence, and that absence, we seem to be told, is represented equally by God (who ends the film) and the word (which will end the world).

Figure 4. List of Scenes and Inter-Scenes

Time is given in (minutes:seconds), rounded off to the second—that is, not counting frames or fractions of seconds.

Title: MISSILE. White letters on black background; the title fades in from black and out to black (0:5); the screen stays black for about 1 second before the first shot of the film; ambient sound of road noises comes up with the title and is continuous throughout the discontinuous shots of the first inter-scene; helicopter sound fades in and out as the helicopter passes overhead.
Inter-Scene 1 (8 shots) (0 minutes:35 seconds)[33] Shots of SAC and Vandenberg AFB signs, cars entering base; entrance guards; sign: Welcome to Space & Missile Country; helicopter passing overhead, pan down to exterior of frame building.
Scene 1: Welcome of trainees by commander, Colonel Jim Ryan. (2:09)
Inter-Scene 2 (5 shots) (0:26) Shots of buildings; cars moving along roads.
Scene 2: Professional responsibility seminar. (15:47)
Inter-Scene 3 (5 shots) (0:25) White frame building, airplane in sky, cars moving, missile on display.
Scene 3: Weapons lecture. "Never point a weapon at anything unless you intend to shoot it." Demonstration and safety lesson for M-15 Smith and Wesson Combat Masterpiece revolver. (3:49)
Scene 4: Firing range. Students practice firing revolvers. (3:25)
Inter-Scene 4 (4 shots) (0:28) Airplane in sky, yellow building, sign, entrance of building as door closes and cut to:
Scene 5: Instruction session. Students introduce themselves. (6:05)
Inter-Scene 5 (2 shots) (0:06) Students in a hallway; babble of voices, young man walking through door into hallway.
Scene 6: Classroom. Instructor tells students to respect safety rules; they are more valuable than the equipment. (0:30)
Inter-Scene 6 (2 shots) (0:06) Corridor, people in background.
Scene 7: Classroom. Instructor explains launch control panel. "It takes two launch modes to launch a missile." (3:19)
Inter-Scene 7 (2 shots) (0:13) Corridor; backlit figures in silhouette.
Scene 8: Classroom. Instructor from scene 6 describes critical errors in simulated flights. (1:52)
Inter-Scene 8 (5 shots) (0:20) Truck; jet plane in sky; cars at intersection; person entering building with a warning sign; interior corridor with muzak playing and red light flashing over a door.
Scene 9: Instructor and flight team; instructor tells two students that they are going to be a team and will be graded as a unit during the program. "That's my philosophy, okay?" (1:31)
Inter-Scene 9 (6 shots) (0:17) Corridor; red lights flashing over doors in medium and close shots; signs; one reads QUIET TRAINING IN PROGRESS.
Scene 10: Clock-reading instruction; in launch control room, instructor teaches two students how to read a clock. (1:41)
Inter-Scene 10 (3 shots) (0:27) Corridor with red light flashing; exterior panning shot of moving helicopter; car passing a building.

Scene 11: Classroom; instructor from scene 7 on weapons safety rules. "That's the only time that a deliberate release is authorized: when we have received emergency war orders that are authorizing the release of nuclear weapons from a competent authority. Then, that's when we'll be able to do it." (1:30)

Inter-Scene 11 (3 shots) (0:15) Red sports car passing building; moving van on highway; plane in flight.

Scene 12: Picnic. (0:52)

Inter-Scene 12 (2 shots) (0:10) Cars driving; sign outside building.

Scene 13: Classroom; instructor from scenes 7 and 11 introduces the personnel reliability program. (1:44)

Inter-Scene 13 (5 shots) (0:21) Building; road with fire engine and passing cars; jogger; cars at intersection; house.

Scene 14: Student review session. Five male students study for exam. (1:42)

Inter-Scene 14 (6 shots) (0:43) [all night shots] Cars driving; cars at entrance gate; woman guard saluting; cars driving.

Scene 15: Threat simulation. White female student gets crank telephone call from supposed terrorist with 25 sticks of dynamite. (1:22)

Inter-Scene 15 (4 shots) (0:23) Corridor, muzak playing, red light flashing; building; garbage truck; yellow building.

Scene 16: Classroom. Instructor from scenes 7, 11, and 13 lectures on the effect of weather on missiles in flight. (2:23)

Inter-Scene 16 (5 shots) (0:19) Missile on display; street sign; cars passing building; man entering building; empty corridor with flashing red light, muzak.

Scene 17: Team simulation. Two students rehearsing procedures as instructor watches. (2:09)

Inter-Scene 17 (6 shots) (0:44) Corridor with flashing red lights; man leaving building; street scene with traffic; missile cone on flat-bed truck; tracking shot of helicopter in sky; classroom building.

Scene 18: Classroom. Instructor from scene 6 and 8 discusses advantages of "soloing" in contrast to leaning on a strong partner. (1:19)

Inter-Scene 18 (4 shots) (1:44 including intermission) Classroom building; INTERMISSION (1:00); street scene with pickup trucks; small building; sign.

Scene 19: Instructors' conference. Jim Ryan discusses a student who has bad debts in his past; Ryan doesn't think he should be pulled from training while the matter is checked; says all of them will have debts sometime in their lives, especially when they send their children to college. (1:33)

Inter-Scene 19 (3 shots) (0:23) Man riding bicycle; plane flying overhead; classroom building.

Scene 20: Young black male instructor advises other instructors [?] about the dangers of fraternization [with female students?]. "It's something the Air Force doesn't like and won't tolerate. . . . We're not going to get paranoid to that point, because it's not worth it. But we do want to make sure that we always be professional." (1:07)

Inter-Scene 20 (4 shots) (0:14) Yellow classroom building; cars on road; signs; small white building.

Scene 21: Officers' conference. Discussion of student who is having trouble in the program; the student will be given a seminar in taking tests. (1:08)

Inter-Scene 21 (4 shots) (0:14) Frame building, man walking by; building; sign on exterior wall; corridor with red lights over doors.

Scene 22: Launching Practice. Same team, same instructor as in scene 17; student still has a hard time reading. The team practices the launching routine by repeating commands exactly; instructor mentions how "very structured" these commands must be: "Hands on keys"; "light on; light off"; "release keys." Instructor: "You have launch indications." (4:29)

Inter-Scene 22 (6 shots) (0:24) Building; long shot of burning building; closer shot of burning building; burning building as jogger passes by, not paying any special attention; cars on the street, a plane passes overhead; yellow classroom building with door closing.

Scene 23: Classroom. Instructor from scenes 7, 11, 13, 16 says, "The only time you're gonna be sending PLCAs is when we're gonna go to war. . . . We have no determination of what kind of war we fight." (2:06)

Inter-Scene 23 (1 shot) (0:02) Corridor.

Scene 24: Corridor; students discuss searching for war artifacts. (5:22)

Scene 25: Classroom. Instructor from scenes 7, 11, 13, 16, 23 describes the inhibit code. (1:23).

Inter-Scene 24 (6 shots) (0:37) Classroom building; cars on street; sign; barracks.

Scene 26: Testing instruction. Instructor gives student "last-ditch tips" on testmanship for multiple-choice exams. (3:37)

Inter-Scene 25 (2 shots) (0:06) Barracks; sign in front of building (dialogue from next scene overlaps).

Scene 27: Student evaluation session. Officers recommend to Jim Ryan that a student be eliminated from the program after failing to come up to performance standards after 107 hours of individual instruction. Ryan: "It's not fair to the crew force to put somebody out there who can't hack it, that's all." (2:11)

Inter-scene 26 (1 shot) (0:03) Small white building; bird call overlaps from inter-scene to:.

Scene 28: Softball game. (1:45)

Inter-Scene 27 (4 shots) (0:16) Truck driving on road; sign; man entering building; corridor with muzak and flashing red light.

Scene 29: Student on telephone reporting a fire. (0:51)

Inter-Scene 28 (2 shots) (0:08) Large building; yellow classroom building.

Scene 30: Classroom. Instructor from scenes 7, 11, 13, 16, 23, 25 lectures on weapons system safety rules. (1:58)

Inter-Scene 29 (3 shots) (0:12) Missile on display; road; frame building.

Scene 31: Conference room. Jim Ryan tells a group of officers that the maintenance branch has complained that instructors are not taking proper care of the place—cigarettes, pop cans, coffee spills. Ryan doesn't "like this kind of unpleasantness." (1:49)

Inter-Scene 30 (3 shots) (0:10) Frame building; jet plane in sky; building.

Scene 32: Examination: Simulated ICBM launch. Two female students practice launching missiles. Key turns. "They're all gone, Debbie. That's it." (4:28)

Scene 33: Evaluation. Two female students are told that their simulated launch was an outstanding performance. "Congratulations. Welcome to the Minuteman Crew Force." (4:11) (This scene starts as a hand-held camera follows the students and instructor down the corridor and into the room where the scene immediately begins, so there is a transition effect but no formally separate inter-scene)

Inter-Scene 31 (5 shots) (0:26) Two people leaving a building; roadside with passing cars; men in group talking; roadside with cars; exterior of chapel.

Scene 34: Chapel. Memorial service for the Challenger disaster of January 28, 1986, which occurred about a week before this scene was filmed. After a minister delivers a eulogy, Jim Ryan leads the singing of a hymn. "Taps" is played; the music continues into the following series of transitional shots. (4:37)

Inter-Scene 32 (4 shots) (0:20) [sound of taps continues through the first three shots] Exterior of chapel; trees by roadside; cars on road; building with sign: Vandenberg Center.

Scene 35: Speech by general. "We do what our elected leaders tell us to do. . . . We give confidence to our national leaders [and to the Soviet Union that] we're not a nation to be treated lightly." "We are a people who are concerned about our God." (6:30)

Inter-Scene 33 (5 shots) (0:21) Center building; sign; school bus traveling down a road; street sounds continue through the credits, which include this screen:

*Thanks to the men and women at*
*the 4315th Combat Crew Training Squadron of*
*the Strategic Air Command*
*at Vandenberg Air Force Base*
*for their cooperation and to*
*Donald Baruch*
*for his assistance over the years.*

<h1 align="center">NOTES</h1>

1 *Missile* was filmed at the 4315th Training Squadron of the Strategic Air Command at Vandenberg Air Force Base, California, in January and February 1986. Wiseman produced, directed, and edited the film; John Davey was cinematographer. *Missile* was shown at the U.S. Film Festival, Park City, Utah, in January 1988. It was first broadcast on the Public Broadcasting System on 31 August 1988. The copyright date of the film is 1987. According to a *Boston Globe* story, a grant of $171,172 from the MacArthur Foundation partially supported the costs of the film with money from a series of grants devoted to projects promoting world peace. The award was announced on 28 January 1988. ("Peace Offerings," *Boston Globe*, 29 January 1988, 2).

2 John 1:1.

3 Julia Kristeva, "Word, Dialogue, and Novel," *Desire in Language: A Semiotic Approach to Literature and Art*, ed. Leon S. Roudiez; trans. Thomas Gora, Alice Jardine, and Leon S. Roudiez (New York: Columbia University Press, 1980), 65-66. On intertextuality, consult Christopher Norris, *Deconstruction: Theory and Practice*, revised (London: Methuen, 1986), 114-115; Peter J. Rabinowitz, *Before Reading: Narrative Conventions and the Politics of Interpretation* (Ithaca: Cornell University Press, 1987), 69-75; Roland Barthes, "From Work to Text," in *Textual Strategies: Perspectives in Post-Structuralist Criticism*, ed. Josue V. Harari (Ithaca: Cornell University Press, 1979), 76-77; Jonathan Culler, *On Deconstruction: Theory and Criticism after Structuralism* (Ithaca: Cornell University Press, 1982); John Fiske, "British Cultural Studies and Television," in *Channels of Discourse*, ed. Robert C. Allen (Chapel Hill: University of North Carolina Press, 1987), 254-289; Mimi White, "Television Genres: Intertextuality," *Journal of Film and Video* 37, no. 3 (1985): 41-47; Mimi White, "Crossing Wavelengths: The Diegetic and Referential Imaginary of American Commercial Television" *Cinema Journal* 25, no. 2 (1986): 51-64; Scott R. Olson, "Meta-television: Popular Postmodernism," *Critical Studies in Mass Communication* 4 (1987): 284-300.

4 There is nothing in the film that would, on its face, offend a patriotic American middle-aged couple with a son or daughter in the Air Force and who happened to be watching public television on an August evening in 1988. In the intertextual view we propose, contrary readings of a text by hypothetical or various actual audiences themselves become part of the intertextual matrix, become part of each others' texts.

5 Kristeva, *Desire in Language*, 65. See M. M. Bakhtin, *The Dialogic Imagination*, ed. Michael Holquist, trans. Caryl Emerson and Michael Holquist (Austin: University of Texas Press, 1981).

6 Figures 1, 2, and 3 are meant to suggest the intertextual relations informing *Missile*. But the limits of graphic representation may make the figures misleading. The boxes are not meant to represent a matrix of separate and exclusive categories so much as sites from which to view the texts sited at the other locations—and the still other texts of which they are merely the examples chosen for representation here. Hence, the reader might begin by imagining the point of view taken in the upper right-hand square of figure 1—the text of *Missile* itself. That text is, at one moment, imagined as self-contained and as bounded with reflecting surfaces that re-present the reader to him or herself and that rebound with ironies and reinterpretations. But this square, and all the others here represented, is permeable as well; depending on the experience and inclination of the viewer, *Missile* is read not only as a text-in-itself, but vertically takes in the text that is the discourse of its characters and the text that is the material culture represented in the film, chiefly the bomb as a text and a part of material culture. Similarly, *Missile* is read horizontally calling (in figure 1) on our knowledge of *Strategic Air Command*, *Dr. Strangelove*, or any of the large body of generically related works or (in figure 2) on any one or all of Wiseman's other works, which *Missile* transforms and re-reads, and which may itself be re-read in the context of Wiseman's earlier (and eventually later) work.

7 Wayne C. Booth, *The Company We Keep: An Ethics of Fiction* (Berkeley: University of California Press, 1988), 10. For a review of other recent work on relations between speaker and audience, see Thomas W. Benson, "Rhetoric As a Way of Being," in *American Rhetoric: Context and*

*Criticism* ed. Thomas W. Benson (Carbondale: Southern Illinois University Press, 1989), 293-322.

8 See Walter R. Fisher, *Human Communication as Narration: Toward a Philosophy of Reason, Value, and Action* (Columbia: University of South Carolina Press, 1987).

9 The term intertextual grid is from Rabinowitz, *Before Reading*.

10 William Rothman, *Hitchcock—The Murderous Gaze* (Cambridge: Harvard University Press, 1982), 113.

11 Rothman, *Hitchcock*, 114.

12 Stanley Cavell, *Pursuits of Happiness: The Hollywood Comedy of Remarriage* (Cambridge: Harvard University Press, 1981).

13 We have borrowed this phrase from Herbert W. Simons, "Genrealizing' About Rhetoric: A Scientific Approach," in *Form and Genre: Shaping Rhetorical Action*, ed., Karlyn Kohrs Campbell and Kathleen Hall Jamieson, eds. (Falls Church, Virginia: Speech Communication Association, n.d.), 33-50. See also Thomas Conley, "The Linnaean Blues: Thoughts on the Genre Approach," in *Form, Genre, and the Study of Political Discourse*, ed. Herbert W. Simons and Aram A. Aghazarian (Columbia: University of South Carolina Press, 1986), 59-78; Bruce E. Gronbeck, "Celluloid Rhetoric: On Genres of Documentary," in Campbell and Jamieson, *Form and Genre*, 139-161; Rick Altman, "A Semantic/Syntactic Approach to Film Genre," *Cinema Journal* 23, no. 3 (1984): 6-18; Rick Altman, *Genre: The Musical: A Reader* (London: Routledge and Kegan Paul, 1981); Stephen Neale, *Genre* (London: British Film Institute, 1980); Barry Keith Grant, ed., *Film Genre Reader* (Austin: University of Texas Press, 1986); Wes D. Gehring, ed., *Handbook of American Film Genres* (Westport, CT: Greenwood Press, 1988); Jeanine Basinger, *The World War II Combat Film: Anatomy of a Genre* (New York: Columbia University Press, 1986).

14 We are obviously using the term genre in its loosest sense here, referring to clusters of films that, by their subject matter, call up associations of nuclear war, nuclear fear, or even, more broadly, the larger categories of "war films" or films cultivating confidence in or fear of government and its agents. For example, films that speculate about nuclear war include *Strategic Air Command, Dr. Strangelove, Fail Safe, Gathering of Eagles, Hiroshima, Mon Amour*, and *On the Beach*. Films such as *The China Syndrome* and *Silkwood* address a more general fear of nuclear accident. The fear of government bungling or plotting that is addressed in *Dr. Strangelove* is extended in such films as *Three Days of the Condor, Seven Days in May, The Spy Who Came in from the Cold*, and many others. The confidence in the military or in bombing upon which *Strategic Air Command* draws was cultivated in *Wings, Mrs. Miniver, Thirty Seconds over Tokyo, The Court Martial of Billy Mitchell*, and extended with such films as *Midway*. These films are certainly of differing subtypes, as are the many documentaries and television miniseries and docudramas that also form part of the fund of images and ideologies they share as inventional resources and the basis for audience response. See Jack G. Shaheen, ed., *Nuclear War Films* (Carbondale: Southern Illinois University Press, 1978); Peter Malone, *Nuclear Films* (Richmond, Australia: Spectrum, 1985); Spencer R. Weart, *Nuclear Fear: A History of Images* (Cambridge: Harvard University Press, 1988).

15 Wiseman's films are widely circulated to colleges and universities, after their showings on public television, by his distributing company, Zipporah Films, One Richdale Avenue, Unit 4, Cambridge, Massachusetts 02140. *Missile*, for example, is advertised in the Fall 1988 Zipporah Films brochure as renting for $150 for a single screening at a college, where no admission is charged. Once, when he was asked why he did not make his films available in videocassette formats at affordable prices, he said "If I sold my films for $19.95 on VHS, I'd be out of business. Who'd rent the films on 16mm at the prices I've got to charge if they could buy or copy a cheap video?" (Wiseman in conversation with students and faculty at a reception following his lecture at Bucknell University; the quotation is based on notes taken by Thomas W. Benson [12 November 1985]). For Wiseman's complaints about his difficulties in gaining funding for *Missile* from PBS, see Paula Span, "Wiseman and the Film of Truth," *Washington Post*, 31 August 1988, C1-3; Frederick Wiseman, "What Public Television Needs: Less Bureaucracy," *New York Times*, 27

November 1988, H35, 42. Although Wiseman has for years controlled the exhibition of his films, the explosive growth of home video recording has drastically challenged the ability of filmmakers to restrict private access to their work. But even though home video recording is possible, that seems like a marginal sort of access when we are discussing an important and complex contribution to a cultural debate about nuclear weapons.

16 For a discussion of the difference between formalist and poststructuralist concerns with meaning and the classical rhetorician's concern with effect, see Jane P. Tompkins, "The Reader in History: The Changing Shape of Literary Response," in *Reader-Response Criticism: From Formalism to Post-Structuralism* ed. Jane P. Tompkins (Baltimore: The Johns Hopkins University Press, 1980), esp. 201-206.

17 Because we will be invoking a version of auteur criticism as part of the context for this essay, we should make it clear that, although we refer to these films, as a matter of convenience, as the work of their directors, the relationship of each director to his films is rather different. The attribution of single authorship may fit best for Fred Wiseman, who employs a cinematographer, but who produces, directs, records sound, edits, and distributes his films, coming very close to a one-man organization. (See Thomas W. Benson and Carolyn Anderson, *Reality Fictions: The Films of Frederick Wiseman* [Carbondale: Southern Illinois University Press, 1989]). Stanley Kubrick is known for his autonomy in comparison with the norms for theatrical filmmaking. With the exception of Spartacus, for which he has disowned any claim to credit in recent years (see Wallace Coyle, *Stanley Kubrick: A Guide to References and Resources* [Boston: G. K. Hall, 1980], 5), each of his film projects has been characterized by Kubrick's personal involvement at every step of production. See Alexander Walker, *Stanley Kubrick Directs* (New York: Harcourt Brace Jovanovich, 1971); Norman Kagan, *The Cinema of Stanley Kubrick* (New York: Holt Rinehart and Winston, 1972); Gene D. Phillips, *Stanley Kubrick: A Film Odyssey* (New York: Popular Library, 1975); Thomas Allen Nelson, *Kubrick: Inside a Film Artist's Maze* (Bloomington: Indiana University Press, 1982); Michel Ciment, *Kubrick*, trans. Gilbert Adair (1980; New York: Holt, Rinehart and Winston, 1983).

Working within the Hollywood studio system, Anthony Mann developed a personal style singular enough to earn him the label of film "auteur" from a number of critics. See, for example, Jeanine Basinger, *Anthony Mann* (Boston: Twayne, 1979); Jim Kitses, *Horizons West* (Bloomington: Indiana University Press, 1969), 28-87. Still, Mann's reputation as an auteur has been based largely on his Westerns. *Strategic Air Command* was one of three films Mann directed in 1955. It was a project that "Mann reportedly did not want to direct but accepted as a favor to his old friend James Stewart" (Basinger, *Anthony Mann*, 29). The careers of Mann and Kubrick intersected when Kubrick replaced Mann as the director of *Spartacus* after Mann "walked off the set after three days of filming" (Coyle, *Stanley Kubrick*, 4). The difficulty of assigning authorship is illustrated by the differing accounts given by other writers. According to Basinger, Mann "directed all the scenes [in *Spartacus*] taking place in the desert, and all those in the school for the gladiators (excepting those with Jean Simmons)" (31). According to Ciment, *Kubrick*, "In 1960 the producer of *Spartacus*, Kirk Douglas (also the star of *Paths of Glory*) asked Kubrick after one week's shooting to replace Anthony Mann, with whom he had serious disagreements (Mann had directed the opening sequence and prepared the gladiatorial bouts)" (36).

The date of release for *Dr. Strangelove* is problematic, and reveals something of the rhetorical context it addressed. Most references to the film list its release date as 1964. According to Coyle, *Stanley Kubrick*, the New York opening of the film occurred on December 3, 1963, but it was released on January 30, 1964 (48). Seth Cagin and Philip Dray claim that "the film's release coincided with the assassination of John F. Kennedy—the opening was actually pushed back after Kennedy's death, from December 1963 to late January 1964, out of a sense of decorum—but there is a historical connection, for the film dramatizes the epochal loss of innocence that Kennedy's murder would come to symbolize" (*Hollywood Films of the Seventies* [New York: Harper & Row, 1984], 6). The issue of directorial autonomy is complicated by the related issue of other constraints upon a film, constraints that, in managing access and image, influence the ideology of the resulting film. A long succession of films have gained access to military hardware by trading with sympathetic portrayals of the military; *Strategic Air Command* is a clear example of this relationship between Hollywood and the Pentagon. Stanley Kubrick got no cooperation for *Dr. Strangelove* because of his unsympathetic treatment of the military. See Lawrence Suid, "The Pentagon and Hollywood: *Dr. Strangelove Or: How I Learned to Stop Worrying and Love the Bomb*" in *American History/American Film: Interpreting the Hollywood Image* (New York: Frederick Ungar, 1979), 219-235. Wiseman worked with the cooperation of the Department of Defense. Paula Span quotes Donald Baruch: "'He brings an honest and true picture to the screen,' says Donald Baruch, the Department of Defense public affairs officer who smoothed Wiseman's course" (Span, "Wiseman and the Film of Truth," C2).

18 To a large extent, the intertextual relations of all of these films to one another may be understood either as an assertion of how audiences may be likely to make use of one text to understand another, or of how a critic may construct a set of relationships among texts for the purposes of discussion—thus creating a new context, and a new text. It may also be argued, for example, that a text may explicitly refer to another text, or that an author may do so, as Fred Wiseman, in an interview about *Missile*, explicitly referred to *Dr. Strangelove* (Span, "Wiseman and the Film of Truth," C2). Or a text may allude to another text, as *Dr. Strangelove* seems to allude to *Strategic Air Command* by opening with a parodistic sequence of mid-air refueling. We have based this understanding of reference and allusion on James H. Coombs, "Allusion Defined and Explained," *Poetics* 13 (1984): 475-488. Reference and allusion are part of the matter of intertextuality, but do not exhaust it as an analytical category.

19 Peter Biskind, *Seeing Is Believing: How Hollywood Taught Us To Stop Worrying and Love the Fifties* (New York: Pantheon, 1983), discusses *Strategic Air Command* (64-69) and compares it to *Dr. Strangelove* (343-346). Biskind argues that *Strategic Air Command* is a "corporate-liberal" film (312), embodying the pluralist ideology of the Eisenhower years (2-6). Biskind and other critics have written of *Dr. Strangelove* as one of several films that exploded one historical period and introduced another. See Nora Sayre, *Running Time: Films of the Cold War* (New York: Dial Press, 1982); Cagin and Dray, *Hollywood Films of the Seventies*. We are forced by considerations of space to leave undeveloped in this chapter the intertextual treatment of women in these films, but it is perhaps worth noting how in *Strategic Air Command* Allyson works as a domestic distraction who partially motivates Stewart's reluctance without, however, exercising the traditional "female" role of ethical resistance to the military. In *Dr. Strangelove*, women appear very little, and primarily as distractions or the objects of sexual fantasy. In *Missile*, women are integrated into the military (although they are not seen as instructors), and in crucial scenes in the movie are shown as particularly enthusiastic and effective participants in the nuclear missile launch process.

20 Robert B. Ray, *A Certain Tendency of the Hollywood Cinema, 1930-1980* (Princeton: Princeton University Press, 1985), 65. For a review of Ray's book that raises some misgivings, see Thomas W. Benson, "Respecting the Reader," *Quarterly Journal of Speech* 72 (1986): 198-200.

21 See Figure 4 for a list of scenes.

22 Richard Rhodes, *The Making of the Atomic Bomb* (New York: Simon and Schuster, 1986), 325.

23 Rhodes, *The Making of the Atomic Bomb*, 378-379. In the months preceding the conference for which this paper was prepared, both the doctrine and the system of control of U.S. nuclear weapons seemed in doubt. The nomination of John Tower as Secretary of Defense, a crucial post that has both policy and operational missions in nuclear strategy, was delayed by a bitter debate over accusations of conflict of interest and personal irresponsibility. At the same time, the Bush administration seemed unsure of its direction in nuclear policy. The Associated Press reported that Brent Scowcroft, President George Bush's National Security Advisor, who was directing a review of nuclear policy, "described Reagan's nuclear policy as one of 'near total ... confusion'" (*Centre Daily Times* [State College, PA], 5 March 1989, A4). At the same time, it was reported that the Department of Energy, admitting to failure in safely disposing of radioactive wastes, was negotiating with private garbage-hauling firms, some of them with records of alleged criminal activities, to take over the job (Michael Zielenziger, "Department of Energy May Turn Nuke Wastes over to Trash Haulers," *Centre Daily Times*, 5 March 1989, A4). On March 25, 1989, the

press reported that Defense Secretary Dick Cheney had severely rebuked, in public, Air Force Chief of Staff General Larry D. Welch for allegedly negotiating with members of Congress about land-based missile strategy. Cheney's "stunning" rebuke was widely seen as an assertion of control by the civilian over the military in nuclear matters. See "Cheney Rebukes Air Force Chief for Arms Talk with Legislators," *New York Times*, 25 March 1989, 1. Events such as these typically occur in a sphere where historical memory is vague but in which public consciousness has been informed by a nuclear paranoia generated by films from *Dr. Strangelove* and *Fail Safe* to *China Syndrome* and *Silkwood*. Dramatic assertions of control such as Cheney's may have the side effect of reinforcing stereotypes of launch-on-impulse generals who are ready to take over nuclear policy for themselves. See also Steven Kull, *Minds at War: Nuclear Reality and the Inner Conflicts of Defense Policymakers* (New York: Basic Books, 1989); McGeorge Bundy, *Danger and Survival: Choices About the Bomb in the First Fifty Years* (New York: Random House, 1988); Paul J. Bracken, *The Command and Control of Nuclear Forces* (New Haven: Yale University Press, 1983).

24 Wiseman does allow himself to suture time and space within a scene at a single location, using cutaways to make fragmented and condensed time seem continuous—a common practice in single-camera documentary—but he appears to do much less of this in *Missile* than in some of his earlier documentaries. See Bill Nichols, *Ideology and the Image: Social Representation in Cinema and Other Media* (Bloomington: Indiana University Press, 1981), 218-233 for an analysis of Wiseman's editing style in *Hospital*. For a discussion of the "system of the suture," see Daniel Dayan, "The Tutor-Code of Classical Cinema," in *Movies and Methods,* ed. Bill Nichols (Berkeley: University of California Press, 1976), 438-451; William Rothman, "Against 'The System of the Suture,'" in Nichols, *Movies and Methods,* 451-459; Kaja Silverman, *The Subject of Semiotics* (New York: Oxford University Press, 1983), 194-235.

25 A similar formal strategy of omniscient and ubiquitous narration is employed in *Fail Safe,* where the President (Henry Fonda) is unable to communicate with his bomber crew, though we can follow the action in both locations.

26 TV displays are used, for example, in *Dr. Strangelove, Fail Safe, Threads,* and *WarGames.*

27 Leo Hurwitz, "One Man's Voyage: Ideas and Films in the 1930's," *Cinema Journal* 15 (Fall 1975), 9; cited in William Alexander, *Film on the Left: American Documentary Film from 1931 to 1942* (Princeton: Princeton University Press, 1981), 17. See also Brian Winston, "The Tradition of the Victim in Griersonian Documentary," in *Image Ethics: The Moral Rights of Subjects in Photographs, Film, and Television,* ed. Larry Gross, John Stuart Katz, and Jay Ruby (New York: Oxford University Press, 1988), 34-57.

28 Wiseman, quoted by Span, "Wiseman and the Film of Truth," C2. Wiseman's flirtation with the texts of existential and absurdist drama includes a passage in *Welfare* in which a disappointed client invokes existentialism's most famous drama of absence, *Waiting for Godot.* For a discussion of that passage, see J. Louis Campbell III, "'All Men Are Created Equal': Waiting for Godot in the Culture of Inequality," *Communication Monographs* 55 (1988): 143-161. See also Benson and Anderson, Reality Fictions, 242, 267-268. The phrase "absent guest" also evokes the notion of the guest of honor at a funeral, the dear departed, the absent one. See, for example, Susan Letzler Cole, *The Absent One: Mourning, Ritual, Tragedy and the Performance of Ambivalence* (University Park: Pennsylvania State University Press, 1985). Daniel Dayan, following Jean-Pierre Oudart, writes of the "absent-one (*l'absent*)" in the context of the shot, reverse-shot sequence of narrative cutting: "The spectator discovers that his possession of space was only partial, illusory. He feels dispossessed of what he is prevented from seeing. He discovers that he is only authorized to see what happens to be in the axis of the glance of another spectator, who is ghostly or absent" (Dayan, "The Tutor-Code," 448). Tzvetan Todorov writes at length about absence as an issue in the tales of Henry James. *Tzvetan Todorov, The Poetics of Prose*, trans. Richard Howard (Ithaca: Cornell University Press, 1977), 143-189. Wiseman shares with James both a fascination with the shape of human conversation and the sense of absence described by Todorov. See also Frederic Jameson: "Every universalizing approach, whether the phenomenological or the semiotic, will from the dialectical point of view be found to conceal its own contradictions and repress its own historicity by strategically framing its perspective so as to omit

the negative, absence, contradiction, repression, the *non-dit,* or the *imense*." Fredric Jameson, *The Political Unconscious: Narrative As a Socially Symbolic Act* (Ithaca: Cornell University Press, 1981), 109-110. It is by invoking absence, we mean to argue, that Wiseman restores to a seemingly merely descriptive documentary the possibility of a dialectical view. There are Wiseman films that seem closer in their agendas to Hurwitz's call for showing the audience "what its eyes were turned away from." In *Titicut Follies, Primate,* and *Meat,* Wiseman seems to provide a sort of window into the forbidden and horrible, but even then the vision is comic, sardonic, aggressively unsentimental—he shocks but does not share our shock.

29 Frederick Wiseman, "What Public TV Needs: Less Bureaucracy," H35.

30 Richard P. Feynman, *What Do You Care What Other People Think?* (New York: W. W. Norton, 1988), 200.

31 David I. Grossvogel, *The Blasphemers: The Theater of Brecht, Ionesco, Beckett, Genet* (1962; rpt. Ithaca: Cornell University Press, 1965), 50. Grossvogel comments at several points on the issues of detachment and engagement; see, for instance, 7-8. For a discussion of Ionesco's *Exit the King* that describes "tragedy as exit-play" see Cole, *The Absent One,* 88. The parallels between *Missile* and *Exit the King,* together with Wiseman's explicit admiration of Ionesco, provide a clue to the intertextual speculations that may have acted as another resource for *Missile.* See also Ronald Hayman, *Eugene Ionesco* (New York: Frederick Ungar, 1976).

32 Paul Rabinow, "Representations Are Social Facts: Modernity and Post-Modernity in Anthropology," in *Writing Culture: The Poetics and Politics of Ethnography,* ed. James Clifford and George E. Marcus (Berkeley: University of California Press, 1986), 241. On perspective by incongruity, see Kenneth Burke, *Permanence and Change: An Anatomy of Purpose,* 3d ed. (Berkeley: University of California Press, 1984), 89-96.

33 Figure 4 describes each separate scene and inter-scene in *Missile.* We have listed as "scenes" what appear to be the major narrative episodes of the film—each of these episodes contains discernible and continuous human speech. We have listed as "inter-scenes" the groups of shots that are placed between almost all the episodes. Typically, the inter-scenes show various settings on the base, with mixed ambient sound and with shots of moving cars, planes, helicopters, and people; the inter-scenes are sometimes indoors and sometimes out of doors. There are places where the assignment of a shot to a scene or inter-scene may be arguable, as in inter-scene 27, leading to scene 28—in this pair, a bird call can be heard in inter-scene 27 and the same bird is heard in scene 28; this could be taken as a formal indication that the two are not separated. Similarly, inter-scene 23 shows a group of students in a corridor; scene 24 shows a group of students, at perhaps the very next moment, having a conversation in a corridor. In these two cases, perhaps the only reason to regard these shots as inter-scenes 23 and 27 is that the form of the film so strongly leads to the expectation that there will be inter-scenes between scenes. Even so, the wall between the two is clearly being treated in Wiseman's cutting as in some ways variable; he creates the form and then varies it for his own rhetorical ends. Some of the inter-scenes lead by conventional indications directly to the following scene, as when inter-scene 22 ends with an exterior shot of a door closing, followed by a cut to a classroom; normal film conventions would indicate that such a pairing indicates that the classroom is in the building shown in inter-scene 22. But the inter-scenes are so different from the dramatic episodes that they seem to require a different designation in a scene list and as part of the intertextual work of the spectator. The inter-scenes are ambiguously "meanwhile," "elsewhere," "nearby"—and they are so much longer, more regularly repeated and formally similar, than merely transitional bridges that they call attention to themselves as requiring interpretation.

In almost all of his films Wiseman has played with "corridor shots" to punctuate and comment on his films; such shots have become a distinctive mark of his work. For example, in *High School,* we see shots of a cleaning woman moving down a corridor, a police guard standing in a corridor and consulting his watch. *Welfare* and *Hospital* are punctuated with portraits of waiting clients. In *Model,* we see frequent shots of "ordinary people," and of New York street scenes that serve a transitional and commentative function. In *Missile,* these "corridor shots" and transitional sequences have been given a very large and formal role in the film, but a role that is clearly, for viewers familiar with Wiseman's films, part of the intertextual transformation of a continuing body of work.

## CONTRIBUTORS

Karen E. Altman (Ph.D., University of Iowa) is Assistant Professor of Communication Arts and Sciences at the University of Southern California, where she also participates in the Program for the Study of Women and Men in Society. Her related work on discourse, gender, and technology appears in the *Journal of Popular Film and Television.*

Carolyn Anderson (Ph.D., University of Massachusetts) is Assistant Professor in the Department of Communication at the University of Massachusetts at Amherst. She teaches film and television history and criticism and is co-author of *Reality Fictions: The Films of Frederick Wiseman.*

Wayne A. Beach (Ph.D., University of Utah) is Associate Professor of Speech Communication at San Diego State University. Among his publications on the conversational organization of casual and institutional discourse is his recently edited *Sequential Organization of Conversational Activities.* He is also working on a book about how judges regulate court traffic.

Thomas W. Benson (Ph.D., Cornell University) is Professor of Speech Communication and Chair of the doctoral program in Mass Communications at Penn State University. He is a former editor of *Communication Quarterly* and *The Quarterly Journal of Speech,* and co-author of *Reality Fictions: The Films of Frederick Wiseman.*

Charmaine Bradley (M.A., Texas A&M University), an Acoma/Navajo, is a doctoral candidate in educational psychology at Texas A&M University. Ms. Bradley presently is co-director of a mentorship project for minority students.

James W. Carey (Ph.D., University of Illinois) is Dean of the College of Communications at the University of Illinois, Urbana. His most recent book is *Communication as Culture: Essays on Media and Society.*

Stanley Deetz (Ph.D., Ohio University) is Associate Professor of Communication at Rutgers University. He is author of two books and over thirty published essays on interpersonal and organizational communication, holds offices in both the Speech Communication Association and the International Communication Association, and is the current editor of *Communication Yearbook.*

Thomas S. Frentz (Ph.D., University of Wisconsin) is Associate Professor of Communication at the University of Arkansas, Fayetteville. He has published numerous articles in areas of communication that include psycholinguistics, cultural dialects, conversational analysis, rhetorical theory, and media criticism.

Alberto Gonzalez (Ph.D., Ohio State University) is an Assistant Professor of Speech Communication at Texas A&M University. His research interests include the study of popular music and cultural myth, and Mexican-American political rhetoric. His work has appeared in *The Quarterly Journal of Speech,* the *Western Journal of Speech Communication,* and the *Southern Speech Communication Journal.*

Bruce E. Gronbeck (Ph.D., University of Iowa) is Professor and Chair of Communication Studies at the University of Iowa. His field is rhetorical studies, with particular interst in cultural studies of media and politics.

Steven Hartwell (J.D., University of Southern California) is Associate Professor of Law at the University of San Diego. He teaches and writes as a clinician, with particular interest in law and social science.

David Henry (Ph.D., Indiana University) is Professor of Speech Communication at California Polytechnic State University, San Luis Obispo. His essays and reviews on political rhetoric, criticism, and argumentation have appeared in the *Quarterly Journal of Speech, Communication Education, Southern Speech Communication Journal,* and eight books.

Robert Hopper (Ph.D., University of Wisconsin) is Charles Sapp Centennial Professor of Communication at the University of Texas at Austin. He has written seven textbooks and over a hundred scholarly papers concerning language and conversation. His essays on telephone conversation appear in recent issues of *Communication Monographs, Western Journal of Speech Communication,* and *Journal of Language and Social Psychology.*

Michael J. Hyde (Ph.D., Purdue University) is an Associate Professor of Communication Studies at Northwestern University and a past National Fellow of the W. K. Kellogg Foundation. His research and critical reviews appear in scholarly journals and texts emphasizing rhetorical and philosophical perspectives on human communication. He is editor of *Communication Philosophy and the Technological Age.*

Martin J. Medhurst (Ph.D., Pennsylvania State University) is Associate Professor of Speech Communication at Texas A&M University. His areas of interest are media criticism, political rhetoric, and civil-religious discourse. He is co-author of the forthcoming book, *Cold War Rhetoric: Strategy, Metaphor, and Ideology* and co-editor of *Rhetorical Dimensions in Media.*

Roger C. Pace (Ph.D., Pennsylvania State University) is an Assistant Professor of Communication Studies at the University of San Diego. He has published articles on the decision-making process in *Communication Monographs,* the *Western Journal of Speech Communication,* and the *Southern Speech Communication Journal.*

Tarla Rai Peterson (Ph.D., Washington State University) is an Assistant Professor in the Department of Speech Communication and Theatre Arts, Texas A&M University.

Janice Hocker Rushing (Ph.D., University of Southern California) is Associate Professor of Communication at the University of Arkansas, Fayetteville. She writes about rhetoric and values in mass-mediated narratives, focusing primarily upon the evolution of dominant American myths.

Micheal R. Vickery (Ph.D., University of Texas) is an Assistant Professor in the Department of Speech Communication and Theatre Arts at Texas A&M University. His research focuses on the ethos of technological society and the rhetoric of technique.

LIVE AND

LEAD —

ACCOUNTABLY!

BEST REGARDS,

# ACHIEVE

## WITH

## ACCOUNTABILITY

# ACHIEVE

## WITH

## ACCOUNTABILITY

Ignite
Engagement,
Ownership,
Perseverance,
Alignment,
& Change

Mike Evans

**WILEY**

*Library of Congress Cataloging-in-Publication Data is available:*

978-1-119-31408-0 (Hardcover)
978-1-119-31410-3 (ePDF)
978-1-119-31409-7 (ePub)

Cover Design: Wiley
Cover Image: © Jose A. Bernat Bacete/Getty Images

Printed in the United States of America

10 9 8 7 6 5 4 3 2 1

*To Zack and Nick, my heroes.*

# Contents

The Author                                                              xi

Resources, Solutions, and Workshops                                    xiii

Preface: Taking Accountability                                          xv

Chapter 1  **Taking Accountability**                                      1
           Positively the Best Decision                                 1
           Take Accountability or Blame? The Stakes
               Are High                                                 5
           The Magic of Taking Accountability                           6
           The Rewards of Taking Accountability                        10
           Accountability Accelerator                                  11
           Shine the Spotlight on the One and Only                     15
           An Unexpected and Illuminating Discovery                    19
           Do They Know the Rules of the Game?                         21

Chapter 2  **Choices**                                                   35
           Decisions Determine Destiny                                 35
           Stay the Course—Avoid Drowning "Below
               the Water"                                              39
           Beware of the Bell                                          43
           Leading Accountability—It Is a Choice,
               and Everyone's Job                                      46
           Where Do They Learn Those Things?                           49
           Words Are Cheap—The Actions We Choose
               Are What Matters                                        52
           The Enemies: Justifications, Explanations,
               Reasons, Rationalizations, Vindications,
               and Excuses                                             54
           Why Did Randy Cross the Road?                               56

**Chapter 3    Accountability—What Is It?**                    **67**

Once Defined, It Flows                                          67
Are You Fully Recognizing Your Realities?                       74
The Eyes Have It                                                76
Have You Accepted Ownership for Your
    Realities?                               78
Yoda—The Accountability Coach                                   81
Lessons from Hyrum                                              83
Solutions Abound                                                86
The Obvious Next Step                                           90

**Chapter 4    Accountability Transcends**                     **97**

Powerful Shift in Perception                                    97
Cement the Focus                                                100
The Magical Mindset Shift                                       103
Lessons from the Marauders                                      106
Accountability Trumps Responsibility                            109
To LAG or to LEAD                                               111

**Chapter 5    The Preeminent Organization**                   **117**

Accountability Is the Catalyst                                  117
Shaping an Accountable Culture                                  120
Ignite Desired Accountable Behaviors                            122
The Power of Genuine Appreciation                               127
Culture: How We Do Things Around Here                           129
Forging an Accountable Culture: Little Things
    Matter—A Lot!                            131
Are the Experiences in Your Culture Nurturing
    Accountability?                          133
The Relationship Triad: Experiences, Beliefs
    and Actions                              135
You Alone Are Accountable for How You Are
    Perceived                                137
Sustaining an Accountable Culture: Tune-Ups
    Required                                 139

Accountability Tip: Do Not Shoulder Their
   Accountabilities                                    142
Do Not Dwell on Why: Keep Moving Forward    144
Accountability in Times of Change: Overcome
   Fear and Denial                                     146
The Pinnacle: The Accountable Organization    150
Live and Lead Accountably                       161

**Index**                                                **163**

# The Author

**Mike Evans** has developed a unique perspective from working alongside a star-studded list of world-renowned thought leaders, including Dr. John Kotter, Dr. Stephen Covey, Tom Peters, Jim Kouzes, Hyrum Smith, Steve Farber, and Chris McChesney.

Mike served in senior leadership and consulting roles with Kotter International, FranklinCovey, and Tom Peters Company.

In addition to unleashing personal and organizational accountability, clients rely upon Mike's solutions to ***Accelerate Change, Shape Their Optimal Culture, Flawlessly Execute Key Strategies, Ignite Leadership Capacity at All Levels, and Amplify Employee Engagement***.

He consults with senior teams, facilitates custom workshops, delivers keynote speeches, and provides coaching and consulting on all of these subject areas.

Mike has worked with organizations in virtually every arena, from the tech sector to financial services, manufacturing, health care, hospitality, entertainment, retail, and the US government. Clients include: Intel, Capital One, Apple, BP, BNY Mellon, Pfizer, The United States Navy, Fidelity Investments, Johnson & Johnson, Symantec, Cigna Corporation, Oracle, Astra Zeneca, Baxter International Inc., Shell Oil, Cargill, American Airlines, DuPont USA, and NASA.

His personal mission is to help individuals, teams, and organizations accelerate their ability to achieve more than they ever believed possible. Clients describe him as inspiring, motivating, and a ball of energy with an unequaled focus and passion for helping them achieve their desired results.

Reach Mike at: www.questmarkcompany.com or mike.evans@ questmarkcompany.com

# Resources, Solutions, and Workshops

There is a reason that "accountability" continues to top the critical needs list of most organizational leaders. These leaders are keenly aware that accountability is the essential ingredient that allows individuals, teams, and organizations to soar and leave the competition in their wake.

*Accountability helps individuals, teams, and organizations*
- Develop the agility, flexibility, and resiliency to adapt to constant change
- Adopt a can-do, solutions focus and resolute mindset in the face of difficult challenges, obstacles, and barriers
- Eradicate the blame-game and excuse-making that stifles peak performance
- Strengthen collaboration and teamwork
- Shed feelings of disarray, discomfort, apathy, entitlement, cynicism, and despair
- Cultivate a sense of control over your circumstances to achieve what matters most
- Establish unshakable trust and credibility
- Unleash voluntary contributions of discretionary performance that is often left untapped in individuals, teams, and organizations
- Stimulate creativity and innovation
- Flawlessly execute crucial strategies
- Galvanize alignment and ownership
- Ignite a desire to learn, grow, and improve

If you are interested in creating a highly accountable team or organization, **Mike and QuestMark offer the following options and solutions:**

**Speaking:** Bring Mike Evans into your organization, your conference, or offsite event to keynote on: Accountability, Change

Leadership, Culture Shaping, Exemplary Leadership, Flawless Execution, or Employee Engagement

**Workshops:** Schedule an interactive, half-day, one-day, or multiple-day onsite custom workshop for your team or organization:

- Cultivate a Highly Accountable Team
- Leading and Accelerating Change
- Creating Your Optimal Culture
- Flawless Execution—Achieve What Matters Most
- Unleash Exemplary Leadership Capacity at All Levels

**Consulting:** Hire Evans and QuestMark to advise, counsel, and coach your team. Integrate proven and pragmatic models, tools, methodologies, and principles into your leadership repertoire to supplement and enhance your current efforts.

To Learn More Visit: www.questmarkcompany.com or contact mike.evans@questmarkcompany.com

# Preface: Taking Accountability

## If Not You, Who?

Accountability continues to be a topic high on the list of interests of executives and organizational leaders. What we find more interesting is that accountability is just as high on the list of what employees want from their peers. Everyone wants more!

Accountability is the catalyst to accelerated change, robust employee engagement, intensified ownership, relentless perseverance, impeccable alignment, and it propels individuals, teams, and organizations to intoxicating heights of achievement and success.

Accountability crumbles silos, boosts teamwork and collaboration, strengthens camaraderie, creativity, resiliency, agility, trust, and communication.

*Achieve with Accountability* presents a recipe for awakening the *belief, resolve, confidence, perseverance, determination, can-do mindset, whatever-it-takes attitude, esprit de corps, drive,* and *creativity* to achieve what matters most to individuals, teams, and organizations. Discover how to transform accountability into a positive, engaging, and forward-looking experience that will secure your position in the new world of work. Learn how to kick-start a revolution to achieve what matters most.

## Why Accountability?

We are in a brawl with few rules, where the fast, flexible, and agile will eat and spit out the slow, over-thinking, and complacent. In the past, change was episodic, transient, and gradual; now, change is constant as business, technology, and society itself continue to evolve at an ever more rapid pace. Individuals, teams, and organizations that fail to adapt will find themselves vulnerable, uncomfortable, and at the mercy of other people, events, or the competition.

With the world coming at you fast and furious every day, it is easy to feel like you have lost control of your life, your team, or your organization. By choosing to *take* and *lead* accountability, you reclaim control and are able to direct your own destiny, as well as the fate of your team or organization. Accountability is a current that feeds into the slipstream of success. Your performance, your decisions, and your results are all within your control; when you operate from that premise, magic happens. You deliver like never before because you are personally invested in the outcome. *Achieve with Accountability* shows you how to nourish that can-do mindset so that you can start achieving what matters most. Accountability helps individuals, teams, and organizations

- Develop the agility, flexibility, and resiliency to adapt to constant change
- Adopt a can-do, solutions focus, and resolute mindset in the face of difficult challenges, obstacles, and barriers
- Eradicate the blame-game and excuse-making that stifles peak performance
- Strengthen collaboration and teamwork
- Shed feelings of disarray, discomfort, apathy, entitlement, cynicism, and despair
- Cultivate a sense of control over your circumstances to achieve what matters most
- Establish unshakable trust and credibility
- Unleash voluntary contributions of discretionary performance that is often left untapped in individuals, teams, and organizations
- Stimulate creativity and innovation
- Flawlessly execute crucial strategies
- Galvanize alignment and ownership
- Ignite a desire to learn, grow, and improve

Individuals, teams, and organizations that are mired in the blame-game, self-pity, excuse-making, cynicism, complacency, and apathy, and that lack alignment, engagement, and ownership to achieve what matters most, are in jeopardy. When you relinquish accountability, you place your future in the hands of fate. It is only by taking accountability and ownership for our circumstances that we can achieve what matters most. Today is the day to start owning your situation and take control—to orchestrate and realize the future you want.

Start believing in you, your team, and your organization. Embrace the accountability fundamentals that have helped individuals, teams, and organizations for years to achieve and reach beyond what they had believed was possible. The race is about to begin. Don't miss the starting gun while wallowing in the blame-game.

## What Is Instigating the Call for More Accountability?

There are myriad reasons individuals, teams, and organizations crave more accountability. Some that are listed below may strike close to home. What would you add to the list? What realities are you, your team, or organization faced with right now that heightened levels of accountability would be the ideal prescription?

| | |
|---|---|
| Increased Competition | Commoditization |
| New Technologies | Increased Shareholder Demand |
| Pressure to Innovate | Shrinking Margins |
| Globalization | Talent Wars |
| Economic Downturns | Penetrating New Markets |
| Nonstop Change | Doing More with Less |
| Regulations/Legislation | Expansion |
| Brand/Reputation Management | Political Uncertainty |

### Winning in the New World of Work

For many of the reasons cited above, Beth and her leadership team of nine were always looking for a competitive advantage. So the opportunity to attend a forum with a world-renown authority was a no-brainer. Almost immediately, this guru captivated Beth's team.

He started with passion and energy, "The interplay of unstoppable forces that are creating the 'Perfect Storm' in the deep waters of commerce will sink organizations and individuals that are ill prepared, or that lose focus on what they have identified as most important.

"The white waters of change are unnerving for many. But the fact is they will only become more turbulent. For organizations and employees who are unprepared, the future will be devastating.

"The need for increased levels of personal and organizational accountability has never been higher. And that need will expand daily.

"Consider:

1.  There are companies and entrepreneurs at this moment looking to reinvent the way your business is done. How many bookstores were wiped out because of Amazon.com? What happened to Kodak? Blockbuster? Sears? Woolworth's? Nokia? Rest assured there are some college students ensconced in a dorm room somewhere about to revolutionize your industry.

2.  How many jobs have been expunged, teams eliminated, and firms driven to extinction due to ERP, SAP, the Internet, or White-Collar Robots (e.g., Cash Machines). Blue-collar robots came and triumphed. Are you, your team, and your company confident that the value you exhibit outweighs these options? What are you doing to take accountability for your future success?

3.  MIT's No. 1 computer guru, Michael Dertuzos, said not long ago that India could easily boost its GDP by a *trillion* dollars in the next few years. How? By performing "backroom" white-collar tasks for Western companies. He estimated that fifty million (white-collar) jobs could be sucked from the West and transplanted to India, at less than half the cost. What are you doing to take accountability to make certain one of those jobs is not yours in the future?

4.  It took thirty-eight years for the radio to get to fifty million users. The Internet got there in four years. Change is coming at us faster than ever before. There is no time to rest on our success. Success often leads to complacency, and complacency kills. What is your plan to take accountability to keep up with the pace of change?

5.  In his book, *Change or Die: The Three Keys to Change at Work and in Life*, Alan Deutschman, cites that only one in nine people will make lifestyle changes (diet, exercise, etc.) after they

are told they could prolong their life, restore their health, and even reverse diabetes, hypertension, and heart disease. Even when people know that these relatively simple changes can potentially save their lives, 90 percent choose to not take accountability to do so. Imagine the gargantuan task leaders face in order to create heightened levels of personal accountability for organizational objectives when only 10 percent of people will make simple changes to save their own lives.

6. Every job being performed by white-collar workers employed in any corporation is also being performed by someone on the outside that can be hired as a consultant to do the same work, probably with higher quality, and at a lower cost. How are you, or your team taking accountability to make sure you or your department survives in the new world of work?

7. Whether individually, as a team, or organizationally if you cannot answer these three questions in a compelling manner, you are in trouble: What is the overt benefit I/we offer? What reasons can I/we point to as to why anyone should believe I/we are the best option? How am I, or are we, dramatically different from all other options?

The speaker paused, surveying his audience before driving home his final point: "Those who will survive and thrive in this new world of work are individuals, teams, and organizations who can maintain high levels of accountability, remain focused, and do whatever it takes to achieve what matters most.

"To take and lead accountability? If not you, who?"

# Chapter 1

## Taking Accountability

### Positively the Best Decision

Frustration and exasperation were escalating among Janet's team. Their attempts to heighten accountability to boost organization performance and achieve better results had negligible impact. In fact, there was a noticeable decline in morale, enthusiasm, and engagement, with a touch of resentment and irritation to boot.

Fierce competition, shrinking margins, pressure to innovate, new government regulations, and a downturn in the economy had Squire Medical on its heels. Squire Medical was losing ground to the competition at an alarming rate.

Clayton offered, "It seems our plan to instill a stronger focus on personal accountability has had little to no affect on achieving more accountability. If anything, we have created stress, anxiety, acrimony, bitterness, and tension among the workforce."

The board of directors viewed Janet as an up-and-coming leadership superstar. They had seen her perform miracles in other roles over the previous two years and had confidence in her abilities to resurrect what was once an industry leader. Complacency, with a hint of arrogance, had thrust Squire Medical into a downward spiral. The competition was intense and the stakes were high. Employee morale and engagement were at an all-time low. Top talent was leaving in droves and those remaining had little hope of things getting better. There was a very real possibility that the plug may be pulled and assets sold off.

"I agree with Clayton. I have observed the same reactions," Janet said. "We must create a culture that is engaged, focused, resilient, innovative, and agile.

"I believe I may have a solution to jump-start that journey. I had an epiphany last night while attending a Miracles for Kids meeting. Within our volunteer group it is commonplace that every member is passionate, engaged, committed, and energetic when asked to get involved with a project. Members leap at the opportunity to participate, take ownership, and accountability. Everyone involved voluntarily chooses to take personal accountability to achieve what matters most. There is an indestructible level of personal ownership among all involved, and nothing will deter the members from accomplishing the desired outcomes.

"Obstacles, challenges, and barriers are viewed as trivial and as minor bumps in the road on our way to attaining our goals. Excuse-making, finger-pointing, blame, and inaction simply do not exist. The group's desire, focus, determination, and can-do mindset are unrivaled. With all of the external challenges we are faced with at this time, that same passion, zeal, and energy are paramount to our future success here."

Not quite sure what Janet meant, Clayton asked for clarification.

"What I discovered," Janet shared, "is that too often accountability is something that is addressed after the fact. Most often after a mistake is made, when somebody drops the ball or when someone screws up. So naturally in those circumstances, accountability is viewed as punitive, historic, focused on blame, and unpleasant. Far too many people hold a negative connotation or perception of accountability because of their past experiences.

"Think about it, when do you typically hear the question being asked, 'Who is accountable?'"

Andrew, dripping with indignation, chimed in, "Usually after somebody makes a mistake."

"Exactly! So, what are people really hearing when that question is asked?"

"Who messed up and who is to blame!" Andrew stated with a tone of disdain.

"And when they are really hearing 'who is to blame,' what do people tend to come up with?"

"A litany of excuses, stories, and reasons," Andrew declared. "People spend more time explaining why something is not their fault than they do on finding a solution."

"And time spent playing the blame-game is not helping any-one," added Janet. "And as we hear more excuses being offered, as leaders we often are mistakenly compelled to ask the question 'who is accountable' even more. Not realizing that we are reinforcing the prevailing perception many hold of accountability being a negative experience. The more we ask the question after the fact, the more we fortify that belief. It can become a hairy monster that cannot be stopped.

"I cringe when I hear someone say the words, 'We need to *make* them accountable.' It even sounds like a punishment. We need to flip this and create positive experiences with accountability.

"How much more effective would we be as an organization if every employee voluntarily chose to *take* accountability rather than being pressured to *be* accountable? We need to engage employees on the front end—before the results are in. Think about it, when accountability is positioned up front, people have the opportunity to get excited about the ability to help while there is still time to influence the outcome.

"Our folks want to succeed in the workplace. They want to make a difference, find meaning in their work, and contribute. For the most part, people are not stupid, lazy, or defiant. People crave to find significance in their work and to feel they are part of a team doing something worthwhile. We have plenty of talented, smart people here at Squire. We need to engage and energize them to take accountability on the front end as we define and declare our desired results. Asking 'Who is accountable?' once the result is in does not change the result. Leadership is about getting our folks to voluntarily choose to take accountability while they can still help shape the outcome.

"With this approach we will begin to create positive, engaging, and forward-looking experiences around accountability, which I believe will result in employees exhibiting accountable behaviors."

Janet challenged her team to create positive experiences around accountability by engaging employees on the front end. Instead of the after-the-fact experience of "Who is accountable for failing to achieve the result?" the question became "Who is accountable to achieve the desired results?" The second format captivates people because they have the ability to choose the appropriate actions that will positively influence and move the team closer toward the desired result. Positioning accountability before the results are in promotes a heightened sense of control, which in turn leads to increased productivity and when people feel they have control and are productive it bolsters self-esteem, morale, and engagement.

This new focus on enlisting employees and creating ownership, energy, and passion preceding a clearly defined "must-achieve desired result" had an immediate and noticeable impact. Janet's team observed a distinct shift in employee enthusiasm and willingness to voluntarily take personal accountability to achieve desired results. Accountability began to be viewed as something that was positive, forward-looking, and energizing.

Engaging, aligning, and enlisting employees up front requires leadership traits that reside inside all of us. Exemplary leaders foster high levels of personal accountability in a variety of ways on which we will elaborate throughout this book.

The cornerstone, and most essential element, to cultivating soaring levels of personal and organizational accountability is communicating top priority desired outcomes so that they are unquestionably clear in the minds of every employee. As we will reveal in later chapters, explicit and precise clarity on desired results or expectations is not as common as many believe, and the implications are severe.

A very close second element is connecting to both the head and the heart of those involved. Individuals must not only understand the logic behind the desired result, they must also recognize how it will benefit them. How will it make their job better or easier? How will it help their team? The organization? Their career? Their family? How will it make their life better or develop them? What is it that will light the fire within and get them to voluntarily choose to engage? What is in it for them? Deprived of compelling answers to these questions, many employees select compliance instead of commitment. To thrive and excel in the new world of work demands leadership create waves of enthusiasm and commitment. Obedience and compliance are ingredients commonly found in failure.

Janet discovered that when leadership changes the way accountability is experienced that employees welcome and embrace the opportunity to contribute fully. She understood that nobody relishes being told, "I am holding you accountable," "I am going to *make* you accountable," "You need to be accountable," or being asked "Who is accountable for this?"

The leadership team at Squire Medical attributed the shift in perception around accountability as a key component of revitalizing their culture. As they described it, they transitioned from a "have-to" to a "want-to" culture. What the team had described as a complacent, slow-moving, apathetic culture had transformed into one dubbed as agile, focused, innovative, and opportunistic. This change ultimately allowed Squire Medical not only to strengthen their competitive position in the market, but also to once again become the recognized leader.

## Take Accountability or Blame? The Stakes Are High

Lack of accountability can lead to dire consequences. Consider the following scenario that played out in the Pacific Northwest back in 1993.

In what many described as the most infamous food-poisoning out-break in history at the time, 732 people became seriously ill. Four children died and 178 others were left with severe injuries, including kidney and brain damage. Panic set in throughout the Pacific Northwest. Investigators quickly discovered that people had become stricken with the *E. coli* virus after consuming food from Jack in the Box.

As the story was reported, Jack in the Box chose to ignore Washington state laws stating to cook hamburgers to a temperature of 155 degrees to completely kill *E. coli*, and instead adhered to a standard of 140 degrees.

The company's almost unforgivable response was, "No com-ment." That soon led to Jack in the Box blaming their meat supplier, Vons Companies, Inc. Vons, evidently, was in no mood to take that sitting down and in a variety of ways pointed the finger of blame at the meat inspectors.

The meat inspectors were all too willing to play the blame-game and some declared that it was the fault of the USDA due to their inadequate number of inspectors. In testimony, some went as far as to blame Congress for not providing sufficient budgets.

On and on it went. No accountability. As a result of their refusal to immediately take accountability, and with their reputation tarnished, it took Jack in the Box years to recover financially.

Jack in the Box eventually became known as an industry leader in safety and health procedures because of changes throughout the company.

Flashback to Chicago a decade earlier. Seven people died after taking capsules of Tylenol that some lunatic had laced with cyanide. Hysteria set in worldwide.

Within 48 hours Johnson & Johnson not only recalled all Tylenol products from around the globe, they incinerated $300 million (estimated in today's dollars) worth of product. This took their market share from over 35 percent to rapidly approaching zero within 48 hours.

Many people would have given Tylenol a break. After all, Tylenol could not reasonably foresee something like this taking place. Instead, Johnson & Johnson chose the approach of taking accountability and ownership.

The company determined and believed they had not done enough to protect consumers. They adopted a steadfast mindset of "what else can we do to protect and keep consumers safe?" Within weeks following the tragedy they developed tamperproof packaging, which is still used today. In fact, if you have a headache today you need to plan three days in advance so you can figure out how to open the darn package. But at the time of this tragedy, they were the innovators with the concept.

Because of their accountable "what else can we do?" mindset Tylenol recaptured over 80 percent of their market only months following their launch of the revolutionary tamperproof packaging.

Take accountability and recover in two months? Play the blame-game and recover in years? The choice is always yours to make.

These stories clearly demonstrate the power of taking accountability. The decision as to whether you take accountability is solely yours, your team's, or your organization's.

## The Magic of Taking Accountability

It is essential and vitally important to understand that nobody can **make** any other person accountable. One can choose to **take** accountability and model it in hopes that others choose to do the same. But again, you absolutely cannot ***make*** anyone accountable. In fact, you cannot ***make*** any other person do anything. The choice to ***take*** accountability is personal. You either will take accountability or you will not.

Vanquish the phrase "make them accountable," and any variation of that, from your vocabulary. You can, however, ***hold*** others accountable. And there are best practices associated with holding others accountable. We will address those later.

Let us experience both choices.

"Good morning. I'm here to pick up the truck I rented for today," I offered to the inattentive employee seated behind the counter. As he grunted and reluctantly peeled himself from his chair and slowly moved toward the counter, I looked at my son who mirrored my expression of concern with what we both believed we were about to experience. "You Evans?" he mumbled, never making eye contact. "Yes. I called last night to confirm my reservation and to let you know what time I would be here to pick up the truck."

"We talked to you five minutes ago to tell you not to show up because we don't got no truck for you." Dumbfounded and with a look of confusion, I silently glanced at my son, wondering if I had missed something. From the expression on his face, it was obvious that Nick and I were on the same page.

Nick was always observant of the experiences employees create for customers. Even as a very young child he would ask me to take him to a specific Burger King over another one that was closer to home because he liked the way the employees treated customers at the store farther down the road. Nick has a penchant for engaged, upbeat, enthusiastic, accountable employees who display a can-do mindset. Never one to openly judge, and always with kindness, fairness, and generosity in his heart, Nick developed a habit of closely examining the experiences employees created for customers. After quickly synthesizing an interaction he would then render a decision on whether or not he would want that employee representing his future company.

Without a word, we communicated to one another, "What on earth is this guy talking about?" We had left the house fifteen minutes ago. Nobody called the house before we left and my mobile phone never rang during the drive to the rental agency.

"I believe there is a misunderstanding. What is your name?" I calmly asked the employee.

"Arnie," he shared.

"Arnie, I did not speak with anybody from your company in the last twelve hours. My last communication was yesterday, with you I believe, to confirm the reservation and the pick-up time. We are ten minutes early. My phone did not ring and I have no message from anyone from your company. So, are you telling me that there is no truck here for me, or just not the one I wanted?"

Arnie, with a hint of disdain, uttered, "My boss talked with you and told you we got no trucks today. We told you not to show up."

Knowing my son was watching me closely, I took a deep breath, looked at Nick, whom I could sense was anxious to see how I would respond. I took a few seconds, attempted to synthesize the series of events, and consider that perhaps I missed something.

Reminding myself to remain calm, I offered to Arnie, "I am 99 percent certain I did not speak with anyone from your company in the last ten minutes. If I did, and was told there was no truck, I would not have shown up. Do you have any trucks at all available today? We are moving a bunch of furniture into an apartment today and we need a truck. I'll take any truck you have."

"Do you see any trucks in the lot?" Arnie sarcastically questioned with a strong dose of indifference.

I must admit, that caught me off guard and maybe in my younger years would have set me off. However, we acquire wisdom, patience, and grace as we age. "No, Arnie. I don't see any trucks in your lot. Is there anything you or your boss can do to help me? You do have other locations in the area, don't you? Can you check with them to see if any have the same type of truck I reserved?"

"We're a franchise. I don't know what other franchise owners have."

"Would you call a few or maybe check on your computer to see if one of the other franchises in the area might have what I need?"

Without a response, acknowledgment, or eye contact, Arnie pulled out his cell phone and walked away. Nick and I looked at each other wondering if he was checking for us, was ordering lunch, or had just given up on us. Nick took on an expression of bewilderment. He jokingly declared that someone might be playing a prank on us. I asked Nick, "Is Arnie the type of employee you would want working for you?"

"My lord no," Nick asserted. "He seems angry that we showed up, has no interest in trying to resolve the situation, has not offered any alternatives, is taking no initiative to accommodate us, and has pretty much called you a liar. He's blamed you and his boss for creating this situation. He's pathetic and I would never use this rental agency in the future. Why on earth the owner of this company would have him interfacing with customers is perplexing."

After watching Arnie whisper into his phone in the back office for fifteen minutes he finally strolled out and ventured back to the counter. "Got a truck for you at the Moon Township location if you want it."

"That is about an hour away, Arnie. Is there anything closer?" Arnie looked as though someone had just asked him to solve world hunger.

"Do you want it or not? It ain't the same size truck, but you can't be choosy."

Reluctantly submitting to Arnie, I said, "I'll take it. We cannot move without a truck."

Arnie blurted into the phone, "He wants it," and walked into the back office never to be seen again.

We all have had similar interactions with the Arnies of the world. At work, while shopping, while getting your car repaired, with our children, while traveling, when dining out—the list is endless. These are instances when you are dealing with someone who has no interest in helping resolve any problems, snags, or hiccups. They are drawn into the trap of blaming others, fashioning excuses, withdrawing from the predicament, or even ignoring the dilemma. Instead of focusing on what they can do to help achieve the desired outcome, they choose to take the easy way out and engage in the unwinnable blame-game.

Apathetic, disengaged, cynical, negative, pessimistic, disinterested, lethargic, aloof, inactive, bitter, indifferent, entitled, confused, lazy, and lacking ownership are just a few of the words that commonly describe folks that renounce accountability and are sucked into the blame-game.

Throughout the remainder of this book we will share pragmatic and memorable principles and natural laws that are proven to generate increased levels of personal, team, and organizational accountability.

## The Rewards of Taking Accountability

About two miles into our drive to Moon Township, Nick suggested that we stop by, or call, a competing truck rental firm not too far from us. We pulled over into a store parking lot and made the call.

What a difference that phone call made.

From the way Chris, the sales rep, answered the phone I sensed this was a great suggestion from Nick.

Chris was filled with the desire and willingness to find a solution for us. His focus, engagement, and energy were contagious.

I had the call on speakerphone, and as Chris was doing his thing, Nick and I were giving one another a thumbs-up. This was a guy we wanted to work with.

Chris suggested we stop by and take a look at what they had available, and that if what they had did not suit our needs, he would make sure he found us something that would.

As we approached the counter to ask for Chris, a pleasant voice wafted from behind, "Mr. Evans, I'm Chris. I am pleased to meet you. I have some trucks lined up outside. What a gorgeous day you two have for moving. Follow me and I'll show you what we have."

"Wow! Thanks so much, Chris. I really appreciate you taking time to help us with such short notice."

"Not a problem at all, sir. That is why I'm here. I wish we had more trucks on hand to show you, but this weekend is really busy. If none of these work for you, I can connect with other lots in the area to make sure you have a truck for your move today. Whatever it takes. I am here to please."

As much as we wanted to work with Chris, he did not have a truck large enough for our needs.

"No problem at all," Chris said with a smile. "Follow me into the office and let me make a few calls." He got on the phone. "Jim, this is Chris at the McKnight site. I need you to help out my friends. Do you have a twenty-four foot truck available today? Great, I am sending Mr. Evans and his son over. Please take good care of them."

Nick and I climbed into the car to go pick up the truck. We looked at each other and said in what anyone observing would have thought was rigorously rehearsed choreographed synchrony, "Hire that man! That was a great experience!" What a difference in mindset, attitude, engagement, and accountability from our stop earlier in the morning.

There was nothing that was going to deter Chris from helping us. No barrier, challenge, or obstacle was going to prevent Chris from creating a very happy customer, and two customers for life. Chris chose to take accountability to achieve what mattered most to him and his company.

It was obvious that Chris was crystal clear on what mattered most to his organization. Creating exceptional client experiences is one of the company's highest desired results. That desired outcome manifested itself in the manner Chris worked with us. There was no need for his manager or owner to tell him how to behave or interact with customers. When Chris found himself in a position to influence that desired result, there was not a challenge, barrier, or obstacle that was going to prevent him from doing so. Chris could have easily used several "reasons" to offer an excuse as to why he could not help us. Instead, he maintained a can-do mindset and focused on finding a solution. He seized complete ownership to create an optimal experience and chose to take personal accountability to ensure that outcome was realized.

## Accountability Accelerator

Implant, embed, and tattoo this into your brain. When your team members or employees possess only a hazy, faint, or general understanding of your must-achieve desired results, it smothers the flames of peak performance and accountability. Worse yet, it fuels confusion, misunderstanding, miscommunication, lack of collaboration, and disenchantment.

Furthermore, when the corresponding metrics of your desired results are nebulous, or inexact, it is no better than employees not knowing them at all. Being "close" may be helpful when playing horseshoes, but not so much in business.

Your desired results (outcomes) must be memorable and internalized by every single employee involved. Equally as important, every employee must know how each of the results is being measured, what the desired metric (goal) is, and where the team or organization stands today (current metric) in relation to the desired outcome. As my former colleague and best-selling author Dr. John Kotter regularly proclaimed, "If employees do not know where the team resides today against the desired outcome, how can they possibly make an informed decision of what actions to take in order to help move the needle in the right direction?" Think about it this way. If the players on a basketball team did not know the score of the game, did not know if they were winning or losing, or how many points they needed to ensure victory, how would that influence the decisions they make during the game? Let's take this a step further. Suppose some of the players believed they had scored enough points to win, while others felt more buckets were needed, how would that affect team play? Robbed of this information, how could the players possibly make informed decisions on how to best execute?

Many supervisors, managers, and leaders assume their employees know the top goals and objectives. Data collected over the past twenty-five years reveals that nearly nine times out of ten, employees are not clear beyond any doubt on what is expected of them, or on their organization's most important goals and objectives. The consequences are grave.

Imagine you are a member of a highly skilled team on a clandestine mission essential to securing peace on earth. Your team is navigating the high seas as you journey to your classified outpost. Once at your desired location the mission's success is all but guaranteed. The captain of your vessel asks you to tend the helm, proceed at full speed, and hold course at 225° SW for six hours. Instead, at full speed, you maintain course at 220° SW, which is close to 225°, for six hours. What ramifications may that have on your mission's success? But you were close to doing what you were asked!

Suppose you are a musher in the Iditarod, and each of your sled dogs was running only a few degrees in different directions. What impact would that have on your desired outcome of winning the competition?

Daresay you were injured and needed an ambulance. Your friend quickly grabs her phone and dials a number "close" to the ambulance service number. How long will you be waiting for your ambulance?

Picture yourself quarterbacking a team. You identify a weakness in your opponent's defense and you call the perfect play to capitalize on it. But some of your teammates thought they heard you call a play that was "close" to the one you specified. What may happen when the ball is snapped to you?

While reading a magazine, you stumble upon what you believe is the perfect cake recipe for your upcoming family holiday event. As you follow the directions, you are pretty "close" to including all of the ingredients, but exclude the flour and sugar. What feedback might you receive from your family about your cake?

The notion of clarity and the impact it has upon high levels of accountability and performance was reinforced in my mind not long ago. I was joining my colleague Al to work with a client. When I arrived at the hotel to meet him I noticed Al rented a convertible sports car for our trip and that stirred up some internal anxiety.

Al typically drove as though he was competing to win the Indianapolis 500. He drove the same way he managed his business—full acceleration, proactive, and enthused. So I was a bit reluctant to sit shotgun for our three-hour drive heading north on California State Route 1, which promised tight winding roadways, slow-moving tourists, treacherous cliff-side overhangs, few passing lanes, not to mention the frequent landslides and erosion. I prepared to white-knuckle it and hope for the best.

We met in the lobby at 5 AM. It was still dark outside and heavy raindrops were pinging off the rooftop. As is typical near the shore, there was also a low-hanging veil of thick fog that limited our vision severely. Not realizing it until we were several minutes into our trip, I discovered that the weather was a gift from above. The elements had come together to protect me from Al's nerve-racking driving.

Without clear vision, direction, or sense of where he was, Al had no alternative but to move slowly, tentatively, and react to what was around him. Both of us were a bit stressed, anxious, apprehensive, nervous, and frustrated because of our circumstances.

I secretly prayed the weather would not change until we arrived at our destination.

Al's cautious driving fortified a crucial business lesson and foundational leadership practice: the importance of a well-articulated and crystal-clear vision of where you are going, that connects to the head and the hearts of those you are relying on to help you get there.

Exemplary leaders paint such a vivid picture of the future that employees can see themselves in it. When employees at all levels can see that future, and what is in it for them, they will voluntarily choose to take appropriate actions to help create it. It is like increasing the speed limit from 30 mph to 75 mph. The premise is simple actually. It is difficult to act inconsistently with how you see yourself in the future.

I am not talking about a long-winded, overly verbose vision that you often may see (but not read) hanging in a corporate lobby.

What we are referring to is extreme clarity on the top two or three most meaningful results that, once accomplished, will place you, your team, or organization in a much better circumstance. Dr. Kotter referred to these desired results as "The Big Opportunities." Another colleague of mine, Chris McChesney, co-author of *The 4 Disciplines of Execution*, calls them "Wildly Important Goals."

Selecting your two or three top priority results is not a simple task. However, it is essential to establishing heightened levels of personal and organizational accountability. At any given time, organizations have so much in play (strategies, efforts, initiatives), almost all of which are important, that asking senior leaders to select just two or three is challenging, yet absolutely vital.

There will always be more good ideas than capacity to execute on them. If every "good idea" and every goal your team or organization is targeting becomes a top priority result then you end up with no focus. If everything becomes a priority, then nothing is a priority and the result is a giant blur that chokes performance and effectiveness. The law of diminishing returns takes over and you will end up with poor to mediocre outcomes with all of your objectives. When employees are overwhelmed with a plethora of priorities, they default to a focus of being "active" rather than focusing on what matters most: your top priority results. Activity alone does not guarantee desired results will be achieved.

The key is to create unmistakable clarity on the top two or three top priority results that exist at this particular point in time that, when accomplished, will bring about the highest return. Once one or more are realized, new top priority results replace them. The top priority results are not static, but rather dynamic based upon external and internal driving forces.

Metrics are essential for focus and guidance, but your desired results should be outcomes from achieving those metrics. The selected outcomes must benefit all stakeholders and members of the team. A groundswell of momentum and waves of passion can be generated for a cause, but almost never for a metric.

Finally, the outcomes must be precise so that everyone involved is clear on what success looks like. Desired results such as "to be the industry leader," or "become world-class," may be admirable and worth pursuing, and not many would argue with them either. But they do not provide the absolute clarity necessary. As an employee, these may sound great to me, but I am probably asking myself what exactly do I need to do to help us be the industry leader, or become world-class? How close are we? How will I know when we get there? What does it look like?

When a vision, outcome, or set of desired results, is foggy, cloudy, hazy, or nebulous, employees will act and feel the way Al and I did during that trip: stressed, anxious, apprehensive, worried, nervous, unsure, and frustrated. They will be reactive, tentative, slow-moving, and lack engagement. Those feelings and actions asphyxiate optimal performance and stifle accountability.

Imagine Al driving on that same road on a clear, sunny day. I did, and I experienced how he behaved and felt: engaged, fast-moving, proactive, empowered, confident, assertive, poised, and with a can-do mindset.

The same thing holds true in the business world. Clarity on must-achieve desired results will generate those same desirable behaviors and feelings.

Close is not good enough. Unmistakable precision and accuracy are imperative.

## Shine the Spotlight on the One and Only

There are times when even the best intentions inadvertently undermine the development of high levels of accountability, engagement, collaboration, and teamwork. The scenario that played out with Cameron Medical is a classic example.

At the annual sales conference, Barbara's opening comments, clear vision of the future, and the benefit it would bring to all in attendance created excitement. The conference was in full swing and the 720 salespeople attending hungered to learn more about the new products as well as the new sales plans and quotas.

Barbara was the newly appointed senior vice president of sales at Cameron Medical. She and her team were confident that the new products and technology now available to their clients had set up Cameron Medical to have perhaps their best year ever.

The sales force was adrenalized to learn that cross-selling and collaborating with sales representatives from other regions would be instituted. This change in approach would allow them to increase individual sales and ultimately their income.

The remainder of the conference agenda included a heavy focus on product-training, consultative-selling skills, and developing regional sales plans.

Nancy, David, and Tony had started with Cameron Medical together seven years prior and had forged a strong friendship over the years. Although in different regions, they supported, encouraged, and helped one another over the years. As the regional teams were assembling to move into their breakout rooms, the three friends committed to meet for a drink following the regional team dinners on day three.

The next two days were a whirlwind. Intense product and sales training were followed by detailed specificity and discussion of regional sales plans. The processes and guidelines for the newly adopted strategies of cross-selling and collaboration with salespeople in other regions were introduced and meticulously covered. Over the three days, excitement about the future possibilities swelled.

The regional vice presidents devoted the final afternoon to focusing on regional revenue goals and individual revenue assignments.

With the conference concluded, it was time for the three friends to meet up. David was the first to arrive and spotted a table in the back corner. He sat down, and was soon joined by Nancy and Tony. The three friends all appeared exhausted, but upbeat. It had been a demanding and intense, but rewarding three days.

The friends spent time catching up on family and talked about the new products and sales processes. All three agreed the new products and processes would set them up to have their best year ever.

The conversation quickly turned to the regional and individual sales contributions their vice presidents introduced during the regional breakout meetings.

"These new products are exceptional, but the increase in my revenue target is substantial and has me more than a little bit worried," Tony admitted to his friends.

"Really?" Nancy replied. "I am feeling pretty good. My assignment is almost identical to what I had last year. Your numbers seem more aggressive than ours."

"I am glad I do not have your number, Tony. I think I might be able to hit 100 percent of mine even though it is more than what I have done in my best years," David said with some tentativeness in his voice.

"We were told that it will be a stretch and will require us to be more innovative and creative, and push us to think outside of the box and take risks," Tony disclosed.

"The message we received was different," Nancy divulged. "We were told to stick with what has worked and not to change our approach dramatically."

"Even with the new products, growing to $1.4 billion will be tough," Tony added.

"What do you mean $1.4 billion?" asked Nancy. "Walt told our team that $1.1 billion was the company target."

"Wow, Maureen made us work with an organizational objective of $1.25 billion when we were developing our plans," David confided. "It seems all of us came away from our breakouts with a different understanding of strategies and the actual organizational goal."

As the friends talked they discovered that each regional vice president introduced different organizational goals and strategies to their regional teams.

Tony's vice president zeroed in on the "stretch" goal that the senior team felt could be achieved if the stars aligned. David's vice president used the goal the senior team felt was more reasonable. Nancy's vice president focused on the number to which the senior team had committed to the board of directors.

Annoyed, Tony claimed, "They do not even agree with what we are supposed to go after. You know what? I am just going to do my job and focus on what I have to do. Based upon what you shared with me, Nancy, the quota they assigned me is 22 percent more than what they really need from me."

David thought aloud, "I am not sure now if my number is enough, or too much. Is it a stretch goal, must achieve, or nice to have?"

"I have made President's Club for the past six years," Nancy remarked. "If your regions are using much higher numbers than ours, how will that impact how my performance is viewed? I am confused."

The reality was Barbara and her regional vice presidents were quite confident that the new cutting-edge products coupled with the new selling strategies would easily produce $1.1 billion in sales. That was a slam dunk in their minds and was the number provided to the board of directors.

In their hearts, they believed that $1.25 billion was more likely and that if everything fell into place perfectly $1.4 billion was possible.

Similar scenarios play out often in organizations of all types, and although planting a stretch goal in the minds of those you are relying upon is well intended, often it triggers negative consequences.

The lack of clarity and complete alignment around one specific desired result for Cameron Medical led to confusion, miscommunication, misunderstandings, mistrust, cynicism, apathy, stress, frustration, and isolation.

The enthusiasm, engagement, passion, and energy infused throughout the sales force was demolished by lack of alignment on the part of the senior executives.

Upon discovering the unintended consequences of their actions, the leadership team at Cameron took measures to wipe out any confusion. Though it took several weeks of focus, attention, resources, and time, the senior team at Cameron eventually mobilized the sales force to rally around one metric, $1.3 billion, and the desired behaviors for flawless execution.

Cameron Medical closed out the year with sales of $1.37 billion. Their best year ever.

## An Unexpected and Illuminating Discovery

My former colleague and mentor, Tom Peters, co-author of *In Search of Excellence* and many more best-selling books, once told me, "If leaders would just get out of the way, most organizations would achieve incredible results." His thinking was that senior leaders should spend more time visioning, strategizing, and removing obstacles, and less time directing.

George S. Patton put it this way, "Never tell people how to do things. Tell. them what to do and they will surprise you with their ingenuity."

We could not agree more with both of them. When leadership (at any level) provides pristine clarity on top desired results, employees will astonish and amaze. Angela experienced it firsthand with her team at Lakeshore Bionics.

Angela was uneasy to the point of being distressed about her impending leave. She confided to me that she had been losing sleep dwelling on what might happen during her absence. In less than three weeks Angela would be taking leave for health-related issues. For the four months of her sabbatical, she would be completely out of touch with anyone from Lakeshore Bionics.

"We are in the midst of implementing new systems and structures that are integral to all functions within Lakeshore Bionics. All eyes are on my team right now. Anything short of awe-inspiring success will be viewed as disappointment.

"There is a lot riding on this. Lakeshore's recently announced growth plans, new strategic partnerships and distribution channels, my team's future, and, as selfish as this may sound, my position here. There are a lot of people counting on my team right now."

Over eighteen years, Angela had climbed her way to a senior leadership position with Lakeshore Bionics. She described her leadership style as a bit controlling, hands-on, directive, and with a heavy dose of managing the actions of her people. Her belief was that without her presiding over their actions it was wishful thinking on her part to expect much.

Having the opportunity to provide coaching and consulting to Angela over the prior two months, I can attest that her self-assessment was spot-on, and perhaps a bit light on the "command" sentiment.

It was obvious why she was feeling anxious, stressed, and apprehensive. Angela clearly did not believe her team could pull this off without her.

I encouraged Angela to spend focused time to visualize what success would look like upon her return.

Angela meticulously crafted her vision of what she was hoping her team would achieve in her absence. She spent two weeks agonizing over and detailing every one of the most crucial results that must be realized, why they were important, and how success would be measured.

A few days prior to her departure, Angela shared her vision of success with a small team of her key leaders and managers. The sparkling clarity of what was expected was unmistakable to everyone. They were aligned and committed. Yet, Angela was still distraught.

Upon her return four months later, Angela was astonished to find her team had far exceeded everything she had hoped for. Although pleased, she was puzzled. How did this happen?

As Angela talked with her team, she learned that because she was so clear in her presabbatical communication detailing her vision of success and specificity in clarifying the top desired outcomes, her people had no problem figuring out how to make them reality. In fact, she was astonished with the level of innovation, creativity, and ingenuity her team displayed during her absence.

The results she needed her team to produce were so precise, measureable, and memorable that each team member knew exactly how they contributed, no matter their role.

She discovered she did not have to waste time, energy, and effort on directing, telling, enforcing, and instructing her people. She stumbled upon a key leadership practice of creating a highly accountable culture. Angela learned that if she was precise and clear on the desired results her team must accomplish, they would figure out how to get it done. Angela found that when what is expected of them is clear, her team would voluntarily tap into their discretionary performance area to give more than what is required to achieve what matters most.

With this crucial learning as motivation, Angela's thirst to learn more was unquenchable. By internalizing, practicing, and implementing the best practices of fostering a high-performing and accountable team, she developed into what her CEO called a model leader that he encouraged others to emulate.

## Do They Know the Rules of the Game?

There is a considerable difference in the degree of engagement, commitment, passion, and personal accountability possessed by employees coming to work every day playing to win, versus those playing not to lose. This is not a subtle issue. Are your employees coming to work every day trying to keep their heads above water, or are they focused on what they can do to further the cause and help the organization move closer to what matters most? Lack of clarity breeds the mindset of playing not to lose and brings about inaction, confusion, misunderstandings, lack of alignment, anxiety, frustration, and ultimately resignation to a "just doing my job" mindset that smothers peak performance and accountability in any organization.

Deprived of the "rules of the game" and immaculate understanding of what matters most (must-achieve desired results), the tsunami of the "day-job" mentality takes over and employees can easily allow themselves to be sucked into the trap of busywork.

The leadership team at CVP International embraced this idea and used it to develop a culture with unsurpassed levels of accountability, ownership, collaboration, and engagement.

Debby, the senior vice president of human resources at CVP International, invited me to meet with the company's senior team of nine. CVP International partners with local and international organizations to promote social and economic change throughout the world. A highly successful firm over the years, they were looking to become even more effective. Debby shared with me that Robert, her CEO, held the topic of accountability near and dear to his heart. Her belief was that Robert would embrace the opportunity to create even higher levels of accountability among all employees.

The firm was considering several options to further develop their personnel in order to leap to the next level of performance. We were offered a window of four hours to meet with the senior leadership team, share a bit about our work, and present the case for enhancing levels of personal accountability as the most viable and beneficial option. Debby coached me in advance, sharing that these team members would be skeptical. They had already been pitched by several other firms, and were growing weary of the dog and pony shows. Debby confided that she had had to do a lot of "internal selling" simply to get this team to submit to yet another consultant's presentation. We were warned that some might choose not to show up.

Prior to our meeting with the team, Debby was able to secure a thirty-minute telephone conversation for us to speak with Robert in order to gain a sense of what was most important to him and what he believed would propel them to the next tier of greatness.

During that conversation, we asked Robert what were the top two or three most important things his organization must achieve over the next twelve to eighteen months. This is not a simple question to answer for any CEO. There is so much that must be done in organizations, and to narrow it down to only two or three top priorities is incredibly difficult. Robert spent the next twenty-five minutes whittling down a list of eighteen important goals down to what he believed were the three "must-achieve no matter what" desired results.

With only a minute left in our telephone appointment, we asked one final question to Robert, "What percentage of your employees know that these are the three most important results that must be achieved?" Without hesitation, Robert confidently responded, "At least 90 percent."

We thanked Robert for his time and told him we looked forward to meeting personally the following week.

Debby escorted us into the boardroom thirty minutes before the meeting. While we were setting up, Karen strolled in and, without introducing herself, grumbled, "How long is this going to take?"

As the next five members of the leadership team arrived, I noticed their excitement about spending four hours with me mirrored, or was substantially less, than Karen's.

Jim arrived and stated that Derek and Ken chose not to attend as their mornings had filled up and they did not have time for this. Robert was the final member to arrive. He had just finished a call with a colleague in Nigeria, and it was obvious from his demeanor that there was an issue weighing on his mind. Based on what appeared to be a common sentiment in the room—not much interest in being there—we decided to alter the planned approach.

Instead of starting with the standard introductions and niceties, along with a quick overview of our work and sharing our understanding of what they were hoping to achieve with whatever firm they chose to partner, we opted for another tactic.

"Before we start, would you open up the materials we developed for you to page six? We know your time is valuable and you all have a lot on your plate. So, let's jump right into this.

"You will see that page six is a blank lined page. Without discussing with your colleagues, write down on that page the answer to the following question, 'If you were sitting in Robert's office and asked him what the top three must-achieve desired results this organization must absolutely realize over the next twelve to eighteen months, how would Robert respond?'"

The facial expressions of the team were priceless. We could only imagine what they were thinking. Seven perplexed faces looked at us and then at each other, wondering what in the world they were doing here.

"Go ahead. Write them down exactly as you believe Robert would respond. We will give you five minutes to think about it."

Robert glanced at us with what seemed to be calm assurance. You will recall that he confidently claimed to us during our telephone discussion a few days earlier that at least 90 percent of all employees, not just his senior leadership team, would be able to answer that question exactly the way he did.

As the other members sat in silent contemplation for two, three, then four minutes without writing anything in the materials, Robert's optimism turned to concern bordering on dismay.

"One more minute to finish your list, team. If you cannot come up with three, write down at least one," we told the group.

"Now that you have your list compiled, we have two more requests for you team. Next to each result that you recorded, please write down two numbers for us. First, what percentage of your sixty-seven hundred employees would answer the question exactly the way you did? We will give you one minute to write down that percentage.

"Secondly, how are you measuring the result? In other words, what is the metric you are using? For example, if you wrote down that we must achieve optimal customer satisfaction, how are you measuring that? What is the metric, or result, that you must achieve, and where does the organization stand against that metric today?

"Okay, so let me capture these on the whiteboard. Karen, share with me just one that you recorded on page six."

"Well, I am not sure this is right, but I wrote down 'employee satisfaction.'"

"Great. How many of you in the room, by show of hands, wrote down 'employee satisfaction'?"

Two hands went up, and Robert's was not one of them.

"So, Karen, what percentage of your sixty-seven hundred employees would have responded exactly as you did?"

Shyly and almost inaudibly, Karen said, "I wrote down 95 percent."

"Okay, great. So let me note that on the whiteboard.

Attempting to be humorous, we suggested to Karen, "So two out of seven members of your senior team had that as a must-achieve desired result, but you believe 95 percent of all employees would have stated that result?

"And finally, Karen, what is the metric you are using to measure that result?"

Karen whispered, "I did not get that far. I am not sure how we measure that."

"No problem at all. Karen. Let's move on to another one. Jim, share just one that you have on your list."

Reluctantly, Jim offered, "Healthy EBITDA."

"Perfect. So how many of you in the room had EBITDA on page six?"

One other hand went up. Yes, of course, it was the chief financial officer.

"Okay Jim, what percentage of your sixty-seven hundred employees would have answered exactly the way you did?"

"Well, I had written down half, but based upon what I am seeing here, I want to adjust that now."

"For now, let's use what you have. What is the EBITDA metric that must be achieved?"

"We need to be at 12.7 or better."

"And Jim, how many employees would know that number, or for that matter, would know what EBITDA is and how they contribute toward improving that metric?"

Humbly, Jim responded with, "Few to none."

This exercise went on for another ten minutes. As the list of top three desired results swelled to nineteen, we could see that Robert had a moment of self-discovery. He allowed the conversation to go on for another five minutes and then asked if he could speak with his team.

As we have discussed to this point, clearly defined desired results are foundational to cultivating high levels of accountability. See figure 1.1 for common consequences when this principle is absent, and for a partial list of some of root causes.

"This has been eye opening for me. I need to take ownership for our lack of alignment around what is most important. I assumed you all knew. I failed you. I was certain that all of you would answer Mike's question with the same three results I identified with him on the telephone last week. I was wrong. If I did not stop this exercise that list may have ballooned to thirty or more. I would like Mike

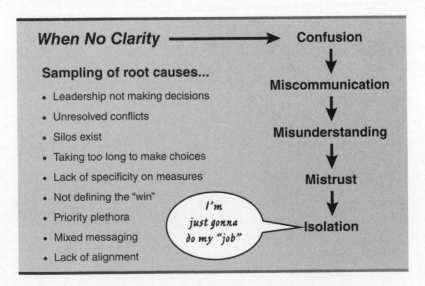

FIGURE 1.1    Lack of Clarity: Consequences and Sampling of Root Causes

to help us come to agreement on and help us create clarity and alignment around the top three must-achieve desired results for this firm. Then we must do what is necessary to make sure all sixty-seven hundred of our employees know those results are what we want them to take accountability to help us achieve. Everything they do should contribute to assuring we achieve those results."

All seven members enthusiastically agreed.

Sandra asked, "I see the value in everyone knowing the most crucial desired results, but I am not sure I get the whole 'metric' thing. Why is developing a metric for each one so important? Is not just knowing what is important effective enough?"

"A very valid question, Sandra," I replied. "And the reason is rather simple. If employees do not know how we are measuring success, and where we stand today in relation to our desired state, how can they possibly show up daily and make a conscious decision about what they can do each day to take accountability to help move the needle in the right direction? It would be like opening up a board game with hundreds of pieces, but no directions detailing how to play and, more important, how to win. If I do not know how to win, how can I possibly make a purposeful decision about what action I can take to do so?

"Another bonus is that employees are at their best, and most engaged in their work, when they believe they are playing a winnable game. The highest standard to which a leader can aspire is if they are creating a winnable game for those they lead. To do so, everyone must know how we are keeping score and what the score is today.

"Suppose this huge whiteboard in the front of this room was instead a window and that we were five floors up looking over a playground basketball court. Close your eyes and create this picture in your mind. On that basketball court are a bunch of kids in the middle of a game. Assume we cannot see any scoreboard, but we know they are keeping score based on the behaviors they're displaying. Tell me what you are seeing that suggests you know they are keeping score?"

The group in the room rapidly began to shout out responses.

- "They are high-fiving each other after a basket."
- "I see them supporting and encouraging each other."
- "They are playing hard, hustling, and giving it their all."

- ◆ "They are cheering for one another."
- ◆ "They are engaged and focused."
- ◆ "They are talking to each other and giving advice and feedback."
- ◆ "Everyone is contributing. Even the subs on the bench are involved and energized."

"That was a simple exercise demonstrating how the power of keeping score ignites a variety of beneficial behaviors. Are those the types of behaviors you believe would contribute to creating a culture of accountability within CVP International?"

The message was received loud and clear.

After six months of concerted effort by Robert and his entire team to adopt and implement this key accountability principle among others, Robert called us with an update. He reported that the board of directors commended him and his team for leading the organization into heights of performance they never believed would be realized. Furthermore, they confided that they sensed a positive difference in the attitude and demeanor of the culture.

Robert went on to share a story about his recently hired new chief of party in Nigeria. After being on board for less than a week, Dikembe knocked on Robert's door and, unsolicited, offered his plan to help the firm achieve the three must-achieve desired results that the team identified several weeks earlier.

"Evidently," Robert exclaimed, "the constant discussion about these three results among all employees within the organization and the fact all meetings begin with a thirty-second reminder of how that meeting will help move us closer to one of the results was eye opening for him. Dikembe divulged that the abundance of visual reminders and employee passion inspired him to proactively seek me out to pledge his alignment and dedication to doing whatever it takes to make sure we achieve them all."

Creating a winnable game for your team or your employees is compulsory to cultivating a highly accountable organization. Employees at every level must be keenly aware of the rules of the game: the desired results; how each of those results is being measured; and where the team or organization stands today in relation to the desired metric (what must be achieved).

Heightened levels of employee engagement, commitment, creativity, perseverance, alignment, collaboration, and morale are additional bonuses resulting from this key principle.

## Key Learnings to Internalize

### One:

There is what some may call a gravitational force that pulls people into the depths of the blame-game. It can feel as if all of your friends are frolicking in a crystal-clear turquoise pool, drinking margaritas, and declaring, "Dive on in, this is great. All of these problems are not our fault and there is nothing we can do about them."

All individuals, teams, and organizations at some point will fall into the blame-game, and may even be justified to play it. That is human nature. We all go there. Some unforeseen event may cause disarray, spawn obstacles, or precipitate chaos. Some person may drop the ball, not follow through on a commitment, or not perform as expected. You may be faced with a challenge, barrier, or obstacle that is more difficult than anticipated or beyond what you feel you are capable of overcoming. A supplier may fail to produce a shipment. A colleague may make a mistake. In these instances, and the list is endless, you can decide to play the blame-game and cling to a notion of "this is not my/our fault and there is nothing I/we can do about it," or you can recognize the reality you are in and press on with a mindset of "okay, here are the realities, what do I/we have to do to achieve what matters most?" The choice made is always directly linked to the degree of ownership you or your team possess for the result. How much do you or your team truly want the result? If ownership for the result is low, an abundance of excuses and blame will come quickly and easily. When ownership is high, solutions will always become apparent.

### Two:

Even when we are justified to blame someone or something for our predicament, the time we spend doing so is not time well spent. It may feel therapeutic to blow off steam, vent, and point the finger of blame. But it is important to be able to recognize the realities of your situation and understand that nothing meaningful, important, or significant is ever accomplished while submerged in the depths of playing the blame-game.

Think about somebody you know in your personal or professional life that seems to consistently achieve or exceed their desired results. I bet they rarely, if ever, let any challenge, barrier, or obstacle deter them from moving closer to achieving what matters most to them. This holds true for top-notch professional athletes, Olympians, musicians, businesspeople, students, parents, engineers, bankers, teachers—you get it—for anyone who has achieved anything significant, important, or meaningful.

By the way, it is not wrong to play the blame-game. We all do it. In the past three weeks, have you played the blame-game? If you replied no, you may not be recognizing reality. Sometimes we need to vent and understand the reasons why we are in our current situation. The key is not to stay there. The difference between highly accountable people, teams, and organizations and those that are not, is that they have the ability to recognize when they are playing the blame-game and possess the internal fortitude to pull themselves out of that losing proposition.

When it comes to recognizing your realities (personal or professional), it can be helpful to consider the following power-ful, but not well-known truth. *Human beings are the only living creatures with the capacity and propensity for self-deception.*

In other words, do not kid yourself! Things may not be as rosy as you are convincing yourself to believe.

Playing the blame-game is easy. It requires no effort. No action. No thought. No energy. No resources. Many people spend a lifetime playing the blame-game. They are always quick with a story, excuse, or explanation why they cannot achieve what they really want. I would venture a guess that you know some of them, or work with some of them. Beware of these

folks. They want company and companionship while wallowing in self-pity, blame, excuses, and denial. They will do all they
can to have you join the dark side. These people can destroy a
team and demolish an organization.

How does it feel to work with somebody who has an
endless supply of excuses, stories, and explanations why they
cannot do anything, or why something was not their fault? It is
draining. These folks can suck the life out of you, your team,
and your organization.

### Three:

Tom Peters once shared with me this golden nugget that I have
internalized and believe with every fiber of my being: *If you
want something badly enough, there is always something you
can do to take a small step closer to achieving it. It is just a
matter of ownership. How badly do you want it?*

This sentiment was further cemented within me while
watching the show *Alone* on the History Channel recently. As
the network describes it:

"Ten people enter the Vancouver Island wilderness carrying only what they can fit in a small backpack. They are alone
in harsh, unforgiving terrain with a single mission—stay alive
as long as they can. These brave men and women must hunt,
build shelters, and fend off predators. They will endure extreme
isolation and psychological distress as they plunge into the
unknown and document the experience themselves. No camera
crew. No producers. It is the ultimate test of human will."

After eight of the ten contestants had "tapped out," one of
the two finalists now vying for $500,000 in prize money, Larry
Roberts, after having spent more than sixty days alone with no
help, no resources, and no camaraderie, was nearing his breaking point. Larry had displayed an iron will throughout, but every
man has a breaking point. Larry began to prepare mentally to
hit the "tap" button on his satellite phone, essentially giving
up. Larry reappears only a short time later. Recording himself
on camera, Larry spoke from the deepest crevice of his heart

delivering a soliloquy that sent chills down my spine. Larry revealed what all of us are capable of, but far too many choose not to endeavor. His short, but inspirational insight:

*"When you are right at that point where you say I did all I can do, there is always a little more you can give. There is always a little more. You can dig always a little bit deeper. I'm digging deeper. Digging as deep as I can possibly go. I'd rather go out swinging."*

Imagine what that level of commitment, ownership, and accountability can do for you, your team, and your organization. The remaining chapters in this book will equip you with principles, tools, and ideas to do just that.

**Four:**

When your team members or employees are not clear on what you want them to achieve, how can they possibly choose to take accountability to help achieve those goals? Consoling yourself by thinking "Most of my employees know, or would be close" is a recipe for failure. Assuming that "everyone knows what is important here because we talk about it all the time" is a blueprint for disaster. Because nearly nine times out of ten, they do not know what the top desired results are and will default to working toward what they decide and believe to be most important. Often, what they choose may not be aligned or congruent with what you identify as most important.

When top priorities are not abundantly clear, leaders abdicate cultivation of a highly accountability team or organization to pure luck. Team members and employees simply cannot choose to take accountability to achieve results that are ambiguous, vague, or foggy.

Bereft of pristine clarity, teams and organizations experience lack of alignment coupled with degrees of confusion, misunderstanding, miscommunication, mistrust, cynicism, inactivity, denial, apathy, stress, frustration, skepticism, and isolation. Ultimately this will give rise to excuse-making, finger-pointing, and a heavy dose of the blame-game. The blame-game breathes life into the toxic emotions of guilt, resentment, and anger.

**Five:**
When every single employee is crystal clear on your top desired results they are more likely to go above and beyond what is "required." Helping their team or organization realize the top priorities becomes viewed by each employee as "my job," which leads to employees voluntarily tapping into their discretionary performance area to contribute more fully.

**Six:**
Clearly defined desired results increase alignment, nurture collaboration, strengthen teamwork, improve communication, instill focus and determination, forge camaraderie, spawn creativity and innovation, and deepen employee engagement and morale. Most important, they breed a can-do mindset versus the destructive thinking brought to life by the blame-game.

**Seven:**
When must-achieve desired results are not known to all employees, variations of the blame-game begin to be played out, such as "I am confused," "I am not sure what you want me to do," "I did not know that was important," "Nobody ever told me that," "That's not my job," "I thought they were supposed to do that," "I was waiting for someone to tell me," "I thought we were tracking okay," and many more.

## Take Accountability Now

**One:**
Identify the top two or three most crucial desired results that your team or organization needs to accomplish. Assign a metric to each one so that every member of your team knows when the desired outcome is reached. Limit yourself to only one metric per must-achieve desired result. Remember that stretch

goals create confusion, misunderstanding, miscommunication, mistrust, lack of alignment, frustration, cynicism, apathy, anxiety, and other negative consequences that fuel the blame-game. If you believe that a stretch goal is truly possible, then make it your desired result. Commit to one outcome.

**Two:**
Once you have identified your top results, go out and ask people to recite them. Ask a lot of people. If fewer than 90 percent can state them exactly as you communicated them, develop your plan to increase clarity, alignment, and understanding. It is only when team members and employees are keenly clear on what must be achieved, and how success is being measured, that they can choose to take accountability to help move the team or organization closer to your desired outcomes.

**Three:**
Develop your communication plan in a way that engages both the head and the heart of all parties and inspires them to voluntarily choose to take accountability on the front end, while they have time to positively affect the result. When accountability is positioned up front, people get excited, enthused, and engaged about the opportunity to help.

**Four:**
Identify three important results, projects, tasks, or assignments where you, your team, or organization have fallen into the unwinnable trap of playing the blame-game. Develop a plan to propel you, your team, or organization out of the blame-game quagmire and to take accountability to achieve what matters most.

In order to move yourself, your team, or your organization to a heightened level of accountability, consider these questions: What are the consequences if you do not achieve the desired result? How much time do you spend blaming other people, things, or events as to why you have not made progress?

How much time do you devote to searching out solutions and exploring alternatives? What is the one specific next step you will implement to make progress today? How much time and energy do you, your team, or organization spend dwelling on reasons why no progress has been made toward achieving what matters most? How much time do you or your team spend rehashing the same old barriers, challenges, and obstacles time and again? What actions have been taken to move forward since the last discussion about these impediments?

There are two additional powerful questions to consider. What is the one thing you have been unable to do, but you know if you could, it would have a fundamentally positive impact on your ability to achieve what matters most? If your job depended on making at least a little bit of progress in the next day toward your most important desired result, what action would you choose?

Waiting for things to get better is not an effective strategy. Ignoring your realities is fruitless. Nobody is going to solve your problems for you. Accountability is a personal choice. People may be well intended and good-natured and many will offer to help. However, your challenges and problems are not theirs. When the going gets tough, even the well-intended find it easy to bow out.

**Five:**
Decide now—at this moment—on one small step you will take to make progress toward your desired result. Detail it clearly, and schedule a time to implement it.

# Chapter 2

## Choices

### Decisions Determine Destiny

When you believe you have no choice, you become a victim. When enveloped in the paralyzing cloak of victimization, individuals, teams, and organizations surrender control of their destiny. The conscious awareness of choice is life changing and instrumental to achieving what matters most.

The choice to take or not take accountability is solely and exclusively yours to make. It requires a perpetual unfiltered willingness to recognize your realities, take ownership of your circumstances—good or bad—create a way forward and take action to achieve what matters most. Let us explore this further.

Tito was so affable, upbeat, and such a pleasure to be with that nearly an hour went by before I noticed that this man driving me to the conference center was missing four and a half fingers and most of his right hand. My driver, I thought to myself, now a bit alarmed, was missing most of his right hand. His sincere charm and obvious love of his work had so captivated me for sixty minutes that I did not notice.

My interest in learning what had happened to him was percolating inside. But, I thought, how do I ask someone I hardly know a personal question like that?

By this time in our journey together, we knew the names of each other's children—he had five—where they went to school, what we believed were their hopes and aspirations, and we praised them all endlessly. Tito was genuinely interested, and so was I. Two of our children attended Villanova at the same time. Tito loved the wings at Winger's as much as my son and me. I learned that he grew up in the Dominican Republic, that his wife, Lovee, was his childhood sweetheart, and we both enjoyed running and cooking.

I convinced myself that we had built enough rapport with one another that asking about his hand was probably not out of line.

"Tito, can I ask you a question that may be a bit personal?" I questioned.

"It was an accident in a manufacturing plant when I was seventeen," Tito volunteered without allowing me even to ask. "I was working to help my mother and father support our family of twelve. I was operating a machine with many gears and pulleys that were not protected in any way. That day I was wearing a long-sleeved shirt. A friend walked by and asked me a question. I lost focus as I stepped to the side of the machine and attempted to lean on it while talking with my friend. My shirt got caught up in the gears and sucked my hand and arm into the machine. I do not remember much after that other than waking up in the hospital. I was told I was lucky to be alive."

I sat and listened as my new friend described this horrific experience. He was smiling as he told the story. His story sent shivers down my spine, yet he was smiling.

"I was pretty depressed for many weeks," Tito admitted to some chagrin.

"My lord, what human being would not be depressed?" I lamented.

Then, Tito went on to share with me that at that time of the accident he was the top-rated basketball player in the Dominican Republic. He was being recruited by some big-time division one colleges in the States. Friends, family, and coaches had painted a picture of a future full of fame, success, money, and accolades. All of that was literally ripped away from Tito in a matter of seconds.

Although this had happened about thirty-two years before I met Tito, my heart broke for this man. His future, his dream, his love was snatched from him in a moment.

"Yeah, man. Best thing that ever happened to me," Tito boasted.

I was completely blindsided. What did he say? Did I hear him correctly? I asked, "Did you say 'best thing that ever happened to you'?"

"Yes. It made me a better person. I was an arrogant, self-centered, pompous, entitled little punk. I thought the world should bow to me. I expected other people to solve my problems and make my life easier.

"After months dealing with my depression, I committed to focus on learning as much as I could about as many subjects as I could. I dedicated my life to learning and I earned several degrees. To this day, I believe that is the primary reason why my children have been so successful. They have seen firsthand what it takes to succeed in life. I am a chemical engineer. I drive this car in the evenings to earn extra money to help my children.

"I learned that in this lifetime, which is a gift from above, what we achieve is a direct result of the choices we make. We make hundreds of choices every day. Some are easy. Some are difficult. Some take commitment, perseverance, resolve, and dedication."

"The toughest choices are when we are confronted with immense obstacles and challenges. I realized that nobody was going to make my life better. Sure, people expressed compassion, and offered support and encouragement, but I came to grips that my future was completely up to me. It was not easy for me. I had setbacks and wanted to give up more times than I care to admit. I felt sorry for myself and wanted pity. But I ultimately found that what I went through is peanuts compared to what others have had to overcome. I just kept reminding myself of how much there was in this life that I wanted to experience.

"You know what, Mike? Two people can be confronted with the identical challenging predicament and how they choose to move through it will almost always differ. Some people wait and hope for help, or that someone else will solve their problems, while life passes them by. Some people accept their realities and choose to make things happen for themselves."

I hugged Tito as he walked me to the conference center entryway. I hired this man to drive me to a destination, and what I got in return was one the most memorable and important life lessons that everyone should internalize.

Tito embraced an incredibly powerful truth that, once embodied, is life changing for many. All people with decision-making capacity are where they are today, both personally and professionally, as a result of every decision they have ever made in their lifetime. Granted, some people are confronted with monstrous and even nightmarish hurdles and problems and have much more difficult decisions to make and roads to travel than others. However, no matter the barrier, challenge, obstacle, or opportunity confronting you, the choice you make as to how to approach it or move through it is 100 percent yours to make. In fact, the choices we make are one of the very few things over which we have complete and total control. We will discuss this in more detail later.

For some, this truth is difficult to accept. Often people will adamantly point to some minor or horrific event or tragedy as the reason they are in their situation. They will convince themselves that it was not their fault and there was nothing they could do to prevent it. And that is the point here. Yes, something or someone may be the reason you are thrust into a difficult situation or circumstance, but how you choose to deal with it is all on you. My personal belief as to why this is difficult for many to accept is that it does not allow any wiggle room to play the blame-game.

Where you are today, personally and professionally, and where you will arrive in the future, is a direct result of the choices you have made and the choices you soon will make.

In the movie *Shawshank Redemption*, after enduring years of pain and misery in prison, Andy Dufresne looks his friend Ellis "Red" Redding in the eye with solemn commitment and says, "I guess it comes down to a simple choice, really. Get busy living or get busy dying." With that decision made, Andy was no longer a victim. When it comes to choices, Josh Shipp puts it this way:

*"You either get **bitter** or you get **better**. It's that simple. You **either** take what has been dealt to you and allow it to make you a better person, or you allow it to tear you down. **The choice** does not belong to fate, it **belongs to you**."*

Tom Peters, the best-selling co-author of *In Search of Excellence* and more than a dozen other books, coined the phrase "Tower Death" years ago. Tom would show his audience a visual of a gravestone etched with the following epitaph:

> John Doe
> 1955–2016
> Would have done a lot of
> cool stuff, but his boss
> would not let him.

His point was that too many people live their lives convinced they are trapped and have no control over their destiny. The reality is you have complete and total control over your destiny.

Likewise, the choice as to how you view your realities or situation is entirely yours. Will you, your team, or your organization choose to concentrate and place emphasis on all of the reasons why you are not making the progress you want to make? Or instead, will you zero in on solutions, opportunities, and possibilities to achieve what matters most? The parable of the two shoe salesman assigned to a remote region in the middle of a third world country is apropos. Once settled into their new locations they report back to their boss with an update:

**Salesman One:** "The situation is hopeless. Nobody here wears shoes."

**Salesman Two:** "The opportunity is unlimited. Nobody here wears shoes."

The next time you find yourself falling into the trap of playing the blame-game, think about Tito, or someone you may know like him, who has faced significant challenges and chose to maintain a mindset of "what can I do to achieve the results I want?"

## Stay the Course—Avoid Drowning "Below the Water"
"We didn't do anything wrong, but somehow we lost."

The CEO of what was once the world's leading mobile phone manufacturer made that statement on November 13, 2015. Read it again and let it sink into the depths and every crevice of your brain.

That statement is perhaps the epitome of the blame-game. It includes absolutely no ownership of why they ended up in their calamitous situation, and no hint of a plan to find a solution.

When I read that statement my mind flashed back to my good friend Captain Hartman.

Captain Hartman described to me a concept his team in the navy called "Below the Water" behavior. This was the term they used to describe civilian and enlisted personnel who would fall into the trap of excuse-making, finger-pointing, or any other destructive facet of the blame-game.

"You surely have dealt with these types of folks, Mike. They exist in every organization," the Captain said with disgust. "They can wreak havoc on a team and devastate a culture. We have no tolerance for their behavior.

"Individuals with a 'Below the Water' mindset are always full of excuses and cling to the futile notion that nothing is ever their fault. Worse yet, they are often insistent there is nothing they can do about the situation. They wallow in self-pity and blame their problems on other people, events, or things. They live their lives snuggly nestled in a cauldron of blame and denial."

Captain Hartman challenged me to think about the last time one of my colleagues, friends, or even family members approached me with an endless list of excuses and reasons as to why they could not achieve an assignment, complete a project, follow through on a commitment, or finish a chore.

Several familiar faces immediately flashed to mind. Some very quickly.

He then asked, "How does it feel when someone always offers excuses why everything is not their fault and consistently complains about their challenges, their boss, company policies, their job?"

He saw by the look on my face that he did not need to wait for an answer. That sort of behavior is exasperating. That behavior, left unaddressed, will spread like wildfire within a team and through-out an organization. It will drag a team down and immobilize an organization.

"Now," he suggested, "think of a time when one of your colleagues took ownership and accountability and said, 'Listen, I know I did not deliver on what was expected. I take ownership for dropping the ball, and here is my plan to make sure it never happens again.'

"That has a completely different feel. That is the mindset supervisors, managers, and leaders around here notice in our people. That's the culture we want.

"People around here know that excuses are not tolerated. No matter how compelling and ingenious the excuses may be, we do not accept them. It is clear that around here we learn from mistakes and accidents. Mistakes, oversights, and accidents happen. We are all human. We do not dwell on reasons why, or excuses, nor do we seek to punish, admonish, and call out who was to blame. Instead we take pause, consider the current realities, and turn our attention to finding solutions to move forward. Achieving what matters most is always foremost on our minds.

"Our folks know that when faced with difficult circumstances, challenges, and hurdles, we expect them to search out solutions and not to look to point the finger of blame.

"That blame-game and excuse-making mindset is contagious. But so is the 'let's find a solution' mindset. So we focus there to nourish that behavior so that it becomes inculcated into our culture.

"It takes self-awareness, and sometimes a kick in the ass, to realize when you are drowning 'Below the Water.' But once people realize they are floundering in the blame-game, they quickly recognize their realities and understand they must take ownership and accountability to figure out how to move forward and make progress toward helping the team to deliver on our objectives.

"We devote a lot of time and energy to creating and sustaining our culture. Bringing attention and awareness to desired behaviors is one of the most important parts of maintaining our desired culture. You would expect high levels of accountability in the military, and we have that for the most part. However, even the majority of our enormous civilian population personnel possess a 'whatever it takes' mentality. That is a result of the importance and focused attention we place on developing our optimal culture.

"You know what I learned, Mike? Perseverance comes from the same part of the brain as decision making and emotion. So to instill that tenacity and determination I push my folks to figure things out on their own, even when it may be easier or quicker for me to solve problems for them.

"I have found that when I am able to spark both the decision-making and emotional parts of the brain that perseverance, determination, and resolve flourish.

"You will find that leadership here is intentional about the words we choose and the experiences we create for our people. Words create pictures. Pictures stimulate emotions. Emotions fuel beliefs. Beliefs drive actions. And most important, actions produce results."

Just then, there was a knock at the door. "Come on in," Captain Hartman shouted. Vinnie, one of the project owners from the civilian workforce, entered the room. Without any pleasantries or chitchat, Vinnie immediately launched into a tirade. With veins popping out of his neck and blood coursing into his beet-red face, he rattled off a laundry list of all the problems, obstructions, impediments, policies, and people that were making his team's job impossible. I felt uncomfortable and awkward as I sat there and observed.

Vinnie was relentless. I turned my attention to the Captain. He appeared calm, reserved, and emotionless, yet he was genuinely listening.

Finally, Vinnie had run out of complaints. Captain Hartman calmly asked Vinnie, "Is there anything else that is holding you back?"

Vinnie, appearing a bit puzzled, thought for a few seconds and then said, "No, that's all of it."

"So, of all that you just shared with me, and correct me if I am wrong, there are seven or eight issues that are out of your control and it sounded like there are two or three that you can control. Is that about right?"

"Yes sir, that sounds about right," Vinnie admitted.

"Vinnie, you know as well as I do how important this project is and what is on the line here. There are lives at stake. We have people depending on us. What is the one thing you can do now to make progress toward the results we must achieve?"

At that point I could see Vinnie's disposition and countenance change. The Captain had skillfully helped Vinnie focus on what he could control instead of what he could not.

In a matter of seconds Vinnie began to rattle off several ideas of how his team could move through the perceived barriers and make progress to ensure the project was not delayed.

Once Vinnie left, Captain Hartman shared, "When Vinnie walked in here, he was 'Below the Water.' He was stuck and frozen in what you like to call the blame-game. I have learned over time that to get people to take accountability, ownership, and move forward, it is best to let them draw their own conclusion to do so. Asking open-ended questions is one of the tactics I have found to be most successful."

# Beware of the Bell

Generating a groundswell of momentum, passion, energy, commitment, and accountability is essential when teams or organizations set their sights on seizing must-achieve desired results. Knowing it is important is common. Executing it is not.

The reality is that there are some employees that will simply never contribute toward your success. Let us call them the "nevers."

An article, "Three Types of People to Fire Immediately," by G. Michael Maddock and Raphael Louis Viton, appeared in *Bloomberg Businessweek* on November 8, 2011. The three types of people identified were "The Victims," "The Nonbelievers," and "The Know-It-Alls." Keep those descriptors in mind as you read the story of Harton Industries.

"We have wrestled with this issue too often," Dan, the CEO of Harton Industries, bemoaned.

Dan was a brilliant strategist and regarded as one the leading experts in his field.

"Too many of our initiatives seem to launch successfully, but ultimately that initial velocity dissipates. The excitement, ownership, and dedication fade into the ether. Over time, and with each new strategy we launch, it has become increasingly more difficult to muster and sustain any enthusiasm."

Dan knew that if they were to launch what they believed was a transformational new strategy crucial to the firm's future success in the same manner as they had the previous three initiatives, the outcome would most assuredly be the same.

Dan and his functional leaders planned to devote time during their leadership retreat to devise a new approach.

The research revealed by the guest speaker on day one of the leadership retreat was timely. It smacked Dan in the face as if he was hit in the head with a bat swung by Miguel Cabrera (First Baseman, Detroit Tigers).

That's it, he thought. We have focused way too much of our time and energy trying to convert the nevers. Why are we wasting our time trying to convert the unconvertible?

*Nevers* was the term the speaker used to describe the 20 percent (on average) of employees in any team or organization that will *never ever* contribute to your success.

"In fact," the speaker shared, "the nevers often proactively sabotage and derail your efforts. These are the folks that show up physically every day, but check out mentally once they arrive."

The speaker described them as often apathetic, cynical, bitter, negative, disenchanted, distrustful, skeptical, disinterested, unconcerned, indifferent, dismissive, argumentative, combative, and simply useless.

The speaker continued, "You know who they are. They stand out like a sore thumb. Most employees hide or go running when they see them. Yet, in many organizations they are not only tolerated, they are appeased. That, my friends, is the recipe for failure.

"The best advice I can offer is do not waste a second of your time or a brain cell on this group. They are called the nevers for a reason. It is entirely fruitless to devote any time or attention attempting to sway or change this group.

"An article by Gallup referred to them as poison. Gallup's research concluded that you are better off paying them *not* to come to work because of the extensive and irreparable damage they engender upon your culture.

"On average 20 percent of your employees reside on the other end of that continuum. We call those your 'models.' These are the employees that will walk on broken glass for you if you ask them. Launch a new strategy and these are the folks that dive right in and do what is asked.

"Have you ever believed you had some real traction with a new initiative that slowly fizzles away? That was your models taking the ball and running with it. But they cannot do it alone.

"The 60 percent in the middle of this bell-shaped curve I call the 'maybes.' They are on the fence and can move in either direction. They are waiting and watching to formulate their decision. What they observe determines whether they decide to move toward your models or your nevers."

As Dan shared his experience with me, I thought back to my time working with Dr. John Kotter. Dr. Kotter referred to the nevers group as the "naysayers." They will actively obstruct, impede, disrupt, and cripple change efforts. His advice was to assign them to a "special project" deep down in the basement of the building way down the road. In other words, keep them very far away from everyone.

My former colleague Tom Peters put it slightly differently. Tom's stance was: "Fire every last one of them immediately. They are a cancer to your culture. If you have to pay the price in lawsuits, so be it. They are doing more damage than that anyway."

We all need to choose the approach that works best for our unique situation.

The mistake too often made is placating to this group of malcontents (nevers). Placation comes in a variety of forms, such as slowing down a new initiative in order to allow folks to adapt or get comfortable, spending inordinate amounts of time attempting to convince them to buy in, asking your models to take on even more responsibility, and, worst of all, allowing select teams or groups not to participate. That is like squirting gasoline onto a fire. It only intensifies their efforts and galvanizes the masses.

More consequential is the reverberation this has on your top employees. When your best folks (models) observe you acquiescing to your poorest performers, complainers, and cynics (nevers), consider the impact that has on their passion, loyalty, morale, and commitment. It is a losing proposition. Do not get sucked into it.

Dan's epiphany was spot-on: "All of our potential is with the maybes. The 60 percent in the middle of the bell-shaped curve is where all the potential resides.

"Our models are already on board. We need to reward that group and do everything it takes to keep them there. We must dedicate our efforts as leaders to leverage the models in order to seduce the maybes to self-select in. The nevers will get none of our attention. In essence that will send the needed message."

Dan always knew that driving a strategy that requires individuals to voluntarily take accountability to change their behavior is the toughest challenge a leader will ever face. There really is not a close second. Nobody likes to be told to change. What he now understood was that he and his team wasted far too much time, resources, attention, energy, and effort trying to change the unchangeable.

Top-performing teams and organizations are always comprised with a higher ratio of maybes and models than poorly performing teams and organizations. Managers and leaders must model the desired behaviors and reward those that do the same.

The remaining chapters of this book will provide you with ideas and tools to foster an even higher-performing culture.

As for Dan and his team, within three months of implementing their "focus on the models and maybes" strategy, leaders, employees, and customers stated they experienced a noticeable change in the culture. Employee engagement and customer service scores that had not budged for years rocketed to unprecedented levels.

Most important to Dan and his team was that the most challenging and complex transformational strategy they had launched to date achieved success to heights never before experienced.

## Leading Accountability—It Is a Choice, and Everyone's Job

Who is the "director of accountability" in your company? Oh, you do not have anyone with that title? Whose job is it then?

Developing heightened levels of individual and organizational accountability is the job of anyone choosing to take it on. You do not need permission, a title, or authority to become an impeccable leader of accountability. You merely have to model the way.

Accountability is like love. The more you give, the more you get.

Consider Ted's experience.

The leadership team at Clearwater Technologies was always looking for ways to boost organizational and individual performance. Resting on their laurels was never an option. From the front lines to the senior team, the culture oozed with an unquenchable thirst to improve. That was one of the many reasons they were cited as one of the most admired and respected companies in their industry.

Rhonda, a renowned leadership expert, was hired by Clearwater Technologies to unleash leadership capacity at every level of the organization. The mandate she received from the senior leadership team was to identify opportunities to bolster an already high-performing culture.

As part of her discovery and analysis process, Rhonda met with and interviewed dozens of employees.

Ted, a mid-level manager who had been with the firm for several years, was considered one of the company's top-performing managers. Rhonda was anxious to learn from Ted what he believed was working well and where he believed opportunity existed to do even better. Given that the topic of accountability was broached in more than 75 percent of the one-on-one interviews to that point, Rhonda decided to explore that further with Ted.

"Ted, I have heard so many great things about you and your team. Spencer speaks glowingly of you and could not say enough good things."

After some small talk and getting to know one another, Rhonda looked Ted in the eye and asked, "Ted, are you an accountable person?"

"Of course I am," Ted replied proudly.

"Do you need more accountability within Clearwater Technologies?" Rhonda asked.

"Don't even get me started, we've been around and around this topic longer than I care to admit. Just last week—"

Rhonda politely interrupted. "Ted, I've asked those same two questions to perhaps ten thousand people in my career. To date, every single person has told me they are highly accountable. Not one person has ever told me they are not accountable.

"But at the same time, 99 percent have confided in me that their organizations need more accountability."

Ted eagerly offered, "Just like us. As I was saying…"

"Ted," Rhonda interrupted again with a big smile on her face. "If every person believes they are highly accountable, why on earth would their team or company need more accountability?

"Say 100 percent believe they are accountable, which is the experience I have had over the past twelve years when asking that question. That's every single employee. So if everyone in the company is accountable, problem solved—right?"

There was silence in the room for several seconds as Ted synthesized the information.

"Whoa, I feel foolish. I fell into the trap," Ted admitted. "I should know better. It's something I do with my kids."

"What is that?" Rhonda asked.

"Model the way. With my children I am very cognizant of the experiences I create around them. I set the example of what I expect from them. I learned that key leadership practice from Jim Kouzes and Barry Posner. It is one of the five practices of exemplary leaders detailed in their book, *The Leadership Challenge*."

"Tell me more," Rhonda said.

"Well, it is human nature to externalize change, to convince yourself that you are doing just fine and that if everyone else would change things would be much better.

"I need to be the change I wish to see in others. I cannot expect anyone to change unless they see it in me. I may believe I am highly accountable, but as you shared with me most of our employees are telling you we need more accountability. Just like I did.

"That poses an important question. Are my fellow managers and I modeling the desired behaviors consistently and visibly so that others choose to emulate them?

"The fact that most employees answer your question stating they believe more accountability is needed leads me to believe we need to do better.

"Our employees are a reflection of us as leaders. For the most part, they think, behave, and perform the way they do as a result of how they are led. The reality is, I have no control over the behavioral choices others make. I can hope to influence those choices, but I cannot make their decisions. I cannot dictate or force anyone to be more accountable. I need to model accountability and set the example."

With the "aha" moment resulting from Rhonda's visit freshly minted in his mind, Ted's team spent time to define the "accountable behaviors" they believed would propel an already high-performing team to even loftier levels of effectiveness. The team sought insight and perspectives from all members of his department. They wanted to know what accountability looked like in the eyes of each team member.

Once their team's specific accountable behaviors were defined, Ted worked with his group to develop a plan to spotlight the behaviors. Sharing the behaviors through a variety of communication channels was one small piece of the plan. The team put primary emphasis on modeling the desired behaviors and making them visible.

Among the accountable behaviors Ted's team discovered to be most effective: more precise, open, and consistent communication; unwavering focus and time devoted on solutions rather than assigning blame and focusing on what cannot be done; rewarding effort rather than chiding mistakes; soliciting ideas and perspectives without judgment; collaborating outside of silos to share resources and best practices; moving beyond boundaries and barriers; not blaming other people or events; and growing and learning from failures and mistakes as opposed to stifling positive intent.

The success Ted and his team had did not go unrecognized. Spencer, the CEO, enlisted Ted and his key team members to implement their "accountability program" throughout all of Clearwater Technologies. Other functional leaders welcomed the opportunity to learn from Ted's team.

With a heightened understanding of what accountability looked like, and with a profusion of "leaders" modeling the way, an already high-performing organization exceeded growth projections the following year by double digits.

If you were to ask an employee of Clearwater Technologies today, "Whose job is it to cultivate accountability?" you would most likely receive the response, "Everybody's."

## Where Do They Learn Those Things?

You are being watched! Do not get paranoid.

People around us are watching closely. At work. At home. At play.

Be aware of the messages you are sending through your words, behaviors, body language, and tone. If you lead a team or an organization, your self-awareness must be even more keenly developed. Every movement you make, every shrug of your shoulder, every roll of your eye, is being thoroughly scrutinized. What you do has far more impact than what you say.

This commonsense principle was forever embedded in the forefront of my mind years ago.

While they were young children, I would observe my kids closely, as most parents do. It struck me one day that it seemed they both played the blame-game more than I wanted. Of course, as a doting father I wanted them to be perfect by age one. Their behavior was absolutely typical for young children. But I was perplexed as to how they had grasped the finer aspects of playing the blame-game at such a young age.

I figured it out. It was from me. Me, and everyone else they observed modeling that behavior.

It was the perfect day for a round of golf for three hackers. My sons may object to that term, but at the time of this incident, they were hackers too. Perhaps just not as accomplished a hacker as me. A few years later, they both became accomplished golfers. I have since mastered the art of "hackership."

The plan was to play eighteen holes and be home by six o'clock to take mom to dinner.

The problem was, we were not even off the golf course by six o'clock.

Upon noticing the time, I stopped looking for my ball in the creek and yelled to the boys, "Hey, we are in big trouble. We are supposed to be home in two minutes."

Wondering why that was a problem, and then pausing to think for a moment or two, Nick shouted back at me as I was sinking deeper into the creek, "We can tell mom that the tee times were running late, that there was a slow foursome in front of us, and that we hit traffic."

My first thought was that he was pretty creative for an eight-year-old. Problem solved in his mind.

Impressed with Nick's plan, I bellowed out the appropriate father response, "Nick, where do you come up with this stuff?" I paused, and then added, "We are going to use those, but we have to talk."

Fast-forward to our arrival at home. Picture an angry wife wanting to know what happened and why she had been waiting fifty-five minutes for us.

With my sons standing by my side in solidarity, I launched in with, "Well, the tee times were behind, and the guys in front of us were playing slow, and when we hit the road..."

Before I could get out another word, I was told in not so many words, in fact she used two choice ones, that my explanations and excuses were not acceptable and that if we truly wanted to be home we would have found a way. Her body language and tone helped drive home the point.

She was right. It is always about ownership. If you want something badly enough you will find a way. We could have left early. We could have played around the slow foursome (if there was one). There are always options—always.

Upon reflection, the worst part of that experience is that I modeled the wrong behaviors for my sons. Unconsciously, as they observed the interaction with their mom, I was teaching them how to play the blame-game.

That was a wake-up call for me. If I wanted my children to learn accountability, ownership, and integrity, I needed to model that behavior.

As I thought through what had just transpired, a vivid childhood memory that is forever etched in my mind cemented my commitment. My father, who was a smoker, would often instruct me to never smoke—as he asked me to bring him his cigarettes. Even as a young child that seemed odd to me.

I made a commitment that day to be extraordinarily aware of the experiences I created in front of my children in order to reinforce desired behaviors.

That awareness, coupled with a lot of other factors, must have helped. Because today, they may be the two most accountable people I know. Never sinking "Below the Water," as Captain Hartman called it, for more than just moments until they realize they own their circumstances.

What those around you observe you do (your actions, tone, and words), leads them to form their beliefs about you. Your actions will either reinforce what they already believe about you, or create new beliefs they will hold about you.

The question you should be considering often is, "Are the beliefs I am developing in others going to help us achieve or slow down and impede the outcomes we desire?"

Being fully aware of your actions—the experiences you are creating for those around you—is a powerful tool. This is true in business as well as in your personal life.

People will choose to take actions based on the beliefs they hold. What beliefs are you developing, sowing, nurturing and instilling in those around you?

## Words Are Cheap—The Actions We Choose Are What Matters

In September 2016, a Senate panel questioned John Stumpf, CEO of Wells Fargo, whose massive bank appropriated customers' information to create millions of bogus accounts.

The bank was to pay $185 million in penalties for acts that dated back to 2011. Fifty-three hundred employees were terminated for reportedly creating false accounts to increase "cross-selling" to build the number of accounts each customer holds.

Why did these more than five thousand employees choose to act in the manner they did? Were their actions consistent with the company's stated Vision and Values Statement? If not, why not? What was the impetus for so many employees choosing to take actions inconsistent with the culture leadership defined in their Vision and Values Statement? What were the repercussions of holding people accountable to a "metric" rather than an outcome? If instead, expectations and clarity on desired accountable behaviors were established, would the outcome have been different?

Senator Elizabeth Warren (D-MA) began her questioning by citing Wells Fargo's Vision and Values Statement. This particular passage was called out, *"If you want to find out how strong a company's ethics are, don't listen to what its people say, watch what they do."*

"So, let's do that," Senator Warren said. She then accused Stumpf of failing to hold himself or any other senior executives accountable for the company's actions. "It's gutless leadership," she said, noting that Stumpf is not resigning, returning any of his earnings, or firing any senior executives.

Important lessons about accountability, leadership, and culture can be drawn from the following partial transcript of the exchange, as reported by C-SPAN September 20, 2016, between Senator Warren and John Stumpf:

> Warren moved on to the subject of cross-selling—calling it a particular focus of Stumpf's tenure as CEO, citing his goal of eight accounts per customer and saying that cross-selling was "one of the main reasons that Wells has become the most valuable bank in the world."
>
> The senator asked Stumpf, "Cross-selling is all about pumping up Wells' stock price, isn't it?"
>
> "No," the executive answered. "Cross-selling is short-hand for deepening relationships," he continued—before Warren cut him off.
>
> She then produced 12 transcripts of Wells Fargo earnings calls Stumpf participated in from 2012 to 2014—"the three full years in which we know this scam was going on," Warren said.
>
> "In all 12 of these calls, you personally cited Wells Fargo's success at cross-selling retail accounts as one of the main reasons to buy more stock in the company," Warren told Stumpf. She went on to quote him from the transcripts, as he touted the company's record growth to more than six accounts per household.
>
> "Here's what really gets me about this, Mr. Stumpf. If one of your tellers took a handful of $20 bills out of the cash drawer, they'd probably be looking at criminal charges for theft. They could end up in prison.
>
> "But you squeezed your employees to the breaking point so they would cheat customers and you could drive up the value of your stock and put hundreds of millions of dollars in your own pocket.
>
> "And when it all blew up, you kept your job, you kept your multi-multimillion-dollar bonuses, and you went on television to blame thousands of $12-an-hour employees who were just trying to meet cross-sell quotas that made you rich.
>
> "This is about accountability. You should resign. You should give back the money that you took while this scam

was going on, and you should be criminally investigated by both the Department of Justice and the Securities and Exchange Commission. This just isn't right."

During his allotted time to question Stumpf, Sen. Patrick Toomey (R-PA) voiced skepticism at the idea that the 5,300 Wells Fargo employees who were fired had all acted independently.

"You state unequivocally that there are no orchestrated effort or scheme, as some have called it, by the company," Toomey told Stumpf. "But when thousands of people conduct the same kind of fraudulent activity, it's a stretch to believe that every one of them independently conjured up this idea of how they would commit this fraud."

Responding to several questions, Stumpf said he lacked the appropriate expertise, declaring himself at various times not to be a lawyer, a compensation expert, or a credit consultant.

On October 12, 2016 John Stumpf resigned as CEO of Wells Fargo. It remains to be seen how this ordeal will play out and how any lasting reverberations, if any, have an impact on the bank. What happened at Wells Fargo is an example of how everything leaders do (actions, words, decisions, implementation of programs, etc.) will influence the actions our employees choose to take.

Are your actions driving accountable behaviors in those around you?

## The Enemies: Justifications, Explanations, Reasons, Rationalizations, Vindications, and Excuses

Captain Hartman asked me if I would like to join him at a working meeting that his colleague Harley was facilitating.

The session was already in progress as we entered the room. We quietly sat down and watched as Harley masterfully facilitated the meeting. She was not only entertaining and captivating, but the lessons she imparted were eye opening for many in the room.

Harley posed a question to the group that some seemed eager to answer with passion and conviction.

"What is the difference between reasons and excuses?"

Susan responded, "Reasons are legitimate and excuses are explanations."

Mark offered, "Reasons are justifiable and excuses are things we use to deflect blame."

Patty shared, "Reasons are factual and excuses are fiction."

Harley allowed her team to voice ideas for five minutes or more until Jessica, who had sat in reflective silence throughout the discussion, raised her hand.

"Yes?" Harley asked.

"There is no difference between a reason and an excuse."

"Tell me more," Harley said.

"Well, I believe that reasons become excuses when you just give up. All businesses, teams, and people face challenges, problems, and impediments that can be viewed as reasons or excuses for not achieving what is necessary.

"If we choose to give up and not overcome those *reasons*, they become excuses. In other words, reasons become excuses when we give up and stop searching for a way to move through the obstacles and closer to what really matters."

Harley, beaming with approval, smiled and said, "That may be the best answer I have ever heard to that question. Highly accountable people do not allow any obstacle, challenge, or barrier to prevent them from achieving what they want. They view those obstructions as bumps in the road. They maintain a mindset of 'okay, here's a challenge I must deal with, so what else can I do to get to the result I want?'

"They don't throw their hands up in the air and say, 'Oh no, look at this barrier. This is not my fault, and there is nothing I can do about it. I give up.' Nobody ever achieved anything worthwhile while playing the blame-game, convincing himself or herself that there is nothing they can do.

"Suppose all of us in this room were actually a small company and our primary competitor launched a new strategy that took a chunk of our market position and caused us to miss our forecast.

"We could choose to view that as an excuse why we did not hit our objective. That would sound something like this, 'You saw what the competition did. You were there. Of course we did not hit the objective. It was not our fault and there is nothing we can do about it.'

"We can also view it as a reason. That sounds a bit different. Perhaps like this, 'You saw what the competition did. You were there. Of course we did not hit the objective. Here is what we learned from it and here is our plan to make sure it never happens again.'

"In the first scenario, we completely gave up and succumbed to the blame-game. In the second we recognized our reality, took ownership, and developed a solution so that it would not happen again.

"The Rio 2016 Summer Olympics just took place. Think about Olympic athletes. Most all of them have faced hurdles they had to overcome and extreme challenges on the way to achieving their goals. No Olympic athlete ever won a medal while playing the blame-game.

"One reality that is critically important to understand is that there will be times in your personal life and in your professional role when someone will drop the ball on you, or there may be an event that is out of your control. In other words, you could be justified to play the blame-game. Just know this, you will never achieve anything worthwhile while clinging on to that mindset."

## Why Did Randy Cross the Road?

Sometimes it is the simple things that produce surprising results. Randy and the team at Tundra Oil discovered this firsthand.

While on the North Slope driving from one campsite to another in the Arctic oil fields, Randy noticed that ten to fifteen vehicles ahead of him were swerving off the road as if to avoid something. As he approached that spot, he noticed that they had been avoiding a large piece of metal in the middle of the road.

He pulled over to the side of the road and safely removed the hazard, placed it into his truck bed, and brought it to a leadership meeting I was attending that morning with a group of twenty-two of his colleagues.

While waiting for everyone to arrive, those present were scattered in small groups throughout the room enjoying conversations about plans they had for their down time, personal projects, family, and, of course, work.

Walking into the room with a fairly large disfigured hunk of metal, Randy caught the attention of several of us in the room. He calmly walked to one of the tables and gently placed the metal on the table in front of his seat.

Seeing that he was about to be bombarded with questions, Randy, unfazed, said, "I will explain this a bit later. Let's get the meeting started."

Well, that was not going to happen. Randy had piqued curiosity among his peers and there was no escaping their demands to know what he was doing.

Randy relented and explained to the team what he had experienced only minutes earlier.

He then asked the group, "What do you suppose they were all thinking as they drove by?"

Everyone in the room, in a tone rife with exasperation, chimed in with some form of, "Not my job. Someone should take care of that."

"Exactly," he replied. "Yet, we tell everyone up here daily that safety is job one. We spend millions of dollars to train people, we have processes and systems in place to keep us safe, we have signage regarding safety everywhere, we talk about it constantly, we begin every day with team safety meetings, so why didn't anyone pick this up?

"Anybody that has been up here for more than a week would know it was a safety hazard."

I watched the faces in the room as they sat in perplexed silence, unable to answer Randy's question. I knew where Randy was going with this. We had talked during the prior night's dinner about a key principle to cultivate an organization with unparalleled levels of accountability to deliver on what matters most.

The gist of the conversation centered on creating shared accountability and focus among all employees on key results instead of what is typically seen in most organizations, which is usually a focus on "my job" or "my responsibilities." Randy understood that when employees come to work every day focused on helping the team achieve highest priority top desired results, they will go above and beyond what is required, or what is listed in their job description.

Also during that evening's conversation, I shared a short story with Randy that struck a chord in him.

I shared with Randy how after a busy day spent cleaning and preparing our home for company the next day, we rewarded our kids by allowing them to choose where they would like to have dinner that night. Upon returning home, we headed up the stairs and into the kitchen with Zack leading the way. When we had gone out hours earlier, we had left the windows open, and apparently a gust of wind had blown dozens of papers all over the kitchen floor.

I watched as Zack maneuvered his way through the kitchen as not to step on any of the stray papers.

"Zack!" I shouted with agitation. "What are you doing? Help pick up these papers."

"That is not my job," was the reply from my twelve-year-old son.

At that moment, I had to fight back exceptionally hard to restrain what I truly wanted to say.

After a few deep breaths I looked Zack in the eye and sternly said, "You are right. It is not written on your birth certificate that you have to help pick up papers that have blown onto the kitchen floor. It is not written on mine either. But you do know we like an organized and clean home. So when you see something like this, it is your job because you are part of the family."

I'm not sure how well that connected or resonated for Zack at that moment, but I can report that after that he always had the most organized room in the house.

For some reason that simple anecdote resonated with Randy. For him, it was a rehash of the drivers that morning avoiding the metal hazard rather than taking accountability to ensure the safety of others.

I elaborated a bit and shared with Randy, "Everyone knowing they are on the team is step one. Having your most-skilled players at every position is even better. But if everyone is not aligned around what matters most, it does not matter how skilled they are.

"Suppose you had the world's absolute best musicians assemble to form the preeminent symphony or orchestra of all time. Yet on opening night, as the director raised his baton, each musician played from different pieces. Do you expect they would receive rave reviews?

"Everyone must be on the same page and know intimately how they contribute to the team successfully achieving its desired result.

By this time, the team in the room was pressing Randy. Not willing to wait any longer for an answer, Randy relented and launched into what we had discussed.

"We spend a lot of time talking about safety, and it is absolutely the most important thing up here. It trumps everything else. This is an inherently dangerous place. High pressure, heavy machinery, sharp objects, chemicals, fires, severe weather, ice, and so much more are poised to hurt someone who is not paying attention.

"We are a team," Randy continued. "We were all hired to help this team of ten thousand achieve our desired results. Why then would more than a dozen people drive by that hazard?

"We all get that safety is without a doubt the top focus and most important desired result on the slope. Yet, how many of the ten thousand people up here view safety as their job?"

Randy had the group's attention. He was on a roll.

"I will tell you how many. Jim (vice president of safety) and the few people on his safety team, and that is the problem.

"How has that worked for us? Our safety numbers have been stagnant and have not improved in seven years. We need everyone up here to know and believe that safety is their job. We all must own safety.

"Sure, I may run a drill on a rig, or be a cook at one of the camps, or drive a truck, but I need to view my job as safety first. I want every person on my team to understand their job is to help us achieve our desired results around safety, and then focus on the other items in their job description.

"We need ten thousand people—our team—to take accountability for our desired results rather than focusing on their job description or simply staying busy. When someone focuses only on their job description, it only makes sense that it leads to the blame-game, finger-pointing, and excuse-making. If safety is not written in their job description, they practically have the right to blame someone else."

Randy had fully bought into the principle we had discussed and, based on the heads nodding yes in the room, others were joining him.

Randy asked the group, "Most of us probably watched the Seahawks last night. If I am a wide receiver, say Doug Baldwin, what is my job description?"

Susan, the biggest Seahawks fan in the room, shouted, "Catch the ball."

"Exactly!" Randy proclaimed. "And if I am a wide receiver on a football team and there is a fumbled football, do I stand around and say, 'Man we should have had our fumble recovery guy in on that play'?"

The group smiled and laughed openly at the suggestion. "That's insane. There is no fumble recovery position."

"So, what motivates that wide receiver to jump on the ball?

Barry declared, "Because he's closest to it."

"That is part of it, Barry, but there is a more important reason. Every player on that team is crystal clear on what is most important, and for a football team that is to win the game.

"They will not let their job description deter them from doing whatever it takes to help the team accomplish that result. Doug Baldwin is not going to stand still and tell his teammates and coaches that it is not his job.

"My guess is recovering fumbled footballs is not an activity or responsibility listed on Doug Baldwin's player contract. Yet he does it anyway.

"It is everyone's job to jump on the ball. No player would let their job description dictate their behavior in that situation. They want that ball because they know it will help their team win. The win for us up here on the North Slope is zero accidents.

"We must instill that mindset here when it comes to safety. Accountability is broader than what is on your job description.

"By the way," Randy offered to the team, "the only reason anyone here, or in any organization, has a job, is that at some point someone decided that by adding that person to the team it would help the team achieve their desired results. We do not hire people simply to be active or stay busy. Activity in and of itself will not necessarily produce desired results."

At that point there was no need for Randy to attempt any more convincing. The group realized that unless they developed an organization-wide mindset of shared accountability—the notion that everybody in the company is accountable for safety—they would never achieve their objective of zero accidents.

This group had full support of the senior leadership team to lead this effort and develop strategies to transform the way in which top priority goals were viewed.

They discovered that simple ideas worked most effectively. For example, whenever there was a need to fill a new position, they adopted a slight change in the manner the postings were worded.

Rather than detailing merely an overview of the job description, or primary roles and activities associated with the job, the new approach was put into effect. The postings took on a focus of how the position would contribute to helping the entire team of ten thousand achieve the identified must-achieve desired results, thus instilling a mindset that "we are in this together."

Another tactic was to begin every meeting with a very brief explanation of how the focus and agenda of that particular meeting would help move the organization toward one or more of their identified priorities. The sentiment became: if we cannot make that simple connection, we need to question why we are having the meeting.

Effort was taken to help employees at every level make the connection between what they do, and how "what they do" had an impact on the desired results. This helped employees make the connection not only to how what they did influenced the desired results, but how their job and the choices they made affected the ability of others to do the same.

Employees shared openly that the concept of shared accountability—where accountability for the desired results was everyone's job—changed the way that individuals, departments, and units worked together.

As opposed to performing activities in isolation, as was the norm in the past, collaboration, cooperation, and resource-sharing were now the norm. Employee engagement, commitment, and passion were on the rise. Territorialism, isolation, and the blame-game were on the decline.

By implementing simple, pragmatic concepts like these, the key safety metrics that had not budged for seven years struck levels never before realized.

## Key Learnings to Internalize

**One:**
Individuals, teams, and organizations inevitably will be, or have been, dealt formidable challenges, barriers, and obstacles at some point. Instinctually, as it is human nature, we often point to other people, teams, or events outside of our control as

the culprit. An inexplicable force tends to suck individuals, teams, and at times entire organizations into the ineffective and unproductive lure of the blame-game. That is a losing proposition.

**Two:**

Often you may be justified to point the finger of blame directly at another person, team, department, or event. Stuff happens. Nobody is perfect. We are human and will drop the ball or make mistakes at times.

It may be therapeutic to drop "Below the Water," as Captain Hartman labels it, and vent and blow off steam. However, no individual, team, or organization has ever achieved anything worthwhile living "Below the Water" engulfed in the blame-game.

Accountable individuals, teams, and organizations that seem to consistently achieve what matters most to them take ownership for their circumstances and embrace a different mindset.

No matter how demanding, problematic, or distressing the challenge—or, more important, what or why they are in that situation—they adopt and cling to a winning proposition. They remain resolute, engaged, determined, proactive, tenacious, aligned, committed, and focused on "What else can I/we do to achieve my/our desired results?" They understand that there is always something you can do to move a bit closer to the results you want. If you truly "own" the outcome, you will find a path forward.

Top-performing organizations and teams possess the ability to help one another maintain personal accountability.

**Three:**

An effective approach to move people to choose to take accountability is to ask pointed open-ended questions. Ask focused questions designed to help others discover a solution on their own. Personal ownership, engagement, and account-ability escalate when individuals choose their path forward.

**Four:**

Leaders can release exponential gains in performance of their teams or organizations by focusing more energy, attention, and resources on the "maybes" that reside in the middle of the bell-shaped curve. Highly accountable teams that consistently realize and exceed must-achieve desired results almost always defy the standard 20/60/20 percent bell-shaped curve. The bell-shaped curve in these organizations is always weighted more heavily to the right, often with 30 to 40 percent of employees residing in the "models" section of the curve.

**Five:**

It is easy and quite common to engage in the futile ploy of externalizing rather than internalizing change. Too frequently we wait and hope that others will change their behaviors. We convince ourselves that when everyone else changes that things will be just peachy, and that we are not part of the problem. Effective individuals and leaders choose to model the way.

**Six:**

We cannot control the decisions other people make. But we do have 100 percent total control over the decisions we make and the behaviors we model for those around us.

We must be cognizant of how well we are modeling accountability, building capability, and energizing those around us to choose to do the same.

**Seven:**

The behaviors we exhibit and the choices we make send sig-nals to those around us. What signals are you sending to those watching you? Do your actions and behaviors suggest you are highly accountable? Do you possess a "what else can I do to move toward my desired result?" mindset?

Are you absolutely certain how you show up in the eyes of your colleagues, your boss, your friends, and your family? The actions and choices you have made have shaped the way others see you.

How much time do you spend "Below the Water" focused on what you cannot control? How much time do you spend focused on what you can control and what you can do to achieve what matters most?

### Eight:

Developing self-awareness about the experiences you create for those around you will help you become a more effective colleague, leader, and person.

### Nine:

Top-performing teams and organizations instill a mindset of shared accountability for desired results. Randy, who stopped and removed the hazard from the road, did not have a clause in his employment contract that he must remove hunks of metal from the roadway. He took accountability to do so because he was clear that safety is everyone's job.

Creating a focus on what matters most, rather than roles, responsibilities, or job descriptions is a simple concept that produces demonstrable results.

## Take Accountability Now

### One:

Reflect over the past month. How much time did you spend "Below the Water" playing the blame-game?

What messages have you been broadcasting to those around you through your words and your actions?

Identify one opportunity to take accountability and visibly demonstrate what you are capable of to those around you.

**Two:**
Choose a colleague, friend, or family member who may currently be "Below the Water." Craft five to seven open-ended questions that will help them to recognize their current reality and to take accountability. A proven starter question, on the house, is: "What's holding you back from achieving what matters most?"

**Three:**
The "nevers" in an organization usually stand out like a sore thumb. They can destroy a culture. Do not waste any of your time or energy with this group. Develop a plan to focus on where all the potential exists—with the "maybes."

What can you do to move this group farther to the right and into the "models of accountability" section of the bell-sharped curve? Develop a plan to involve and leverage your models of accountability to help you enhance your culture.

**Four:**
Today, write down at least three accountability behaviors that you and your team can use to create and model the high-accountability culture you want to see in your organization. Plan to model and spotlight these behaviors.

**Five:**
Identify one area personally and professionally where you may have been sucked into playing the blame-game. Establish a plan of action to overcome any obstacles and barriers that have impeded your progress toward your desired outcome. What will you do to move even a small step closer to what matters most to you?

**Six:**

Ask a cross-section of your colleagues, "What is your job?" Do they mention the organization's top two or three most important must-achieve desired results? Do they recite what is listed in their job description?

Devise a plan to create a shift in their mindset from a focus on their job to a focus on what matters most.

Help them make the link between what they do and how what they do contributes to the entire team or organization achieving your most crucial results.

# Chapter 3

## Accountability—What Is It?

### Once Defined, It Flows

While participating in a workshop, the facilitator made eye contact with me and, as many of us do in that situation (or at least I do), I glanced away hoping not be asked a question. My strategy did not work. Linda walked toward me and asked, "Mike, what is time?"

Immediately I thought to myself, Wow, what an easy question. I can answer this. I am glad she did not ask something more challenging.

As I opened my mouth to respond, nothing came out. My confidence quickly declined into dazed befuddlement. With the eyes of the other participants in the room cast upon me, I could feel the blood flowing to my face and the palms of my hands getting damp, as my mind scrambled uncontrollably. What the hell is time? I thought to myself.

"It is how we track our day," was what fell out from my lips. What a stupid answer, I thought to myself, feeling embarrassed. I knew it was ridiculous even as I was forming the words before I spoke.

"That is a common answer and I hear it from lots of people," Linda said with grace as she purposely helped save me from myself.

Linda continued to ask six or seven people to take a stab at answering the same question. She received six or seven different replies. Not one person nailed the answer. I must admit that made me feel better.

The funny thing is that most of the folks who raised their hands to answer seemed self-assured they would nail it. None did.

Linda eventually shared the definition of time, "The indefinite continued progress of existence and events in the past, present, and future regarded as a whole."

If all thirty-two of us in that room spent days working collectively together we would never have arrived at that answer.

Ask ten of your colleagues to define "accountability," and you are most likely to experience something similar.

The team at Mountainview Enterprises pioneered a solution to creatively tackle this dilemma.

Jim gazed at his colleagues with exasperation and snapped, "We need more accountability around here! What does that really mean? Seriously."

"This is frustrating," Jim sighed. "It seems that everyone has his or her own idea about what accountability is and how to define it. Until we get clear on this, how can any of us take accountability or expect it from others?"

After a prolonged contemplative silence of two or three minutes, Lori asked if she could propose an idea.

"May I offer a suggestion?" Lori shyly chimed in. "Why don't we just define it ourselves? We all know people and have professional and personal acquaintances who seem to always achieve what they set out to accomplish. We have folks on our own teams who we often point to as models of accountability. Let's identify their behaviors and use what we discover to establish our definition of accountability."

"I love that idea!" trumpeted Karl. "Other than the definition in the dictionary, which I do not like, the only other models or books on accountability I am familiar with seem to focus with arming individuals with terms or phrases to 'call someone out' when they have failed, missed a commitment, or made a mistake.

"Personally, I do not find that helpful. In fact, I believe it causes anxiety and stress. That approach can create an environment where employees are apprehensive and reluctant to take action for fear of being called out. Think about it. If we foster a culture where employees are basically being told they are not capable, not worthy, and not accountable, how is that going to help us? I would prefer a more positive approach that ignites ownership, enthusiasm, energy, engagement, and trust.

"We are all human, we all make mistakes. Nobody can be 100 percent accountable all the time. Stigmatizing someone for not being accountable does not help. I believe it does more damage than good.

"We can develop our own definition and make sure everyone in the organization has an unwaveringly deep comprehensible grasp of what it is. Once we establish explicit awareness and understanding of what behaviors are expected in this culture, we can lead the way, model those behaviors, and create positive experiences to reinforce them."

Three weeks after the team excitedly accepted Lori's challenge, they gathered to share what each of them had identified. They were all thirsting to learn what the data revealed.

Jim invited me to join the team meeting.

"So," Jim begged, "I am anxious to learn what everyone's observations revealed so that we can we finally provide clarity."

I observed an excitement and energy in the room that I had not noticed previously.

An enormous wealth of information had been compiled. The team spent hours sharing their observations, insights, learnings, spreadsheets they developed, as well as a synopsis of employee evaluations dating back four years.

What had initially seemed to be a simple project turned out to be much more expansive than they had imagined. They sifted through the data for two days in an attempt to compile, synthesize, and simplify the findings.

After two exhaustive days, while looking at the data projected on the screen and beaming with delight, Jim gushed, "You know we have struggled with this for a long time and I now understand why. We've made this much more complicated than it is.

"When I sit back and reflect on these accountable behaviors, they are common sense. Every one of the key behaviors we have identified and determined should be included in our definition of accountability falls into one of the broader categories we have all agreed on.

"I recommend we make it simple and memorable. Why complicate things? I believe we can reduce all we have up there into just a few categories."

The team was energized by Jim's suggested approach. They spent the next few hours paring down all of the information they had collected and discussed and developed their four key categories of accountability. Each category included several of the accountable behaviors their data revealed that were displayed consistently by top performing individuals, teams, and organizations. They spent considerable time debating the flow and order, and settled on what they believed worked best.

With a sense of pride and accomplishment, Jim went on to quote Mark Twain, "The problem with common sense is that it is not that common." He concluded, "Now that I know what accountability truly is, I can model it and clearly articulate it to my team."

"Now," Mary Ann proudly offered, "when any employee is asked, or when we are asked, without batting an eye we will be able to concisely define accountability so that everyone know's what it is and what is expected."

Jim and the team developed a definition of accountability and list of behaviors that worked for them. It was simple, elegant, and memorable.

"Accountability must be rooted in awareness," Jim stated. "It starts with an unbiased, unfiltered and complete recognition of current realities."

Jim was right. It is common within an accountable organization to observe free-flowing communication to assure all perspectives are considered. As Jim put it, "How can we take accountability for things we do not know about, or that people are not addressing?"

"No one person, no matter how smart, can see every opportunity, possibility, challenge, problem, and complication. We need to create open dialogue with a positive intent. We have seen it in the past that when problems, issues, and difficulties are left unacknowledged, those very same roadblocks fester and grow."

Jim continued, "Of course, once an individual, team, or organization recognizes their realities, they must accept ownership. That is the next category.

"I have learned throughout my career," Jim shared firmly, "that it is not good enough to simply recognize your realities. Recognition is just the starting point. To achieve success in life or in business, you need to have a steely unrelenting ownership of your situation. If you do not, who will?

"Without whopping levels of organizational ownership, the best we will achieve is mediocrity. Think about it. If our employees lack ownership for our circumstances, if they simply do no care or are indifferent, it only follows that they will not take appropriate and necessary actions to improve them."

I was impressed with Jim's visible passion about the topic. I later learned how he applied these very concepts to his own life to achieve some incredible goals.

"Once you identify and own your circumstances, the only way forward is to find a solution," Jim explained. "I have seen far too many teams and people waste enormous amounts of time discussing what went wrong, who was to blame, and all the reasons they believe prevent them from achieving success. I say, learn from your mistakes, and then quickly move on and figure out a plan forward.

"Effective and accountable people view complications, challenges, and impediments as minor bumps in the road on the way to success. Think about any world-class athlete or anyone with a singular focus on a goal they truly own. They maintain a solution-focused mindset. They refuse to participate in the blame-game. Creating solutions and a plan to move forward is the third category.

With near contempt, Jim continued, "I read recently that 92 percent of people that purchase a gym membership and 88 percent of people that own exercise equipment do not use them. Imagine that. These people have recognized their realities, accepted ownership of their situation, and went as far as figuring out a solution. But a solution is useless if you do not implement it.

"Taking action and following through with your solution is the final category. Relentless follow-through on commitments must become the norm around here. It must be commonplace for leaders to model the desired behaviors they expect from their teams. Employees must do what they say they will do.

"We must commit to highly visible ongoing alignment, engagement, and support. I believe trust, collaboration, and commitment will snowball when individuals and teams create accountable experiences for one another.

"The order is paramount," Jim explained. "If an individual, team or organization does not recognize their realities, they cannot possibly choose to accept ownership and take accountability to change a reality they ignore, refuse, or have not acknowledged.

"And to recognize their realities and voluntarily accept ownership is pointless without creating a solution or remedy.

"Assume they do take complete ownership for their situation and commit to improving it. Generating and creating solutions without putting those plans into action, well, is just insane."

Jim's elegant, simple, and eloquently described template made perfect sense to the team.

As Jim walked the team through his four categories, I thought back to a day earlier that week. A colleague of mine had e-mailed me a picture that captured the essence of what Jim explained to me. I hope some day to use the picture with a leadership team that is stuck in the trap of viewing their world through rose-tinted glasses.

Envision a middle-aged portly, balding man of generous proportions, wearing a tiny bathing suit and standing in front of a very large full-length mirror. This man could very well have played NFL nose tackle at one time, and "let himself go" for fifteen years since retiring from the league. Rather than staying fit and healthy, he appears to have spent the past fifteen years attempting to set a world record for eating.

As he looks at himself in the mirror, he is admiring what the reflection shows of a very handsome, physically fit man that some might describe as "Chippendales-like."

The point being, if this man were told to "get in shape" or "lose some weight," his response would be, "Why?" In his eyes, he is the picture of fitness and health. Think about that the next time you ask your team members to change, or take ownership to achieve a result they believe is not necessary. If they believe (their reality or perception) that things as they exist today are perfectly acceptable, why would they choose to change or to take ownership to alter a reality they believe is wonderful?

Recognizing your realities truly is foundational to fostering heightened levels of personal and organizational accountability.

"Before we adjourn this meeting," said Jim, "I am asking each of us to commit to find ways to consistently demonstrate these behaviors. We must be the models."

Recognize Entirely Your Current Realities.

Accept and Embrace Ownership for Your Circumstances.

Determine Solutions and Devise a Path Forward.

Take Action and Relentlessly Follow Through.

FIGURE **3.1**    Mountainview Enterprise's 4 Categories of Accountability

The four categories of accountability the team at Mountainview Enterprise's identified and adopted are depicted in figure 3.1. As Jim and his colleagues discovered, creating a common language and understanding of what accountability is, was foundational in their efforts to establish a culture where accountability was embraced, internalized, and demonstrated.

Among the commitments made by the group: more precise, open, and aligned communication; unwavering focus on solutions versus problems; focusing on top priorities; rewarding effort and admonishing inactivity; genuine openness to new ideas and thinking; asking for opinions and insights from others; staying engaged rather than dwelling on blame; and to grow and learn from failures rather than punishing and stifling positive intent.

Through a coordinated effort devised by the leadership group, their definition of accountability was instilled within every employee and team throughout Mountainview Enterprises.

With a common precise understanding of what accountability looked like, as well as what was expected, the senior leadership team detected a noticeable swell in the behaviors they had identified. To the surprise of nobody, within a period of only six months the key metrics related to organizational performance and effectiveness climbed comparably.

## Are You Fully Recognizing Your Realities?

Are you open to perspectives of others? How tightly do you cling, defend, and hold onto only the way you see things? When provided an opposing point of view, is your initial inclination or response to argue for or protect your position? Do you sincerely attempt to see the world through the eyes of the other person?

If you only see the world through your own lens, you are setting yourself up to fail. No two people see the world exactly the same. No one person can see all there is to see. Effective people understand and accept that other folks may see things that they do not. They realize this helps them make better decisions.

"Agree to disagree." We have all heard that statement at some point. When that quip is dissected, it is revealed to be a selfish and useless statement. In essence, "agree to disagree" suggests that you disagree with them. It says that you believe they are wrong, you disagree with their perspective, and that you are right. How do you like to be told you are wrong? What if instead you are both right and you both accept that truth?

People attempting to convince others to see things their way waste a lot of time and energy. You can save yourself from a lot of lost energy, as well as frustration, stress, and potential grief when you embrace the notion that everyone sees the world differently. When you do so, you'll accelerate your personal growth and your ability to learn.

Dudley Field Malone, an attorney, politician, activist, and actor simplified the crux of this with the following:

> "I have never in my life
> learned anything from any
> man who agreed with me."

Actively seeking and being open to others' perspectives pays dividends. Mark Twain, with his wit and humor, captured this perfectly when he said:

> "It ain't what you know that gets you
> into trouble. It's what you know for
> sure that just ain't so."

When it comes to recognizing realities, Willie Degal, an admired New York City restaurateur, puts it this way,

"You cannot fix the
problem if you do not
see the problem."

Gail, a middle manager of a Pac-North Building Supplies, understood and embraced this concept fully.

Gail pulled me aside during a break in a workshop. She shared with me that while attending a recent company leadership event she forced herself to apply what she recently learned and approach as many of her colleagues as possible to seek perspectives on what can be done to improve operations. This was not something she customarily had done in the past.

One junior-level supervisor shared information about a minor glitch in their warehousing operations. This glitch was quickly addressed with a minor investment. However, if it had not been brought to light, it could have ended up being a multimillion-dollar quagmire.

How often do you, your supervisors, managers, and leaders proactively seek perspectives from a diverse cross-section of employees? How often do you seek insights from those closest to your customers? That is where the work is being done. That is where the best ideas usually reside.

The drawing of the woman that was developed by Harvard Business School and appears in Stephen R. Covey's book *The 7 Habits Of Highly Effective People* drives this point home for many. Many people who view the drawing see a *young girl*. Many others see an *old woman*. With that mind, go back to the statement, "Agree to disagree."

You may be willing to politely "agree to disagree," but in doing so you are missing out on all of what is there by dismissing and not working to see the world through the other person's eyes. If I choose to cling only to what I see and not accept and attempt to see what you see, I am not seeing all there is to see. In the example of the drawing described above, I would end up missing out on 50 percent of reality. What opportunities, possibilities, snags, and glitches are you potentially missing by not earnestly working to see the world through the lens of others?

How important is it as a leader, manager, or supervisor to see the entire reality that exists for your team or organization? How much effort and energy are you putting forth to see that entire reality?

Being open to others' perspectives allows us to see and recognize more than we can see or recognize on our own. When we are open to others, we are able to be aware and recognize more snags barriers, obstacles, hurdles, opportunities, and possibilities. Doing so allows us to recognize a larger reality, thereby allowing us to make better decisions. If you only see what you alone see, you are not seeing all there is to see.

When we accept this, we no longer fall into the trap of trying to convince others they must agree with everything we believe. We are all different, and when we accept this fundamental truth, it helps us achieve more.

## The Eyes Have It

A simple yet effective exercise to help demonstrate the value of being open to what others see was a result of a child helping his father.

Peter, a plant manager for W. B. Babcock, happily shared the story about the evolution of this simple exercise.

"I told my son to look somewhere else for my contact lens. I had just spent a solid twenty minutes searching that area thoroughly and did not want him to waste his time. But while I was focused in another area, Teddy went right to the spot I had searched."

"Dad, I found it!" Teddy yelled out as he handed me the recovered contact lens.

"I swear I looked in that exact spot for thirty minutes. How did I not see that?"

"What an 'aha' moment that was for me!" Peter exclaimed. "The lens was in plain sight for Teddy, but never showed up on my radar."

"We have been attempting to develop even higher safety awareness in our plant. Our employees start out every day with focused intent to be safe. However, with all of the machinery, distractions, and activity going on it is tough to maintain that focus all shift long. We need everyone looking out for one another.

"Getting these folks to feel comfortable intervening when someone may be in an unsafe or risky situation is not easily done. They need to be willing to do it, and amenable to someone calling attention to it.

"The incident with Teddy gave me an idea and it has worked well. Join me at the lunch and learn meeting at noon and you can judge for yourself."

Once the folks in the cafeteria settled in, Peter asked all twenty-two of them to look at the projection screen he had placed in the front of the room.

He flashed onto the screen an aerial photograph of the plant floor he had taken weeks prior. In advance of taking the photograph, Peter and his team had strategically placed a variety of objects throughout the entire shop floor. All were objects not normally found in the plant. The objects ranged in size, color, and visibility. Included were items such as stuffed animals, toys, posters, signage, people in costume, and furniture.

After what seemed to be about five seconds, the photograph disappeared from the screen.

Peter walked around the cafeteria from person to person asking them to name only one item they saw on the screen that normally would not be on the shop floor.

As each person responded, Peter asked how many others noticed the same object. In some instances, the identified object in the photograph was seen by 50 percent or more of those seated in the cafeteria. Some of the objects were seen by as few as only one person.

Peter continued to ask ten or so people what they saw.

Peter then asked the group if anyone could name every unusual item in the photograph. The group almost laughed at the suggestion—that was impossible!

"No, not impossible," Peter pointed out. "You were able to identify every object in the picture—as a team.

"So, if no one person in this room was able to see all of the unusual objects, but as a team we spotted them all, you tell me the key lesson from this exercise."

This simple activity had a major impact on the group. Many sat in quiet reflection.

J. T., a respected veteran of the team and chief machinist turned to his teammates in the room and said sternly, "You can't see it all on your own. That's why we need to have each other's back and keep each other safe. With all that goes on around here we are bound to lose focus and let our minds drift away from safety. That's when we need each other the most. Remember that the next time someone brings a safety concern to your attention. They're only seeing something that you are not seeing. They are trying to help you."

Larry, the foreman for that shift, agreed and offered more, "I have been my own worst enemy leading my team and I now know why. I don't have all the answers. I need to leverage the diverse skills, experiences and viewpoints that my teammates bring to the table. As a team we will certainly have more answers and ideas."

Later, as Peter and I were walking back to his office, he shared with me, "The outcome is the same every time I use that exercise. As with most learning experiences that create lasting behavior change, it is best to allow the learner to draw their own conclusions. This fast and fun activity has improved communications, camaraderie, and trust by boosting the willingness to be open to the perspectives and viewpoints of others."

## Have You Accepted Ownership for Your Realities?

Recognizing and acknowledging realities is foundational and a necessity on the journey to accountability. But it is only the start. Taking ownership—unconditional ownership—is the next part of the journey. True ownership is identifying and accepting what you have done or what you have not done that has played a part in your being in your current reality. Note that the word "you" is used in that last sentence. Other people or events may have precipitated and created challenges, barriers, hiccups, and obstacles, but how you, your team, or organization chose to respond is what is crucial. That response is what you must own.

There is a symbiotic relationship between ownership and accountability, as the team from Cambria Ventures discovered.

Kate had captured the attention of even the most hardened of the leadership team. As much as Charles, the ex-marine, wanted to challenge and fight the notion, even he saw the light.

The revelation was that leaders cannot mandate ownership. When you dictate and demand, at best you may obtain compliance. More often this approach results in resistance, resentment, anger, and subversion.

For Kate, the relationship between ownership and accountability was apparent. She had observed firsthand throughout her career that when employees truly owned a desired result, they would figure out a way to make progress and would choose to take accountability to achieve it.

"Think about your past experiences," Kate implored her colleagues. "Nothing will stop a team that is committed to a goal. Nothing will deter a team that truly covets a targeted result. When the magnitude of ownership is low, even minor problems tend to become huge barriers that choke and stall progress.

"I'd like for you to consider your team or those who report to you," Kate suggested to the group. "What percentage of your employees would you insert into each of the four quadrants depicted on this whiteboard?"

As Kate shared with her team, the degree of ownership possessed for targeted desired results and key strategies will vary among teams and employees. Figure 3.2 highlights and describes the four distinct categories of ownership

On the left side of the diagram Kate had drawn, there was an arrow running vertically indicating Low Participation on the bottom to High Participation at the top. The horizontal arrow represented Low Buy-In on the left and High Buy-In on the right.

Kate partitioned the matrix into four quadrants.

The words inside the bottom left quadrant (1) were: Dispute, Complain, Object, Dissent.

The bottom right quadrant (2): Excused, Exempt, Excluded, Absolved.

The top left (3): Concede, Cooperate, Conform, Acquiesce.

The top right (4): Commit, Dedicate, Embrace, Pledge.

"Agreement and buy-in is good, but we need as many employees as possible to choose to participate and to be involved in helping us accomplish our most crucial objectives. I have too many people on my team that consistently buy in and agree with me but do little to nothing to help us move forward."

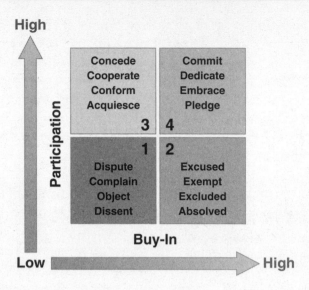

**High**

**Participation**

| Concede | Commit |
| Cooperate | Dedicate |
| Conform | Embrace |
| Acquiesce | Pledge |
| **3** | **4** |
| **1** | **2** |
| Dispute | Excused |
| Complain | Exempt |
| Object | Excluded |
| Dissent | Absolved |

**Buy-In**

**Low** ➔ **High**

FIGURE 3.2    Four Distinct Categories of Ownership

"Okay Kate, we get it," George announced after only a few seconds. "We have too many employees that nod their heads in agreement, but are not involved in any way to help us achieve what is necessary. They live in quadrant 2. So, how do we increase ownership with that group so their accountability soars, and what tricks are up your sleeve to help us maintain ownership with those that already have it?"

"Here's one idea that can help us with both of those, George," answered Kate. "People are much more apt to buy-in and have ownership when they feel they are included and involved in the decision-making process. Nobody likes to have things thrust upon them when they had no input or when their perspective was not sought or considered."

Mark added, "I agree with Kate that involving and including employees boosts ownership. I have also found that when we effectively communicate the business case as to why we have landed upon new strategies and initiatives, ownership escalates."

"This is important," George chimed in. "We need to develop a list of useful approaches we can all use."

The rest of the meeting was devoted to brainstorming additional ideas.

The group unanimously agreed to incorporate Kate's idea to invite more inclusion and involvement from employees prior to pushing out new objectives and goals.

Mark's suggestion to create more understanding about the external and internal drivers necessitating new directives was included in their arsenal as well.

In addition, the team added a few more approaches each of them had successfully used in the past:

1. **Celebrate Wins:** Ownership increases when progress toward a desired outcome is extolled.
2. **Storytelling:** Unlike mandating, when stories are told listeners draw their own conclusion about what must be done and will take action accordingly. Stories also create pictures, and pictures play to emotions.
3. **WIIFM (What is in it for me):** Appeal to and help folks see how achieving the outcome will make their job easier or better, develop them professionally or personally, or will make their life better.
4. **Openness and Transparency:** Both are indispensable to establishing trust, commitment, and buy-in.
5. **Appreciation and Recognition:** When desired behaviors are exhibited, call them out and reward them. Nobody complains they receive too much appreciation.
6. **Encourage Creativity and Innovation:** Invigorate engagement to release passion.

Nine months of concentrated effort by the Cambria Ventures leadership team and their reports paid off handsomely. Their top must-achieve desired results were exceeded, and their next employee survey showed positive spikes in employee engagement, ownership, and commitment.

## Yoda—The Accountability Coach

During his junior year in college, my son sent me a text late one evening. So, as a typical father with a son living in a house with four buddies in the heart of "the place to live" off campus, I had a modicum of concern when I saw his name on my phone.

I should have known from past experiences that there was no need to worry. In fact, the text exchange meant so much to me that I printed and saved it.

"Dad, my roommates and I have been watching all of the *Star Wars* episodes in advance of the new one coming out. Yoda had the best line I have ever heard.

"He was teaching Luke to tap into his abilities with the Force and Luke replied using the word you told me to never ever use. Luke told Yoda he will 'try.'

"Yoda replied by saying, 'No. Try not. Do. Or do not. There is no try.'"

That line from Yoda has been used by people from all walks of life to help others believe in themselves and to truly commit to whatever task, goal, or desired outcome they undertake.

As parents, we strive to teach and instill within our children key life lessons that will help them effectively navigate through life's opportunities and challenges.

That simple but profound statement from Yoda was a lesson deliberately planted in the fertile and accepting soil of my children's minds beginning when they were infants in hopes that it would become rooted into their subconscious.

Talk about making my day. Nick made my year. Knowing that years of teaching and coaching paid off was incredibly rewarding.

As my former colleague and friend Hyrum Smith would often share, "Wisdom is knowledge rightly applied."

Yoda's counsel to Luke was elegant in its simplicity, and indisputably magical in its power.

Why would anyone ever "try" anything? When you use the word "try" you are in fact setting yourself up to fail. When you "try" you are almost conceding defeat before you start by subliminally arming yourself with the excuse that, "I tried."

Yoda's counsel puts 100 percent ownership and accountability on you, the individual. That is one reason some find it difficult and uncomfortable to accept. It offers no leeway for excuses. You either will, or you will not.

The extent to which we believe we can achieve a desired outcome has immense impact on our ability to do so. How many Olympic athletes, top performing artists, world-class musicians, or accomplished professionals launch into attaining what matters most to them with the mindset, "I will 'try'"? How many embark knowing they "will?" As humans we tend to think and act as we envision ourselves in the future. It is difficult to think and behave inconsistently with how we view ourselves.

"There is no 'try'" is the mindset of highly accountable people, teams, and organizations. Once they fully recognize their realities, they accept complete ownership no matter what forces, events, people, or teams placed them there.

Consider this question, "If the majority of your team members or employees 'believe' your most crucial desired results are unrealistic or unattainable, will they commit fully to help make them a reality, or will they 'try' to help?" Based upon what you have learned to this point, what must you do to move the majority of your employees to possess relentless, unstoppable, rampant commitment?

## Lessons from Hyrum

Simple, memorable, practical lessons have the biggest and longest-lasting impact. One that stands out and is closely linked to accountability is about control.

My friend Hyrum's passion, energy, charisma, and desire to teach and share his knowledge filled the room at the Franklin International Institute. I was like a sponge absorbing as much as I could. Being fairly new to the business and to the Franklin International Institute (now known as FranklinCovey) team, my objective was to learn as much as I could during this day with the founder of the firm.

I soaked up more than a dozen key learnings that day, and many more over subsequent years. One model Hyrum shared that day is extraordinarily important as it relates to accountability. Hyrum calls it the "Control Continuum."

More than one hundred of us watched as Hyrum nearly sprinted to the wall on the left side of the room.

He shouted, "This wall represents zero control! When you are at this wall you have absolutely no control at all. Got it?"

Then, jogging the distance across the huge room toward the right wall, he exclaimed, "When you are at this wall, you have total and complete control!"

Hustling back to the center of the room, he announced, as he pointed first to the left wall and then to the right wall, "Everything between there and there is partially controllable stuff. From a little bit of control over here, then as we move away from the left wall, to a lot of control as we get closer to the right wall. Got it?"

The audience was on board and shaking their heads acknowledging the set-up.

Standing smack up against the left wall, with his hand placed on it, Hyrum asked the group, "Name some things that exist on this wall. Stuff we have absolutely no control over at all. None."

The hands went up and the first few responses came quickly. The first three—weather, gravity, and nature—were offered confidently and at lightning speed. But then there was a long period of silence.

Hyrum pushed the group to think of more.

Someone jokingly shouted out, "I cannot control my husband."

After a loud and fun chuckle from the crowd, Hyrum reeled the group back in to focus and had the complete attention of the room.

"That is it, folks. Other than Mother Nature and the laws of physics, you have absolutely no control over the decisions and choices other people make. Sure you can attempt to influence their decisions, but you cannot control what they decide.

"How does it feel to try to control something over which you have no, or very little control?"

The answers from the group flowed with fervor and enthusiasm. Among the many responses included the following descriptors: stressful, anxious, frustrating, useless, time-wasting, taxing, difficult, hard, and agonizing.

"Exactly. Have you ever been stuck in traffic and tried to control it? Even better, have you ever observed people in cars near you with veins bursting from their neck, mouth wide open yelling, as they weave from lane to lane attempting to orchestrate or direct traffic? Is that time well spent?

"Yet, how often do we find ourselves in the workplace trying to control things over which we have little or no control? Think about it. The key is to carefully consider and examine the situation or circumstance and then to adapt accordingly.

"Sounds simple, but it is not easy. We all know the prescription to good health, right? Eat right and exercise. Sounds simple, but it is not easy. There is a lot that goes into doing it well. Knowing what you have to do, or should do, is not enough.

Hyrum pointed to the right wall and calmly asked almost in a whisper, "What are some things over which we have total and complete control?"

In unison, using a variety of words meaning the same thing, nearly everyone replied with, "The choices and decisions I make."

"Precisely," Hyrum confirmed as he took time to let that sink into the minds of everyone in the room.

"If you take nothing else from our time together today other than the profundity of what I am about to share with you, and internalize and act on it, it can change your life for the better.

"There are effects to conditioning.

"The first is that we sometimes believe we can control things over which we have no control. We just talked about how this is not time well spent, as well as the emotional ramifications and turmoil it generates.

"The second, and more important effect of conditioning, is that we at times convince ourselves that we have no control, when in fact we do. It is simply a matter of ownership and how much you want the desired result."

That statement, Hyrum's second effect of conditioning, is paramount as it relates to high levels of personal, team, and organizational accountability.

The irrefutable truth is that if you truly want to achieve a desired result, and have complete and demonstrable ownership of it, there is always, always, always, always some small step you can take to move closer to achieving it. No matter the challenge, barrier, or obstacle. When ownership of the result is high, you will consistently and habitually find a way to move closer to it.

If the degree of ownership of the identified desired result is low, meager, negligible, or nonexistent, watch out. Without ownership, people will almost always revert to reasons, stories, justifications, and excuses as to why they cannot achieve the outcome. They will convince themselves there is "nothing I can do" and most assuredly will fall into the trap of playing the blame-game. Why would they do otherwise if they do not truly "own" the result?

If you have positional authority as a leader, manager, or supervisor, you could attempt to force or coerce team members or employees to take ownership, but what you typically end up with is compliance and not commitment. There is a colossal and consequential difference between results achieved by compliant versus committed teams and organizations.

In today's hyper-competitive, perpetually changing business landscape there is no time to be wasted demanding or enforcing compliance. The need for change is thrust upon us too fast. Cultivating a highly accountable and resilient organization is a monstrous competitive advantage.

## Solutions Abound

With realities recognized and ownership accepted, solutions tend to flow as plentifully and rapidly as the waters falling over Niagara.

Some years ago, a participant of a workshop at Beltway Rail eagerly approached me during a break. It was obvious based on his body language and facial expression that he was excited to share something with me. He looked like my kids used to look on their way downstairs on Christmas morning. He was nearly jumping out of his skin.

"This is great stuff and everyone here needs this," Bruce began. I have a story about perseverance, tenacity, and determination that you should use in your workshops.

"Do you know who Rudy Ruettiger is?" Bruce asked.

"Sure I do. I love that movie. I think most guys do. It is such an inspirational story. The fact that it is true makes it even more powerful."

"I roomed with Rudy in college," Bruce shared. "What he was able to achieve against such incredible odds is astonishing."

When the participants gathered back in the room following the break, I asked Bruce to share the story. Although many in the room had heard it before, they were captivated as Bruce re-created the story in their minds.

"Rudy grew up in a working-class family with fourteen children. As a kid he loved football. His dream was to one day play for Notre Dame. Members of his own family would rid his mind of that silly ambition. He was told he would graduate from high school and go to work in the power plant like his father and brothers.

"Following high school he did a stint in the Navy and then, as his family had demanded years earlier, went to work in the power plant. But he never lost sight of his dream to some day attend Notre Dame. His dream was not only to attend the school, but to play football for the Fighting Irish as well. In the early 1970s Notre Dame was one of the most powerful football programs in the country.

"By the way, Rudy was five feet six and 165 pounds. Notre Dame was not recruiting Rudy.

"During Rudy's time at the power plant a friend of his was severely injured and that incident changed Rudy's life. He quickly determined that working twenty-five to thirty years in a power plant was not the life he wanted. No longer was he going to let other people or events determine his future. He was going to take control. He decided to go after his dream.

"He applied to Notre Dame but was not accepted. His grades were marginal and not to their standard. He discovered later that he was dyslexic, which had contributed to his difficulty with school. Rather than give up on his dream, he instead took on the mindset of 'what else can I do to achieve what I want?' That led to Rudy enrolling to Holy Cross College where he worked his tail off to get his grades up to the standard where Notre Dame would accept him.

"His dedication and persistence paid off. Rudy was accepted to Notre Dame—after his fourth try. He then had to find ways to pay the tuition. Always possessing a mindset of 'whatever it takes.'

"Getting into Notre Dame was only part of his dream. Playing football for the Fighting Irish was the ultimate goal. At his size, his chances were close to nonexistent.

"Back then, Ara Parseghian, the head coach, encouraged kids in the student body to try out or 'walk-on.' That invitation was all Rudy needed.

"Just picture that," Bruce implored his colleagues in the room. "Picture little Rudy going up against the biggest, baddest, and best players in the entire country. He was annihilated, bloodied, and bruised. But he never complained. He kept his dream in the forefront of his mind and never gave up. He maintained his 'whatever it takes' mindset and ultimately triumphed.

"The players and some coaches were so impressed with Rudy's grit and doggedness that they awarded him a spot on the scout team. That basically meant that the real players got to beat him up all week as they prepared for games.

"That was okay with Rudy. For him it was the price he had to pay to achieve his dream.

"Over the next two years Rudy showed more resolve, determination, and perseverance than most people do over a lifetime. He had overcome so many hurdles and challenges on his way to achieving his dream and he'd never complained. He was accepted to Notre Dame, he made the scout team, but his dream was not fully realized. He had never dressed for or played in a game.

"During the last game of his senior year the coaches decided to reward Rudy's dedication and commitment and allowed him to dress for the game and run onto the field with the team.

"That was close to the dream, but not entirely complete. The dream was to actually play in a game.

"With only a few seconds left in the game against Georgia Tech, Notre Dame scored a touchdown to take a larger lead late in the fourth quarter. Dan Devine, the head coach, told Rudy to enter the game and join the kickoff team. Rudy sprinted onto the field to the applause of the fans and adulation of his teammates.

"Following the kickoff and with only seconds remaining in the game the coaches motioned to Rudy to stay on the field. Rudy, at five feet six and 165 pounds, lined up at defensive end. The ball was snapped and the Georgia Tech quarterback, Rudy Allen, threw an incomplete pass.

"With time left for one more play, Allen dropped back again to pass, but this time was sacked by Rudy.

"Rudy was one of only two players in the rich history of Notre Dame football to be carried off the field by his teammates. Dream achieved."

Hearing the story come from someone who knew Rudy made it even more compelling. Bruce's colleagues in the room were entranced.

The demeanor, mindset, and attitude that allowed Rudy to achieve his dreams are commonplace among individuals, teams, and organizations that possess high degrees of ownership. That spirit sparks creativity and passion to discover, uncover, and pinpoint solutions to overcome any barrier or challenge that arises.

The team at Cambria Ventures discovered that once desire and ownership were attained, creativity and innovation to unearth solutions to achieve what matters most soared. Over time they collected and shared ideas, tips, and guidelines for the creative solutions they ascertained to work best:

1. **Rapid-fire brainstorming without analysis:** Whether as a team or on your own, quickly and without evaluating responses, develop an extensive list.

2. **Invite everyone:** The more perspectives you seek, the more options you will have from which to choose. Remember, $5 + 4 = 9$, but so does $6 + 3$.

3. **Do not assume:** Resist the urge to assume things will not work.

4. **Challenge the status quo:** Explore why things have been done the way they have always been done.

5. **Picture success:** Envision yourself or team in your desired state. Then repeatedly ask and answer the question, "What happened before that?" until you are back at your current reality.

6. **Reverse thinking:** Brainstorm or mind-map a list of everything you should not do. This often leads to new idea generation of what you can do.

When it comes to searching for solutions and finding a way forward, internalize this:

"The successful_____
Is willing to do that which
The unsuccessful_____
Is unwilling to do."

The blanks can be filled in with any occupation, team, or organization. What is it that you, your team, or organization will do to reach your most meaningful, worthwhile, and consequential goals?

## The Obvious Next Step

"It is often easier to act your way into a new way of thinking than it is to think your way into a new way of acting."

A friend shared that statement with me years ago. It can be helpful to consider once you, your team, or organization have moved from recognizing your realties, taking ownership for your circumstances, and developing a solution or path forward, to what is ultimately most important—taking action.

One of the reasons many people resist change is that it will require them to do something different, or simply require them to take action. Most of us are wired to stay in our comfort zones. We become creatures of habit. We find what is familiar and known to be comfortable.

Yet, to improve and achieve better results requires action.

Going after a freshly identified noteworthy goal (to become more effective and perform at higher levels) inherently requires new behaviors and, in many cases, new habits.

I was once asked, "Can you be partially accountable?"

I gave that some thought for maybe three seconds. My response was, "Absolutely you can. It is also called being unaccountable."

Accountability is a journey. As with an actual journey, if you stop at any point into your trip and decide to travel no further you will not reach your chosen desired destination.

The same logic holds true with accountability. You must completely understand and own your circumstance, develop a solution or path forward, and, most important, take action and put the plan in play to achieve what matters most.

You may make mistakes and even fail. But accountability is a forgiving friend. It rewards action. Learn from mistakes and failures and continue the journey by maintaining a "what else can I do to achieve what matters most?" mindset.

Once armed with your solutions, put them into action. Block out time on your schedule solely dedicated to taking the actions your solution requires. This time must be considered sacred and must only include actions linked directly to your solution.

As you develop your plan to implement your solution it can be helpful to consider the following:

1. Of all that you could do, what is the one thing you can do now that will have the most significant impact?
2. What roadblocks may pop up along the way?
3. What impediments or barriers must be removed?
4. What obstacles will require assistance from others?
5. What realities might you not be acknowledging?
6. From what teams or other people do you need to gain more support?
7. What self-imposed restraints exist?
8. What assumptions have you made that have not been tested?
9. What mechanism will you put in place to measure progress?
10. What leading indicators can help assure progress?
11. What has worked well in the past? How could it have worked better?

Finally, accountability is not a place at which you arrive and you are finished. The journey must be repeated regularly over time as you, your team, or your organization continually identify new destinations that once reached, will help you thrive in the new world of work.

## Key Learnings to Internalize

**One:**
The four categories pinpointed by the team at Mountainview Enterprises are undeniably visible and present in highly accountable people, teams, and organizations.

The sequence is important. There is a logical flow from one to the next. As Jim emphasized, accountability must begin with a willingness and desire to fully understand and recognize current realties. Only when those realities are clear can ownership be established and solutions to move forward be developed and put into action.

**Two:**

Being open to the perspectives of others allows us to see and recognize more than we can see or recognize on our own. When we are open to others, we can recognize more barriers, obstacles, challenges, opportunities, and possibilities.

It allows us to recognize a larger reality, thereby allowing us to make better decisions. Being open to others' perspectives does not mean you must agree with what they see or believe. Rather, that you are willing to learn from others.

**Three:**

Unwavering unmistakable commitment—an undeniable belief and confidence—is readily apparent in highly accountable people, teams, and organizations.

To achieve what matters most requires high doses of commitment. Involving and including employees often moves individuals from compliance to commitment. When employees feel that their insights and views have been heard and genuinely considered, buy-in, engagement, morale, and ownership climb.

**Four:**

We often have more control over our destiny than what we have conditioned ourselves to believe. There is always some small step you can take to move closer toward your desired result if you truly own the outcome.

**Five:**

When commitment and ownership for a desired result is high, you will see a surge in strategies, ideas, and solutions to move closer to what matter most.

**Six:**

Accountability is a journey. As with an actual journey, if you stop at any point in your trip and decide to travel no further, you will not reach your chosen desired destination.

## Take Accountability Now

**One:**

As you reflect on the four categories of accountability pinpointed by the team at Mountainview Enterprises, what are the specific aspects within each that your team is doing well? Which can you begin to do even better? What can you do more often? Remember that achieving a highly accountable culture starts with recognizing realities. Focus there first.

**Two:**

Today, go and proactively seek the perspectives of two or three other people.

Consider the current challenges that face you as well as the results you wish to achieve. Then identify friends, colleagues, peers, or leaders who may be able to offer you varying perspectives that will help you make more informed choices.

**Three:**

Consider the approaches to increase ownership developed by the team at Cambria Ventures. How often are you creating these experiences? Develop your plan to implement one or more of these methods within the next two days. Be specific. Develop a plan today to include and involve your colleagues stuck in quadrant 2 to develop in them a higher level of ownership and accountability.

- ◆ **Celebrate Wins:** Ownership increases when progress toward a desired outcome is extolled.

- **Storytelling:** Unlike mandating, when stories are told listeners draws their own conclusion about what must be done. Stories also create pictures, and pictures play to emotions.
- **WIIFM:** Appeal to and help folks see how achieving the outcome will make their job easier or better, develop them professionally or personally, or will make their life better.
- **Openness and Transparency:** Both are indispensable to establishing trust, commitment, and buy-in.
- **Appreciation and Recognition:** When desired behaviors are exhibited, call them out and reward them.
- **Encourage Creativity and Innovation:** Invigorate engagement to release passion.

**Four:**

Identify today one perceived barrier or obstacle over which you have ceded control, but which is holding you back from attaining your must-achieve desired result.

Then ask yourself, "If my job depended on it and I had to do something to resolve or begin to overcome this challenge, what could I do?"

**Five:**

To stimulate more ideas and solutions that will move you or your team closer to your must-achieve desired results, deploy one or more of the following tactics:

- **Rapid-fire brainstorming without analysis:** Whether as a team or on your own, quickly and without evaluating responses, develop an extensive list.
- **Invite everyone:** The more perspectives you seek, the more options you will have from which to choose. Remember, $5 + 4 = 9$, but so does $6 + 3$.
- **Do not assume:** Resist the urge to assume things will not work.

- **Challenge the status quo:** Explore why things have been done the way they have always been done.
- **Picture success:** Envision yourself or team in your desired state. Then repeatedly ask and answer the question, "What happened before that?" until you are back at your current reality.
- **Reverse thinking:** Brainstorm or mind-map a list of everything you should not do. This often leads to new idea generation of what you can do.

**Six:**

Schedule time today to develop and execute on your plan to ensure that your must-achieve desired results become reality.

# Chapter 4

# Accountability Transcends

## Powerful Shift in Perception

Ask a random sampling of ten or more reports, peers, individual contributors, supervisors, managers, directors, and senior leaders, "What is your job?" and listen closely to their replies.

How many of the responses centered predominantly on your "must-achieve desired results"? What percentage of the answers chiefly zoomed in on a list of activities and duties you often will find included in a job description?

Teams and organizations derive considerable benefits when employees at all levels view and approach their work with a staunch focus on must-achieve desired results rather than activity, duties, and roles.

Think back to a time in your professional career when you were at your "personal best" and most engaged in your work.

For most of us, those personal bests were during times when we were part of a team, on a mission to achieve something special, and believed we were playing a game we could win. It is during those moments that individuals voluntarily choose to take accountability and to go above and beyond what is "required" to help the team achieve what matters most.

A focus on top desired results ingrains that shift in thinking and provokes beneficial changes in the way employees go about their work. Once grounded in a focus on what matters most and feeling that they are part of something bigger, meaningful, and important, employees are more aware and conscious of how their work has an impact on other people and teams.

With an organization-wide mindset of "the must-achieve desired results are my job," alignment around what matters most is never an issue. With this subtle mindset embraced, collaboration swells, teamwork is strengthened, determination intensifies, camaraderie blooms, creativity and innovation flourish, and employee engagement and morale rocket.

Just as every single player on a football team is focused on doing whatever it takes to help the team win, every employee on your team must embrace that same notion. Getting there requires focused leadership, as Craig the CEO of Harvest Financial discovered.

"So, my job is to make sure everyone in the organization is aligned and clear on those results," Craig said out loud while approaching me as his team was dispersing.

As the CEO of a mid-sized bank with $15.3 billion in assets, Craig was under extreme pressure from the board of directors to capture additional market share and stave off competition.

Craig's leadership team of nine, coupled with twelve additional colleagues identified as rising stars, had just spent ten intense hours working together to identify, define, and determine precise metrics of the top four goals for the upcoming fiscal year.

The day included plenty of passion, debate, conviction, disagreement, contention, and perhaps a touch of exasperation. But in the end these twenty-two individuals coalesced around a set of four critically important outcomes they believed must be accomplished.

Every member declared their alignment, ownership, and personal accountability to ensure that those results would be realized. They agreed that with these four results achieved the firm would be positioned for ongoing success in the future.

Craig continued toward me, smiling with a sense of pride and satisfaction. I smiled back at Craig and asserted, "No, Craig. That is not your job."

He looked at me, puzzled. I continued, "Achieving those four results is your job. Just like every other employee of this bank."

I asked Craig, "At the end of the year is your board of directors going to ask you if you communicated your top desired results to your employees, or are they going to want to know if you achieved them? And if you do not achieve them, whose tail is on the line?"

What Craig instantly embraced and internalized at that moment is vitally important. The desired results are everyone's job.

"What we are looking for, Craig, is to forge an organization-wide mindset among every single person that achieving those four results is their job."

I asked Craig, "Did you ever hear the story about the janitor at NASA?"

"No, not that I can recall."

"Well, I am not sure if the version I have heard dozens of times is accurate, but let me share the crux of the story.

"When John Kennedy made the declaration back in the 1960s that we would put a man on the moon and bring him back safely to earth by the end of the decade, NASA was under the gun. JFK identified a must-achieve desired result that became the job of every single person associated with NASA.

"During a visit to NASA, Kennedy was touring a facility and surrounded by NASA personnel. One of the individuals standing near him was a member of the building maintenance staff.

"Kennedy turned to him and asked, 'What is your job here, young man?'

"Without skipping a beat, the kid's response was, 'I am here to help put a man on the moon.'

"This kid, a janitor at the facility, viewed his job as helping to put a man on the moon.

"Imagine the impact to an organization when every single member views their job as making certain that the absolute most important results are realized?"

Craig could not have been more inaccurate by believing that simply identifying and *communicating* the must-achieve desired results was his job. After all, he was the leader. Leaders are paid to achieve desired results. Not to communicate them.

With his initial thinking in place that his job was simply to communicate what must be achieved, the team of thousands of employees that Craig was leading at Harvest Financial could fail to achieve those results and he could still feel he did his job. That could not be further from the truth, and especially not acceptable for the CEO.

## Cement the Focus

Have you ever had an overwhelmingly hectic and chaotic work week and, upon reflection when the weekend finally arrives, discovered you had accomplished little to nothing important?

It is easy to let the tsunami of "the work" sidetrack and derail us from focusing on what matters most. Top-performing individuals, teams, and organizations avoid being sucked into the trap of allowing the "urgent" to sabotage what is most important.

As a leader of a team or organization, given the choice, would you prefer to have your employees operate at peak productivity or optimal effectiveness?

Does being active and busy ensure success? Consider the following dialogue between two salespeople.

Salesman One proudly boasts to Salesman Two, "I had a lot of good conversations today." Salesman Two responds, "Yeah, I did not make any sales either."

If Salesman One believes the measure of success is staying active and busy, he had a marvelous day.

If he were to string together a series of several hundred identical days, would he without question be a significant, vital, and positive contributor to his employer's success? Would his actions absolutely and totally ensure a triumphant career for him?

Salesman One was immensely productive, yet did not make a sale. If his performance was graded based on activity, he would have every right to go have a beer with his buddies and glowingly brag about what an amazing salesperson he is.

However, if his performance played out this way over time, his determined and vigorous focus on activity alone would not necessarily produce desired results for him or his employer.

I recall sitting in a workshop once when the facilitator of the session asked the group, "Does all activity produce results?" Nearly everyone in the room immediately and confidently replied, "No."

The instructor stayed silent for a few seconds, smiled, and then slyly said, "You are all wrong. Even if you do nothing, you will achieve a result. It just will not be the result you want."

His point was that a singular focus on being busy or staying active does not assure you, your team, or your organization that your desired results will be realized.

The only reason companies hire employees is that at some point someone in a managerial or leadership position makes a decision that if we hire these people and add them to the team it will help us achieve our desired outcomes.

Companies do not hire people to simply show up to be active and busy. Yet, oftentimes during and following the hiring process there is so much attention paid to job descriptions, roles, and duties that companies inadvertently and unintentionally create a focus on activity versus results.

In the highest-performing teams and organizations, employees focus on what they must achieve rather than what they must do. They pay attention to outcomes instead of tasks. They are cognizant of the difference between a means to an end and the end itself.

There are negative repercussions when a majority of team members or employees maintain and hold an intense focus on only their role or job.

When individuals and teams concentrate solely on their roles, it quickly leads to elements of the blame-game being played. On a team or in an environment like that you will often hear comments such as, "That is not my job," "They were supposed to do that," "I did not know that I was supposed to ...," "Nobody told me that," "I am waiting for them to ...," "I do not know how to ...," "I did not know that was important," "If they would just do what they are supposed to do"—and the epitome of the blame-game, "That is not my fault and there is nothing I can do about it."

Those elements of the blame-game foster dangerous, destructive, and toxic emotions of guilt, resentment, isolation, and mistrust. Such emotions often consciously and subconsciously lead to silos being forged, breakdowns in communication, hoarding of resources, unwillingness to share ideas, and separation.

There is enormous benefit derived when a simple and subtle mindset shift is adopted, embraced, and internalized. I speak of the mindset where every member of the team and every employee views and defines their job as helping the team accomplish the must-achieve desired results. That is the only reason we hire them, anyway.

Therefore, from the very first interview with a new candidate your top desired goals should be clearly explained and detailed. New hires must learn from their first interaction with your team or organization what is most important—the must-achieve desired results.

Sure, they may be given a title and perhaps even a list of tasks they must execute, but they must understand that their job is not done until the team or organization collectively attains what matters most.

Leadership must spearhead the endeavor to create and instill that same mindset shift within current team members and employees. This effort cannot be delegated. The highest level of accountability can only be reached when every single employee is crystal clear on the must-achieve desired results, and defines and accepts their job as helping the team or organization secure them.

Robert, the CEO of CVP International, and his team that we mentioned in chapter 1, fully embraced this simple practice and attributes it as one of the key reasons alignment and ownership around their must-achieve desired results are unprecedented. Barriers, such as lack of collaboration, hoarding resources, poor communication, and the blame-game, that in the past had slowed optimal organizational effectiveness, are rarely experienced now. The CVP team uses their strategic plan and clearly defined desired results to indoctrinate new hires, and to create alignment, ownership, and accountability from day one.

In addition to the obvious benefits, there is a significant and compelling by-product resulting when this mindset is nurtured. Employees begin to find more meaning and purpose in their work. Levels of engagement, morale, loyalty, and creativity tend to escalate when employees recognize they are part of something bigger than themselves.

The following parable demonstrates this.

Two men carrying heavy stones down a dusty road are stopped by a stranger who asks, "What are you doing?"

The first man answers, "Can you not see? I am carrying stones."

The second man answers, "I am helping to build a cathedral."

Which of the two would you choose to be part of your team?

With this newly acquired mindset shift in place, silos begin to crumble, collaboration escalates, alignment intensifies, creativity soars, resiliency surges, camaraderie snowballs, boundaries fade, perseverance balloons, and trust blossoms.

When a team or organization embraces the notion that we are all in this together, it immediately eradicates the common blame-game favorites of, "It is not my job," "I did not know," and "I am confused." When the most crucial desired results have been clearly defined, communicated, and positioned in a manner that they are everyone's job, those blame-game elements become absurd to even consider.

Furthermore, with this new shift in thinking, we vanquish inactivity and waiting, and spawn a can-do or "what else can I do to help the team?" attitude. Searching for solutions to overcome complications, problems, and obstacles becomes the norm even when folks may be justified to play the blame-game.

Once this shift in thinking is in place, team members and employees at all levels recognize that they are accountable for more than what is on their job description.

In an organization like this, Salesman One, whom we met earlier, is no longer eager to boast about his busy day.

## The Magical Mindset Shift

Magic happens when we are crystal clear that the must-achieve desired results are everyone's job and forge the mindset that we are all in this together.

With new cafés opening weekly across the country, the leadership team of Café Delight held a shared sense of optimism about the future. Well, except for Audrey and Russell.

Audrey and Russell were puzzled why the performance of the organization's more than eight hundred current stores had plateaued. The fresh innovations to format, tweaks to the menu, and additional staff training had produced unexciting and pedestrian results.

Both believed that simply to continue opening more stores would not be enough to secure and sustain a foothold in this ultra-competitive market. Their belief was that in addition to launching new locations, it was essential to boost performance and profits from the existing stores. It seemed everything they attempted fell short of expectations.

Russell thought aloud, "We've got to find a solution. Our margins are tight and competition is fierce. Sure, we have new stores opening and that is exciting, but my concern is that staying the course and maintaining status quo at current locations is a dangerous proposition."

What great timing, Neal thought. He had just finished reading a report from Acadia, the "secret shopper" service he contracted as part of his effort to more fully understand why there was a substantial gap in performance measures from region to region.

The report illustrated consistent experiences in the top-performing regions that were not observed in others.

"I have been poring through this report that Acadia sent last week," said Neal. "There are some insights and observations in this report that I found fascinating and, to be honest, a bit revolutionary."

Before Neal could ask, the team begged to learn what the report included. Neal was eager to share some of the most compelling and thought-provoking nuggets with the team.

Adrenalized by his colleagues' interest, Neal suggested, "Let me read directly from the report."

"When secret shopper Patty entered a store in the top-performing region she was politely met and immediately seated.

"Patty asked, 'What's your job here?'

"The greeter politely volunteered, 'My job is to make sure all of our customers have an incredible experience here and to help our store achieve a net profit of 12 percent.'"

Neal shared with the team that Patty made a notation on that page that read, "Wow, that is an unusual answer."

Neal continued to read from the report.

"Patty indicated that shortly after being seated she was glancing at the menu and approached by a smiling, cheerful young man who introduced himself.

"Jamie asked if Patty had been there before and if she had questions about the menu. He also familiarized her with the specials that day. Jamie asked if she was ready to order or if she would like more time.

"Patty decided on one of the specials and, after placing her order, she asked the waiter, 'What is your job here?'

"'My job is to make every customer's visit here memorable and help our store achieve a net profit of 12 percent.'

"'This is weird,' Patty wrote plainly in her notes. In fact," Neal added, "Patty made a notation that she was beginning to wonder if somebody was playing a prank.

"Her notes go on to show that while she was waiting for her order she noticed a young man clearing a table nearby. Patty gained his attention and called him to her table. She asked him the same question she asked the other two team members in the store, 'What is your job here?'

"Again, she received nearly the identical answer.

"Patty commented here in her notes that she was at a loss. You can see in bold letters she wrote in the margin that she had never experienced anything like this.

"Patty asked the third team member, 'Tell me more, I am not sure I understand.'"

Neal, in an effort to gain the group's focus and attention, asserted, "Listen to what Patty wrote in her notes, 'The response he offered is unusual and unlike any I have heard.'

"Patty writes here that she pressed the busser for a further explanation.

"She noted that he said to her enthusiastically, 'Sure, every member of the team here knows that for us to succeed together, we need this store to achieve a profit margin of 12 percent or more. My job is to do whatever it takes to help the team and the store achieve that goal. Everyone on our team believes that for us to achieve that goal we need to create a reason for our customers to want to come back. So, we work together to do that.'"

The Acadia report indicated that unlike the other regions where this secret shopper was assigned, in this region she observed no peer-to-peer gossiping or complaining, no blaming or finger-pointing, and no unwillingness to support and help one another. Rather, she observed high levels of morale, engagement, support, and passion.

The Café Delight team was astonished by the information in the report and agreed that this was the level of collaboration, esprit de corps, ownership, and personal accountability they wanted to instill at every store. They sought out the director of that region to lead the effort nationally.

Eleven months following the launch of this new effort, combined with ongoing innovations, Café Delight achieved a significant and positive bump in profit margin numbers that had been stagnant for years.

## Lessons from the Marauders

Peak performance for teams and organizations has more to do with how well each member understands the possibilities, opportunities, and must-achieve desired results than it does with how well the members like each other. Couple that impeccable clarity and individual understanding with a belief that achieving that result will make their life or job better, and voilà—get out of their way and let the magic begin.

Doreen was shouting out names as she read from the small pieces of paper being pulled from the hat. At previous company picnics, each department had selected who would play for their volleyball team. Not this year.

DigiTech was three months into their initiative to improve organizational performance through strengthening teamwork and collaboration. So it only made sense to use the picnic as an opportunity to leverage that effort. The sales department dominated the volleyball tournament every year and there was some unconcealed angst, bordering on resentment, about that.

Max, perhaps the most competitive salesperson employed by DigiTech, eagerly listened to discover who would have the privilege of becoming a member of his team. Inwardly hoping to be joined by at least two or three of his sales colleagues, he openly showed disappointment as he learned this would not be the case. Instead, he was teamed with two folks from engineering who he did not know, one person from human resources with whom he had run-ins in the past, two people from manufacturing he had never met and three people hired the week prior to work in customer service.

Of his eight team members, he knew one of them, and they despised each other. More concerning to Max was that only three of his teammates appeared to possess athletic prowess. Team Max's Marauders was now assembled.

As Max scrutinized his team, he thought to himself that his team was a microcosm of DigiTech. Like the company, this was a group of people that hardly knew one another and were now tasked with morphing into a high-performing team.

He reflected momentarily on the team-building activities he and others had participated in over the past three months. Max, like many others asked to participate, had found those efforts to be an annoyance and disruptive to his schedule. However, once immersed in the activities, he did find some of them enjoyable.

In the hours leading up to the volleyball tournament, Max spent time talking with friends and thinking about work.

During a moment of solitude, he thought, It is undeniably evident more needs to be done to boost teamwork and collaboration. The lessons learned from the teambuilding experiences have not translated back on the job. We continue to operate in silos focused on our own goals with minimal cross-functional communication, scant sharing of resources, and negligible interdepartmental support of one another.

Following lunch, and with his mind again focused on his team, Max was pleasantly surprised to find that although not as physically or athletically gifted as he, most of his teammates had a strong desire to win. After all, each member of the winning team would receive a gift card to the highly touted new restaurant in town.

With eyes on the prize, the Marauders quickly aligned around winning the spoils of victory. Game after game, and set after set, the Marauders banded together with ferocity.

Their communication was constant, upbeat, encouraging, and focused. They supported one another when things were not going well. When a teammate was out of position another would be more than willing to step up and cover for them. They challenged one another in a reassuring manner. They listened intently to one another. When mistakes were made there was no blame. They leveraged the abilities each member possessed. The players all pushed themselves to levels further than they felt they were capable. The Marauders were on a roll and enjoying every moment of the ride.

No longer was this team just Max and a bunch of people he did not know. This was now the Marauders. A tightly bound team that was focused, resolute, and determined to achieve their desired result.

After dispatching the first three teams they faced, Max and his new friends found themselves in the championship game. The prized trophy, and dinner certificates they coveted, were within grasp.

By this point they had developed into a well-oiled machine. They easily trounced their competition in the final and collected their prizes. The Marauders gloated a bit throughout the rest of the day and enjoyed the camaraderie. Even Max and his nemesis from human resources forged an improved relationship.

Two weeks later Max was at lunch with one of his pals from marketing. Sheila asked Max, "So how are all of your Marauder buddies doing?"

"Not sure. I have not talked with any of them since the picnic."

Over the same two weeks the rest of the Marauders had similar conversations with friends and colleagues.

Why did the qualities, attributes, best practices, traits, and characteristics of a high-performing team that the Marauders had developed not translate back to the workplace?

The Marauders galvanized together quickly into a top-performing team. They did so without any trust falls, scavenger hunts, cooking classes, speed networking, egg drops, offsite retreats, solving a mystery, or building paper boats.

This group of strangers united together within minutes and demonstrated characteristics, qualities, and attributes of top-performing teams. Yet, they hardly knew one another. What is it that fuses a group of individuals into a high-performing team?

Accountability for a mutually desired outcome or desired result is the most important ingredient. That deep degree of accountability must begin with clarity, alignment, and ownership around an outcome or result that benefits every member of the team.

Knowing that Billy, Mary, and Susan will catch me during a trust fall is not what motivates me to do whatever it takes to win. Yes, it is nice to know they have my back, but that is not what truly inspires me. What impels and energizes me is my understanding of the desired result and how it benefits my team, my organization, and me.

Even though members of the Marauders were from different departments and knew little about one another, they exhibited many of the practices and qualities we crave to see on our team or in our organization.

They were focused, aligned, resolute, and rallied around a common desired result. They communicated openly and candidly, sought out feedback, and encouraged and challenged one another to perform at peak levels. They were inclusive and leveraged the unique capabilities of each member of the team. They shared knowledge and pointers to develop one another. They focused on the desired outcome rather than their individual roles. They remained engaged in good times and bad and never wasted time or energy blaming or sulking. They enjoyed working together and had fun.

Why do these highly desirable practices and qualities, quickly developed at a company picnic in this case, often not carry over once back in the workplace?

As in the case of the Marauders, oftentimes employees will go back to their department, business unit, or team where the focus is commonly on a distinct set of results only relevant and meaningful to that department, business unit, or team.

When departments, business units, or teams are singularly focused on only their unique set of desired results it generally leads to an internal focus, an us-versus-them mentality, minimal cross-functional communication and collaboration, protection of resources, and separation and isolation.

The combination of these forces commonly fuels the blame-game, which in turn impedes or stifles optimal organizational effectiveness.

## Accountability Trumps Responsibility

Being responsible for your job is commendable. Taking accountability to do whatever is necessary to help your team achieve what matters most is priceless.

Noticing that Ted was stumbling and would be out of position to get to the ball that was in midflight heading back to his spot, Mary bolted from her back-line position and made a remarkable pass that saved the point for the Marauders.

When Mary darted from her back-line position to cover the front-line spot for Ted, who had tripped and fallen, what was the catalyst that spurred her? After all, her job and responsibility was the left spot on the back line.

It was her personal accountability for the mutually owned desired result—to win the game that triggered Mary to take on more than what she was responsible for.

Mary did not let her job description and responsibility dictate the action she chose. She instead chose to take accountability for the desired result and went above and beyond what was required of her job. She voluntarily tapped into her discretionary performance in order to help the team achieve what mattered most.

If she so chose, she could have remained in her back-line position as the ball fell to the ground and exclaim to her teammates that she "did her job" by covering her area around the back line. More important, she would have been justified and accurate to say so. How would that have affected team camaraderie?

There is a noticeable difference when surrounded by colleagues that take full responsibility for their job versus colleagues that choose to take personal accountability for your desired results.

There are subtle but powerful differences prevalent and on display in responsible versus accountable teams and organizations. They include: a focus on activity versus a focus on desired results; a focus on completing tasks others have assigned versus a focus on choosing highest payoff activities that propel us toward our desired results; primary attention to what is best for me versus primary attention to what is best for the team; operating in isolation versus operating as part of a team; compliance around what I must do versus commitment to achieving what matters most; doing what I am asked versus doing whatever it takes; and a "tell me what to do" mindset versus an "I know how I can help" mindset.

To capsulize, "responsibility" is most often what is detailed in a job description. Accomplishing the tasks listed is my responsibility. When employees are focused on their responsibilities it diminishes the attention being paid to the must-achieve desired results established by the team or organization.

This mindset also leads to gaps and cracks within an organization through which things can fall. Employees can be fully responsible for their jobs and miss many opportunities to contribute more fully in order to catapult the organization closer to what matters most.

When people choose to take accountability for the must-achieve desired results, they are more likely to look for ways outside of their defined job description to thrust the organization toward a better state.

## To LAG or to LEAD

If you ever felt that luck has played a role in your professional or personal success, you have probably been focused on LAG measures.

Individuals, teams, and organizations that consistently achieve and exceed what matters most take accountability and control by zeroing in on LEAD measures. They focus on the behaviors and actions that they can control and that are likely to hurl them toward their must-achieve desired results.

Kevin is a regional director of Superior Electronics. His region consistently outperformed all fourteen others.

During our conversation as we walked to his team meeting, he offered to share with me why he believed his team consistently excelled.

"We have discovered over time a simple but proven approach that allows us to maintain high levels of accountability to achieve what is necessary," he boasted.

Wow, I thought to myself. This is awesome. I am going to learn from the master the secret to accountability and high performance.

"I am a big fan of simple, pragmatic, and memorable concepts," Kevin declared. "This one is not rocket science. In fact it is quite simple. We focus on LEAD measures and not LAG measures."

Kevin could see by my facial expression that I wanted him to elaborate.

"We learned early on that when the focus, attention, and discussion are centered on LAG measures, an after-the-fact metric, you are basically relying on a wishful thinking strategy. Think about it. Once the result is in, there is nothing you can do about it. You must live with that result.

"However, when the focus is on LEAD measures—the behaviors and actions that are both *controllable* and *likely* to move you toward your desired results—you are able to gauge progress regularly. Tracking frequency of these behaviors is vital.

"These LEAD measures must be tracked regularly and everyone on the team needs to have access to them at all times. They must be visible. Everyone can glance at them and immediately learn if we are winning or losing. We ask ourselves daily, are these LEAD measures moving us toward our desired results? That allows us to correct course as needed to ensure we achieve the desired outcome.

"This approach provides considerable flexibility. By tracking the LEAD measures on a regular basis, it is easy to determine if those specific actions and behaviors are advancing us toward our desired results. If they are not, we have built-in agility to change on the fly and choose new LEAD measures before the final result is in. Or we may decide the frequency must be adjusted.

"If we determine the behaviors selected (LEADs), are not working as well as we would like, the team can select new actions and behaviors that they can control and that are likely to ensure success. If the new LEAD measures do not appear to be producing, we then identify yet another set again until the data shows we have selected the best LEAD measures.

"Let me share an example. Suppose you have a goal of losing twenty pounds. The LAG measure in this scenario would be weight lost, or twenty pounds. Most often, someone on a diet will focus primarily on that number. The LAG measure of twenty pounds becomes the goal or result they hope to achieve.

"Sure, they may decide to exercise daily and watch what they eat as part of their strategy to achieve the desired weight loss, which is terrific. But the focus and attention is on losing twenty pounds. Of course exercise and diet are important, and they do meet the two criteria of 'controllable' and 'likely.'

"However, without tracking and measuring progress regularly, there is no way to know for sure if the LAG measure of twenty pounds lost will be achieved. The critical component of the entire process is to regularly measure the impact of your diet and exercise—the LEAD measures.

"In this case it becomes a question of *how much* exercise (calories out) and your diet (calories in). Do you need to run ten miles a week, or fifteen? Should you be consuming twenty-four hundred calories per day, or two thousand? If you are not paying close attention to the impact these LEAD measures are having, you cannot possibly be sure they are working or if your desired result will be achieved.

"We have discovered two immense and valuable by-products resulting from this approach. I would like to claim it was part of my plan, but they were happy accidents.

"By allowing the team to choose the LEAD measures, their degree of commitment, engagement, ownership, collaboration, and passion has skyrocketed.

"Secondly, this heightened sense of control has led to increased levels in self-confidence, and that has our performance numbers soaring."

Kevin's insights enkindled lessons learned from thought-leaders and firsthand experiences over the course of my lifetime. The feeling that when one has control, in the case of Kevin's team members it was control over the LAGs, it facilitates increased levels of self-esteem. The combination of the two, being in control and feeling good about yourself, leads to heightened levels of productivity.

## Key Learnings to Internalize

**One:**

Whether a team of five or an organization of five thousand, the must-achieve desired results are everyone's job. Employees are most often at their personal best (accountable, engaged, effective) when they are part of a team seeking to achieve something important, and when they believe they are playing a game they can win.

Leaders must accept the challenge to set that stage.

**Two:**

Activity alone does not guarantee must-achieve desired results will be attained.

There is enormous benefit derived from shifting employee focus from activity to a focus on the desired results. When employees embrace the mindset shift of "my job is our desired results," they begin to go above and beyond what is in their job description and will voluntarily tap into their discretionary performance area to offer more than what is required.

**Three:**

A mutually owned set of desired results spawns team members to choose behaviors often visible in high-performing teams.

**Four:**

Employees are most engaged in their work when they believe they feel they have control and are playing a game they can win.

Once leadership is clear on the must-achieve desired results (LAGs), involving your team to choose the specific actions and behaviors (LEADs) that are *controllable* and *likely* to produce desired outcomes heightens engagement, ownership, and accountability.

Tracking progress and keeping score along the way is essential. The team must be able to know if the LEAD measures they chose are suggesting the desired results will be achieved. In other words, they need to know if the team is winning or losing so that adjustments can be made as needed.

## Take Accountability Now

**One:**

Ask the question, "What is your job here?" to a colleague or two today. If their answers do not mention the team's must-achieve desired results, and how what they do contributes to team's ability to achieve those results, help your colleague make the connection.

**Two:**

Identify opportunities to begin shifting your team's attention and mindset from activities, roles, and job duties to a laser-like focus on what matters most.

What processes or systems are in place now that can be tweaked to germinate this new mindset? How can you lead differently to orchestrate this shift?

Craft one idea today and implement it immediately.

**Three:**

Include your team members in the process to consider and choose the specific actions and behaviors (LEAD measures) that will ensure that your most important desired results (LAG measures) are achieved.

Develop a mechanism for your team to track the frequency and corresponding impact that these LEAD measures are contributing toward moving your team closer to your desired outcomes.

# Chapter 5

# The Preeminent Organization

## Accountability Is the Catalyst

In the "new world of work," ideas, talent, speed, and leadership capacity are what separate top-performing organizations, teams, and individuals from the laggards.

Driving transformational change and strategies that require shifts in human behavior may be the most difficult challenges a leader can face. What do top leaders do to gain voluntary contributions of discretionary performance from those they lead? How do they create alignment, engagement, ownership, and accountability for desired results at every level of an organization? How do they ensure that key strategies are not slowly suffocated but instead supported to produce exponential results?

We have touched on these questions in earlier chapters. We provide more ideas to consider in this chapter.

Lasting change and flawless execution of key strategies to achieve what matters most require voluntary participation from a diverse and large number of personnel. Clarity, transparency, inclusion, alignment, collaboration, empowerment, and trust are essential. When these critical elements are in place, employees will naturally become motivated, passionate, resolute, and accountable to achieve what matters most.

The mission of forging a highly accountable culture must begin with leaders creating a sparklingly lucid and scintillating picture of the exhilarating opportunities and intoxicating possibilities that exist—the "must-achieve desired results." Leaders must take ferocious ownership to ensure that every employee is able to envision how that new future positively affects them.

How will it make their job better or easier? How does it help them? How will it benefit their team? How, in any way, does it make their personal life better? A "want-to" culture trumps a "have-to" culture.

Team members must be able to make the connection between "what they do" and how that contributes to their organization's or team's ability to successfully realize the desired outcomes. With that translation established, the desired results begin to manifest in the manner employees go about performing their work.

Those foundational steps—(1) clearly defining and communicating the must-achieve desired results, (2) inculcating the "what's in it for me?" to everyone involved, and (3) revealing the translation—are paramount to cultivating a culture with soaring levels of accountability. When these steps are engineered poorly, mediocrity wins the day. When they are executed flawlessly, accountability and peak performance flourish.

A highly accountable culture is always a top-performing culture. With deliberate and intentional focus you can shape your optimal culture.

Your culture—the way your folks think and act—is your "brand." The way your organization or team is perceived is a direct result of the experiences your employees and team members create for your clients and customers (internal and external).

Employees and team members think and act the way they do primarily for one reason. That reason is the manner in which they are being led. For the most part, the actions and behaviors that team members and employees choose to exhibit is a reflection of the culture that has been established, fostered, nurtured, perpetuated, and, at times, tolerated by leadership.

An organization's culture is either an engine propelling it toward its most crucial desired results, or an anchor impeding progress and slowing performance.

Corporate culture is the convergence of the cumulative way your employees think and act, and the beliefs they hold that drive the actions they choose. Every organization and every team has a default culture. That culture is producing the results you are achieving now—whether exceptional or poor.

As a leader, you can let your culture lead you, or you can define and shape your optimal culture. This applies to those leading teams of five to ten people as much as it applies to those leading thousands. Through conscious focus, you can ignite extraordinary performance from your team or organization by shaping your optimal culture.

An important question to answer as you begin is, "If the results you must achieve in the future are loftier, more challenging, or just different from those you are achieving today, can you expect to attain them if your culture continues to think, act, and behave the way it always has?"

What is the tongue-in-cheek definition of insanity?

Doing the same thing over and over and expecting a different result.

Accountability is the catalyst to creating your optimal culture. A culture where employees and team members voluntarily choose to think, act, and behave in a manner that will armor your organization or team with a competitive advantage in the new world of work. An accountable culture is the most effective culture.

Leaders must own the process of developing and shaping their optimal culture. For some, it may feel awkward, or perhaps they may believe it is something that should be handed off to human resources. After all, many leaders have been educated, developed, and have spent most of their careers on the strategic side of running organizations. The "people stuff" is most often not their forte.

No matter the multitude of business savvy and skilled human resources professionals you may have on staff, delegating culture shaping is always the death knell to lasting culture change.

Human resources, along with every other department, should be involved. But the senior leadership team must possess the same degree of ownership, rigor, and passion around creating the optimal organizational culture that it holds for devising and launching the most important strategic initiatives. After all, culture trumps strategy every time.

As a leader, you are being thoroughly scrutinized at all times. Your team members and employees are watching you very closely. They observe every signal you send through your body language, the tone of your voice, and, most important, the congruency between what you say and what you do.

The quickest way to derail something as crucial as shaping your optimal culture is to choose not be involved. Think about the message that sends to those around you. Leaders must be involved and highly visible throughout the process.

A second and equally powerful culture-shaping saboteur is to behave in a manner that does not mirror the message. Employees will tolerate what you say; they will act on what they see you do. Leaders must be sentient and keenly aware of the experiences they are creating for those they lead. Those experiences will either reinforce the beliefs your employees already hold, or create new beliefs. When leaders are cognizant of this, they are more likely to create experiences that develop beliefs that will stimulate the desired actions they wish to see from employees. One slip-up—yes, just one—can throw a culture-shaping effort into a state of unrecoverable disarray.

## Shaping an Accountable Culture

Through her observations and analysis over time, Alex captured the essence of an exemplary leader and packaged it in a concise, useful manner that was easily understood and practical.

Many of the managers and leaders with whom I spent time at Gencova suggested she was likely to become the future CEO.

While speaking with Alex, she divulged that as early as her teenage years she was determined to be the model leader she felt she was capable of being. Her aspiration was clearly apparent based on the first few days that I spent with her and her team. As the youngest business unit leader in Gencova, she understood that all eyes were on her.

I was dazzled by the display of exemplary leadership practices for a person considered by some as "junior." Her team of more than three hundred was spearheading a massive $425 million project that was integral to the success of the entire firm.

They were ahead of schedule, under budget, and clearly functioning as a focused, engaged, passionate team on a mission. When I spoke with her employees there was a consistent sense of purpose, pride, engagement, commitment, and accountability. It was glaringly obvious that nothing would deter this team from accomplishing their mission.

I was interested to learn how she had developed and mastered her exceptional leadership skills.

At a team dinner Alex shared with me what she believed was most influential to her development as a leader.

"I am convinced that what benefited me most was my hunger to learn. Throughout my career I was like a sponge and soaked in and internalized the principles exhibited by top leaders I had worked with and those I observed. I wanted to learn and understand what it was they did that inspired, motivated, and engaged employees. What did they do to bring out the best in their people?

"I have always been intrigued with and fascinated by human behavior. My father was influential on my craving to learn. He ingrained in my mind that I could learn from anyone, anywhere, and anytime. All it required was a willingness and openness to learn. You can watch children playing on a playground and learn. You can observe top leaders or poor leaders and learn. You can be surrounded by strangers and learn.

"Since my aspiration was to develop my leadership abilities I focused on what it was people did in personal and professional settings that resulted in others voluntarily choosing to follow them.

"When teams excelled, I would analyze and dissect what those leaders did to set the stage for success. When things did not go so well, I would reflect, contemplate, and consider what those leaders might have done to achieve better results.

"I wanted to know what it was that truly energized employees. What was it that moved colleagues from compliance to commitment, and, more important, to *voluntarily choose to take personal accountability* to help a team succeed?

"I found it fascinating to learn why one particular team would perform exceptionally under the leadership of one person, yet so poorly under another.

"Over time, I was able to uncover several themes. I believe some of the key attributes of the composite ideal leader include:

- **Provide Clarity:** When employees possess a peerless picture of success they will astonish you with their abilities. Without clarity you will create confusion and employees will experience stress, anxiety, and frustration.
- **Practice Excellent Communication:** Model open and candid communication, actively seek perspectives of others, offer

meaningful informal feedback, and be willing to discuss the elephant in the room. And above all, listen empathically.

*   **Focus on the Few:** There will always be more great ideas than capacity to execute on all of them. The law of diminishing returns is an absolute truth. Choose the opportunities with the highest payoff and focus all your attention there.

*   **Include and Involve:** When employees are not involved it is silly to think they will be engaged and committed. No involvement equates to no commitment. People want to be heard. When you deny them that, expect no more than compliance.

*   **Leadership Alignment:** Once decisions are made, all leaders must demonstrate absolute alignment, ownership, and personal accountability. Lack of alignment by key leaders is readily recognized by employees and breeds confusion, mistrust, cynicism, apathy, and indifference.

*   **Silo-Free:** Nurture the mindset that we are a team and we only win when the entire team wins.

*   **Role Model:** As a leader it is imperative that you model the behaviors you expect from others. If leaders' actions are inconsistent with their messaging, it will destroy their credibility.

*   **Head and Heart:** Logic alone is not enough to create a groundswell of momentum. In addition to the 'what' that must be achieved, employees must be equipped with the 'why.'"

What I learned as I got to know Alex over time was that she exemplified those leadership traits through word and deed. Her focus and attention, as well as that of her leadership team, to these practices were central to shaping a highly accountable top-performing culture.

## Ignite Desired Accountable Behaviors

Robust levels of accountability to achieve what matters most are rarely reached and sustained through an approach of command and control. Enforcing compliance around policies and procedures and demanding specific actions seldom results in lasting behavior change.

Telling people what to do, when to do it, and how to do it is time consuming, energy draining, and futile. Because when you are not hovering over employees enforcing compliance, they almost always will revert back to doing things the way they "believe" things should be done, or the way they have always done things.

However, when employees freely choose accountable behaviors that will power your team or organization closer toward your established desired results, the outcome is much more effective—and lasting.

The mandate declared from corporate headquarters in the UK was explicit. The directive was to "dramatically reduce injuries, recordables, and near misses. Develop more ownership and accountability to foster a culture of safety, or the operation may be shut down."

The numbers exposed an alarming reality. Employees in that region were nearly four times more likely to be injured than in any other region worldwide.

This was not news to the leadership team. They had been attempting to drive the safety numbers in the right direction for seven years. Over that time, they had invested millions of dollars and thousands of man-hours to train employees, develop new guidelines, write new policies, provide superior equipment, produce safety videos, and display signage nearly everywhere. Safety was discussed in team huddles every day. Yet those efforts had yielded little to no improvement on the safety numbers over seven years.

From the first workshop, and several months into a large culture change project including every company involved in the multibillion-dollar operation, participants consistently raved about Graff Engineering. Folks regularly would offer comments suggesting that if every firm operated like Graff Engineering there would never be a hiccup.

When employees from Graff were included in the workshops, other participants would beg them to share their secret. Everyone wanted to know, "How did you guys create such a high-performing team and how have you sustained it?" The Graff employees would proudly smile.

A few dozen workshops into our work with the teams, Graff employees shared a variety of stories about how they transformed their company from one that was on the verge of disappearing to one that was now perceived as the model organization. In fact, they were the envy of every other firm.

We had been asked to work with the leaders of every firm involved in this project to develop and design an integrated solution to improve accountability for safety throughout the culture. As part of the overall solution, a variety of workshops were scheduled and every employee from every firm involved would attend. Knowing what was at stake, senior leaders of every single firm volunteered to participate and cofacilitate each workshop. Lauren, the CEO of Graff Engineering, would be joining me in an upcoming session focused on culture change.

As part of the process and preparation, Lauren and I spoke on the phone a few days in advance of the workshop. I had heard numerous accounts of how Graff Engineering had transformed, but now I would be able to learn from the person who had led the feat.

Following some small talk and review of logistics for the upcoming session, I shared with Lauren the consistent praise, accolades, and tributes that Graff Engineering received from employees from every firm represented in the workshops.

"Let me tell you, Mike. If you had been here three years ago, you would have been hearing about our impending doom."

"That is what I have heard, Lauren. If you have a few minutes, I would love to hear from you what happened and how you and your colleagues morphed an organization on the brink of collapse into what is now viewed as the model firm on the project."

"Happy to share. I will give you the condensed version.

"Five years ago we were ecstatic to have won the contract to be part of this enormous undertaking. We were certain we had the technical abilities and know-how to succeed. What I did not realize at the time was that being the most technically qualified does not guarantee success.

"Just a year into the project, I knew the numbers we were generating were unacceptable in several areas. Our client called our poor performance to my attention, and my team began to apply pressure on our staff to improve. Six months later not much changed. Around that time I received a call from the client asking me to jump on a plane and meet with the CEO and her team.

"The short and sweet is I was told that unless a significant improvement was quickly made, Graff Engineering was in jeopardy of losing our $175 million contract. I learned during that meeting that of the few dozen contract companies connected with this multibillion-dollar project, Graff was the poorest performer in the area of safety. We were put on notice that our client was prepared to exercise the cancellation option in our contract. That would have been a crushing blow to Graff and most certainly would have led to bankruptcy.

"To be perfectly honest with you, I left that meeting unsure of what to do next. I was paralyzed with a flood of emotions. My concern was first for our employees and their families, and then for the firm itself.

"As I sat in my study the following morning and reflected on the situation, I noticed my four teenage children in the other room. I thought about how proud I was that they always seemed to act and behave in a manner that made me proud. They consistently and consciously make choices that shape them more and more into the young adults my husband and I hope they will become. With my focus on my kids and in the comfort of my home, the alarming news I received the previous day drifted away momentarily.

"Nearly in a daze, I had an epiphany. Bill and I had always been clear with our kids about ideal behaviors that will move them toward the outcomes they want, and what we want for them. We taught them the importance of being fully aware of their actions and the consequences of those behaviors. Our goal was to get them to filter every action they took with the simple question, 'Will the behavior I am displaying now help me or prevent me from achieving what is most important to me?'

"In fact, we discovered that when we consistently recognized and rewarded the desired behaviors, the kids demonstrated those behaviors even more often. My husband likes to tell me that we need to 'water what we want to see grow.' He is really big on calling out and acknowledging desired behaviors and letting the kids know when they are doing the right thing. You know, catching them making good decisions."

"There truly is power in appreciative feedback, Lauren," I submitted. When you offer someone genuine appreciative feedback on a desired behavior, they will exhibit that behavior more often. The need to feel appreciated is one of the deepest cravings human beings have."

Lauren added, "The power of this principle was reinforced for me just yesterday. A few weeks ago, while waiting for members of my staff to arrive for a meeting, I thanked my COO, Maureen, for always being on time, being organized, and contributing during meetings. So, when she was the first to arrive to yesterday's meeting, guess what she said? 'I know you like it when I am on time, organized, and prepared to contribute.'"

"That is a perfect example," I said to Lauren. "I had a personal experience demonstrating this just several weeks ago with the person in our firm who processes our expense vouchers. She mailed me a check with a small note attached that said, 'Mike, thanks for being so meticulous with your expense vouchers. They are always on time, neat, complete, well organized, and easy to process.'

"So now, every time I am filling out one of those things, I am thinking that I need to make sure this is perfect—Kimberly really likes that. The irony is, I do not especially like completing those vouchers, but I do it anyway."

Lauren excitedly shared, "I believe if leaders completely understood the power of genuine appreciative feedback they would be all over it. It is such a simple principle that has tremendous lasting impact.

"When I think back to the top-performing teams I have been part of throughout my career, the leaders always made clear and called attention to desired behaviors. I vividly recall moments when leaders showed appreciation for specific behaviors, and how meaningful and memorable those experiences were. There was always a clear understanding of the behaviors that were expected.

"So you ask what was the most important strategy that transformed Graff? We simply committed to apply what Bill and I had been doing with our children for years, and what those top leaders did to cultivate teams where employees always seemed to act and behave in a manner that produced desired results.

"The week following my personal revelation, my staff and I launched into the effort. Initially we spent considerable time pinpointing the results that we absolutely had to attain to not only satisfy our client's expectations, but to outperform every other company involved in this operation.

"We then considered and answered some simple questions.

"If the majority of our employees continued to think, act, and behave in the manner they had been, would we achieve those results?

"What shifts are needed in the way our employees currently think and act that would accelerate change and thrust us toward our desired outcomes?

"What are the actions and beliefs that currently exist that are impeding progress?

"What actions and beliefs must we abandon, and what actions and beliefs must we adopt?

"We spent considerable time imagining and describing in detail how our employees would think, act, and behave in our ideal culture.

"After several iterations we landed on a set of nine behaviors that we were convinced would be abundant in our ideal culture.

"We were steadfast in our commitment to create awareness at every level of the organization to those nine behaviors. We were devoted to shining the light on people displaying these by offering appreciative feedback. From there it snowballed. Departments and teams embraced the effort to the point where they developed unique ways to generate even more focus.

"Today, every employee within Graff is able to recite all nine. So not only do folks know what they are, they are also out there looking for them. We have acute awareness and alignment around these behaviors. When someone exhibits one of these behaviors, you will most often hear someone sharing appreciate feedback to the person displaying the desired action.

"Within six months of focusing on creating an optimal culture by calling attention to these desired behaviors, our performance improved exponentially and I was informed that our contract would be renewed. Our employees are passionate about maintaining our culture. They will not tolerate naysayers or cynics. They understand how important these behaviors are to our ongoing success and ultimately their livelihoods."

## The Power of Genuine Appreciation

Creating focus and awareness on desired behaviors has a profound impact on developing and shaping your optimal accountable culture.

Awareness and alignment around desired behaviors will positively contribute to blasting you toward desired goals and objectives.

There is extraordinary power in sincere genuine appreciation. It can be life changing.

Bob Danzig lived his childhood moving from one foster home to another, at one point living in the unlighted attic of a foster family that used to leave his dinner on the bottom step leading to the attic. When a foster family no longer wanted him and it was time to move to a new family, his two sets of clothes would fit easily into a wrinkled garbage bag.

Bob left the foster care system when he graduated from high school at sixteen. He did not graduate early due to academic superiority, but rather to a paperwork error that allowed him to mistakenly slide through some grades. So at sixteen, with no family, no friends, and no close connection to anyone, he found himself on his own.

With extraordinary odds against him, Bob Danzig rose to become the CEO of the Hearst Newspaper Group, a multibillion-dollar company. Bob Danzig writes of two experiences that played a pivotal role in his life and contributed immensely to his future.

While Bob was in the foster care system, a social worker, Mae Morse, would meet with him from time to time when he moved from one foster care family to another. He writes that at the end of each visit, Mae would grasp his hands, look him in the eye, and tell him, *"Bobby, don't ever forget this. You are worthwhile."*

After Bob was fired from a job for a supposed mistake shortly after leaving the foster care system, an acquaintance made him aware of a possible job available as office boy at the *Albany Times Union* newspaper. He was hired to fill the job of office boy.

The office manager that hired Bob called him into her office a few months after he had started. She started by saying, "I've been watching you." Bob's first thought was, Oh no, I am going to be fired again. Instead, the office manager said, *"I just want to tell you that I believe you are filled with promise."*

Bob climbed from office boy to advertising salesman to head of the advertising department and ultimately became the publisher of the *Albany Times Union*. Bob eventually rose to become CEO of the entire Hearst Newspaper Group.

After becoming publisher of the *Albany Times Herald*, Bob tracked down Mae Morse, the social worker from years earlier. In his book, *Conversations with Bobby* (2007), this is how he describes the meeting:

> They had set her up in the parlor chair of the nursing home. She beamed when I walked in. I can see her so clearly, her knit shawl hung over her shoulders. I walked over to her and put my hands in hers. Before I could utter a word, she said to me, "Didn't I always tell you that you are worthwhile?" I was in awe. I told her how I looked forward to this day—the day when I could share with her my gratitude for the confidence and value she placed in me. I said to her, "In a life stuck in the shadows, you, Mae Morse, gave me my first shining moment that penetrated the darkness."

Bob Danzig's story is a living testimonial to the power of genuine appreciation. Bob attributes much of what he achieved in life to those two statements. My friend and former colleague Hyrum Smith developed a menu of "natural laws of successful time and life management," one of which is, "We become what we think about."

Help your employees internalize and manifest the accountable beliefs and behaviors that will ensure your success by calling attention to those behaviors with genuine sincere appreciative feedback.


## Culture: How We Do Things Around Here

Employees, new team members, and especially newer employees all watch others closely to "see how things are done around here." What they observe is far more influential and powerful than anything they may have been told or read in any policy binder. Words are cheap. Employees will watch and observe and then take action based on what they see.

Being swayed to behave in a specific manner happens swiftly, even when we are not sure why we are choosing that behavior.

All of the advice, stories, and counsel I received prior to actually stepping foot in the Arctic did not prepare me fully for the experience.

I was excited and at the same time intimidated to begin the project that would last twelve to eighteen months.

The flight from Anchorage to the North Slope would be departing soon. I did my best to fit in among the diverse group of men and women who were gathering in the gate area and accustomed to the trip that transported them to work every other week. Most worked twelve-hour shifts of two weeks on, and then had two weeks off. I was told folks either loved the lifestyle or hated it. This was not the typical corporate crowd with which I was familiar. My attempt to inconspicuously observe those around me was easily spotted and I was quickly pegged as a first-timer.

With my seat assignment in hand, I sat nervously waiting to board and wondered what I had chosen to get involved with for the next twelve to eighteen months. Those two hours of waiting seemed like an eternity.

With ninety minutes left until boarding, many of those waiting for the flight began to form a line. I thought this to be a bit unusual as we all had seat assignments. And, because most everyone's luggage was checked, there would be plenty of storage space in the overhead bins.

This group of experienced, Arctic-tested employees and contractors continued to form a line that was winding from the gate door all the way down the terminal corridor. I could not see the end of the line. My estimate was the line was easily two to three hundred feet long. Nearly everyone was in line. This was ninety minutes before we were to board. The gate agent had yet to even show up.

I thought to myself, I guess this is the process. So not quite certain why standing in line was necessary, I grabbed my bags and joined the line.

We stood in line for forty-five or fifty minutes until the gate agent, Tony, arrived. Tony knew many of the folks in line and enjoyed quick hellos as he made his way to the gate. He grabbed the microphone and announced the flight would be delayed by an hour due to severe weather on the slope.

I thought, Great, we can sit down and relax. However, to my chagrin, nobody strayed from the line.

Over the next thirty to sixty minutes of standing in line I had the opportunity to strike up conversations with my line mates. As I got to know these folks better, I summoned the courage to ask, "Why are we standing in line so long before boarding time?"

"Just the way we always do it," Stan shared with me.

Feeling I might be pushing my luck, I questioned further, "We all have a seat assignment, right? Why not simply sit comfortably and wait until the plane lands, the passengers deplane and the crew is ready to have us board?"

"Just the way we do things around here," Stan reiterated.

Okay then, I thought to myself. I guess I am going to stand here in line.

Over the next year and thirty plus flights from Anchorage to Dead-horse, Alaska, I would sit in that gate area, watch the line-forming phenomenon occur and join them—because it was the way they did things around there and I wanted to be accountable for how I fit into the culture.

## Forging an Accountable Culture: Little Things Matter—A Lot!

Arriving in Deadhorse I thought to myself, This must be what it would be like to land on the moon.

What I had been told in advance was accurate. There really was nothing there. No hotels, no restaurants, no stores, no trees, no plant life, no taxis, and no nightlife. There was snow and ice as far as the eye could see. And for a place referred to as the North Slope, there was no slope at all. People came here to work—period.

The buildings that were scattered about existed to support the operation in some manner. Fighting the wind that was whipping me across the tarmac, I could feel the bitter cold in my bones. Ninety below with the wind chill. Welcome to Deadhorse.

My plan was simple. I would watch what everyone else did and mirror them. I wanted to fit into the culture.

We walked through what was called an airport, but would be more accurately described as a hut, then continued back outside and climbed into a bus. The bus would drop folks off at various campsites where they would reside throughout their stay.

During the bus ride I reviewed in my mind all of the tips colleagues had offered prior to my trip. Number one, and without any question at all, safety trumps everything else on the slope. Number two, safety is the most important thing on the slope. And finally, safety is absolutely and positively the top focus on the slope.

Okay, that was easy to remember, but what did it really mean? What was I supposed to do? How do I know if I am being safe? What might I do that would cause concern? What might I do that would be viewed as stupid, foolish, or asinine? What happens if I do something that is considered unsafe? My intention was to take personal accountability to be safe and support the safety mandate. I wanted to fit in. However, I was a bit foggy on what that actually looked like.

A few safety practices were drilled into my head quickly. Wear your "spikeys" (slip prevention devices on your boots) any time you are outside, always hold handrails when on stairs, and wash your hands anytime you enter a cafeteria. Washing hands was important because everyone lived together in close quarters. Germs could spread quickly in these camps. Staying healthy was critical to keeping the operation humming. Sanitizing stations were placed just about everywhere you looked. There was a forest of sanitizing stations at the entrance of every cafeteria.

I was assigned a room, which reminded me of my college dorm. My roommate, I immediately discovered, was also very serious about germ prevention. Sanitizing wipes were everywhere.

It had been a twenty-eight-hour journey from the East Coast until I finally was hunkered down in Deadhorse. I was exhausted and hungry. Walking toward the cafeteria I spotted a sanitizing station. Aha, I thought to myself, a chance to demonstrate my knowledge and understanding of the culture.

I stopped and deliberately took more time than needed in washing my hands. I wanted to make sure that as many of my camp mates as possible observed me demonstrating my ability to fit in.

It was pizza night and it looked delicious. After grabbing several slices, I looked around the enormous cafeteria for a place to sit. I spotted a seat in the corner where, in my weary state, I could hide out, eat, and head back to my tiny room.

Before eating, I walked to the window to glance outside. I was curious. Was there anything out there other than snow? There was not. By the time I got back to the table and sat down, a group of five veterans of the slope anchored themselves at the far end of my table. They were in deep discussion about their work project and did not even notice I was there.

I was about to shovel a slice of pizza down my gullet when by chance I glimpsed at the five guys at my table. I had left my glasses in the room, so I had to look twice. But yes, what I thought I saw was accurate. All five of these guys were eating their pizza with a knife and fork. That seemed a bit bizarre to me. I had never done that. In fact, I do not think I had ever noticed anyone doing that.

Immediately, I thought, Maybe it has something to do with the focus on sanitary conditions. That must be it. When I had attempted to place my laptop on the table earlier, I was instantly scolded and directed to remove that "unsanitary thing" from the table.

Within ten seconds I convinced myself that while on the slope you must eat your pizza with a knife and fork to prevent spreading germs. So, even though it felt ridiculous to me, I grabbed my knife and fork and began to eat my pizza. The entire time I was feeling quite silly.

On my way back to my room, I stopped by the desk where rooms were assigned. I asked Peggy, "Why do we have to eat pizza with a knife and fork? How does it help prevent germs from being spread?"

I will never forget the look on Peggy's face at that moment. She appeared bewildered, baffled, amused, and dumbfounded.

"What on earth are you talking about?" she said, almost mysti-fied. "You can eat your pizza however you would like to eat your pizza."

That experience demonstrates the most important facet of how cultures are developed. We watch to see what others are doing. We choose our actions based on what we see happening around us.

## Are the Experiences in Your Culture Nurturing Accountability?

I later shared the pizza story with a diverse group of thirty-five employees who were attending a culture change workshop.

Brianna, a brilliant engineer who was three months into her dream job, raised her hand and asked to share an experience.

"This may not be exactly what you are talking about, but I would like to share it and get everyone's opinion. I am a little nervous about bringing this up because I am new, but I believe it is important.

"As you all know, before we start here we all go through a pretty extensive and rigorous onboarding program. I am a team player and have every intention to be highly accountable and to do what is expected.

"I was told maybe a hundred times over the past three months that when you are on the slope you absolutely must wear and display your badge at all times.

"This is my first time on the slope, and I am wearing my badge and it is visible. But, as I look around this room, I see maybe 25 percent of you wearing your badge. Some of you are in very senior positions and I do not see your badges. So, is this a policy, or is it not? I was told many times it was. Yet I do not see it happening. Is it mandatory? Is it just a suggestion?

"And, if this policy that was ingrained into my head during orientation is not really practiced, what else was I told in orientation that is not really true? How much of what I was told is really worth knowing and how much is not? I am confused."

I watched the faces in the room as Brianna politely and sincerely told her story and asked her questions. Senior leaders in the room instantly had "aha" moments.

If they were saying one thing and demonstrating another, what message was that sending?

Tom, who had twenty-nine years experience, stood up, looked at Justine, one of the senior leaders in the room, and stated, "Thank you, Brianna. You are absolutely right. We are creating too much confusion up here. We need to do something about this. If it is a policy, then we need to make it known and enforce it. If it is not a policy, then get rid of it. And while we are at it, let's look at every other policy that is in the books and not enforced."

One of the most important lessons drawn from that story is how one simple experience can alter a well-intentioned employee's beliefs, create confusion, misalignment, and drive undesired behaviors.

In order to develop a highly accountable culture leaders must be certain that the experiences created are consistent with the messaging.

## The Relationship Triad: Experiences, Beliefs and Actions

Every day there are the hundreds, more likely thousands, of observable experiences that are created throughout your culture. Are those experiences germinating beliefs and actions that will spur employees to voluntarily choose to take personal accountability to achieve what matters most?

The pizza story and the badge story are symbolic of thousands of experiences that take place every day in every organization around the globe.

Exemplary leaders take control and influence the experiences that are being created. They understand the enormous benefit realized when those experiences conjure beliefs and actions that develop, enhance, support, and sustain their optimal culture.

When leaders, managers, and supervisors are conscious and intentional about those experiences, it accelerates culture change. When there is no purposeful effort placed on leading and managing those experiences, the resulting consequences can be devastating and nearly impossible to overcome. Either you will lead and nurture your culture, or you culture will control you.

In an article in the *Washington Free Beacon*, written by Elizabeth Harrington on July 27, 2016, she shared the following:

> The Veterans Affairs administration spent $20 million on expensive artwork and sculptures amidst the health-care scandal, where thousands of veterans died waiting to see doctors. The taxpayer watchdog group Open the Books teamed up with COX Media Washington, D.C., for an oversight report on spending at the VA, finding numerous frivolous expenditures on artwork, including six-figure dollar sculptures at facilities for the blind.
>
> "In the now-infamous VA scandal of 2012-2015, the nation was appalled to learn that 1,000 veterans died

while waiting to see a doctor," wrote Adam Andrzejewski, the founder and CEO of Open the Books, in an editorial for *Forbes*. "Tragically, many calls to the suicide assistance hotline were answered by voicemail. The health claim appeals process was known as 'the hamster wheel' and the appointment books were cooked in seven of every ten clinics."

"Yet, in the midst of these horrific failings the VA managed to spend $20 million on high-end art over the last ten years—with $16 million spent during the Obama years," Andrzejewski said.

The VA spent $21,000 for a 27 foot fake Christmas tree; $32,000 for 62 "local image" pictures for the San Francisco VA; and $115,600 for "art consultants" for the Palo Alto facility.

A "rock sculpture" cost taxpayers $482,960, and more than a half a million dollars were spent for sculptures for veterans that could not see them.

"In an ironic vignette, at a healthcare facility dedicated to serving blind veterans—the new Palo Alto Polytrauma and Blind Rehabilitation Center—the agency wasted $670,000 on two sculptures no blind veteran can even see," Andrzejewski said. "The 'Helmick Sculpture' cost $385,000 (2014) and a parking garage exterior wall façade by King Ray Studio for the 'design, fabrication, and installation of the public artwork' cost $285,000 (2014).

"Blind veterans can't see fancy sculptures, and all veterans would be happier if they could just see a doctor," he said.

Even if well intended, the experiences detailed here surely implanted a variety of deeply held beliefs in the minds of veterans, the general public, and employees of the Veterans Affairs Administration.

The key point of consideration is, "Will those beliefs drive actions among the Veterans, the general public, and the employees that will help or hinder the VA in its efforts to achieve what matters most?"

## You Alone Are Accountable for How You Are Perceived

Each of us is entirely accountable for how we are perceived. Fortunately, we all have complete control over influencing those perceptions.

Lucas was concerned and anxious. Like many of the employees within SeaTech, he was wondering if he would be one of the 30 percent of the workforce that was about to be eliminated. Senior leadership had some difficult decisions to make as the company was facing external pressures and challenges, coupled with a significant drop in product development.

Throughout the day Lucas was highly engaged and I was thrilled to have his zeal and desire to learn on display. Lucas approached me during a break and seemed excited to share his thoughts.

"With everything that is going on around here, I am finding this topic of personal accountability to be calming."

Lucas shared with me that what was discussed in the group earlier that day struck him, specifically the statement, "You are accountable and have complete control over how people view you.

"I was not sure I agreed with it initially," Lucas admitted. "But you are absolutely right.

"The advice I found most illuminating was how the words you speak and, more important, the actions you choose to exhibit over time develop your personal brand, and ultimately the way people view you.

"It is simple logic that others will form their beliefs about us based on what we say, and more important, by what they see us do. I do have complete and total control over what I say and what I do. For some reason, embracing that this morning has been like being covered with a warm blanket. I feel more in control now.

"I want to be crystal clear on how I show up in the eyes of my colleagues. I want to know what my boss thinks about me. I want to be absolutely certain what family and friends think about me. If I do not know what they think, there is nothing I can do to reinforce or change what they think. If I do not ask them, I do not know if they carry any beliefs I do not want them to hold.

"I learned from my last boss that the only way I can be certain about my 'brand,' or how others see me, is to ask for their feedback. For me, feedback is the secret potion to personal accountability."

Lucas knew that asking for feedback allowed him to know for certain how he was perceived. He understood the feedback others had for him was simply the "beliefs" they held about him. And the experiences he created for others was the primary reason for the beliefs they held. He understood that his actions created his brand.

"Some of the beliefs others hold about me are exactly what I am hoping for and helping me achieve my goals. Some of the beliefs others hold are inconsistent with what I want them to believe. In those cases, I must take ownership and accountability to create a new experience, or multiple experiences, for them to hopefully change what they believe.

"What I find frustrating at times is that people may misinterpret my actions or words and form a belief I do not want them to possess. But that does not excuse me from owning what they believe. They will continue believing what they believe unless I choose to create a new experience for them. The choice is totally up to me. I am accountable for the beliefs others hold about me. But I am in control."

Lucas committed to maintaining awareness of the experiences he was creating for those around him, including his colleagues, his boss, family, and friends.

When colleagues would fall into the trap of playing the blame-game, complaining about all that was wrong, making excuses, finger-pointing, and being frozen from taking action, Lucas would choose to focus on what he could control and challenge himself to discover solutions.

When others possessed a mindset of "there is nothing we can do, and things will never change here," Lucas instead maintained a mindset of "what else can I do to achieve the results that we must achieve, and how can I contribute?"

When I bumped into Lucas a few weeks later, after the layoffs, he shared with me that he had consistently been asking himself, "Are the signals I am sending through my actions developing the brand I want? Are my actions suggesting I bring value?"

This awareness helped him develop the brand he wanted others to see. In the eyes of leadership, he was the prototype employee. They believed that with a collection of Lucas-type employees, the sky was the limit to what they could achieve. He was viewed as one of the future leaders of the firm.

## Sustaining an Accountable Culture: Tune-Ups Required

The path to shaping a highly accountable culture does not lead to a final destination. It is an unending journey that requires relentless attention, focus, adaptation, and fine-tuning. Make the mistake that you have arrived and can relax and you may suffer severe consequences. That applies to individuals and teams as much as it does to organizations. Relax, become complacent—or worse yet, arrogant—and you risk disaster.

"No matter how good you are, you can always get better. You find ways. Believe me. I'll do anything to be successful. I hate not being successful."

That is a quotation from Sidney Crosby who many consider to be the top hockey player in the world. Imagine if every single member of your team or organization brought that mindset with him or her to work every day.

Examples of organizations that were admired and enjoyed noteworthy degrees of success, but are no longer with us or have stumbled badly for a variety of reasons, are plentiful.

Blockbuster, Nokia, Borders, Eastman Kodak, EF Hutton, RCA, General Foods, Compaq, Circuit City, Woolworths, Foot Locker, Tower Records, to name a few.

No matter how well you, your team, or organization is performing at this time, relentless desire to improve is imperative to sustain continued success. There will always be somebody else or some other organization—perhaps one you do not even know exists, that is looking to snap up your market share or reinvent the way your business is done.

Consistent fine-tuning and alignment of your culture yields high rewards.

No other region throughout all of Haliboo Dynamics produced like Troy's team—except for Allison's. Troy, a competitive man by nature, was so obsessed with overtaking Allison's team that he could not focus on anything else on his drive to work.

My people are every bit as talented as hers, he thought as he nearly missed the highway exit. They are smart, motivated, collaborative, and accountable. So why does she beat us every quarter? What am I overlooking?

Finally at the office and excited to meet with his team, Troy jumped out of his car and headed to the lobby. "Wait," he said aloud. "My phone!" He darted back to the car. As he approached, he noticed the tires he purchased just weeks ago looked worn.

Troy had splurged on top-of-the-line, high-performance tires for his car. After all, investing in premier equipment would get him the most out of his investment. Yet there were his new tires, practically bald.

He said aloud, "How can this be?"

Later that night at the dealership, Gus, the mechanic, scolded Troy. "The wear on the tires is because the car is out of alignment. You passed on getting the alignment when you bought these. You cannot expect to get continued peak performance out of even premier tires when they are even slightly out of alignment."

With those words, something clicked in Troy's mind.

"That's it!" Troy gushed, turning to Gus, tires no longer on his mind. "I have premier top-notch, skilled, capable people. They are the best. Just like these tires. And like these tires, I can get the most out of my team when we are completely aligned. Only then will we be at our best.

"I have allowed my team to drift off course ever so slightly. Over time that tiny deviation in direction has caused our gap of separation to grow. We are not as aligned as we were just a few months ago. I have let my team lose sight of what matters most. We have strayed off course slightly and we are not as aligned as we need to be."

Troy realized that even highly accountable, top-performing people and teams need alignment adjustments regularly so that everyone is moving precisely and exactly toward the same desired outcome.

With his revelation fresh on his mind, Troy made plans to assemble his team in order to reinvigorate and capture the alignment they had possessed just months earlier.

During our meeting, Troy shared with me his rather simple but effective process he developed to create and sustain alignment.

1. **Perspectives and Insights Sought:** The team is involved to help ensure the best decisions are made by offering input and ideas. Ask, "What are the external and internal drivers at this time that we must consider?" Teams do not make decisions, leaders do. But the most effective leaders seek perspectives first.

2. **Open and Candid Discussion:** No holds barred. Now is the time to voice your perspectives and thoughts—not after the decision is made. Leaders must be sure all team members have had an opportunity to speak their mind.

3. **Commitment Articulated:** All the team members commit to each other their ownership and accountability for the decision. This strengthens camaraderie, collaboration, and engagement.

4. **Consistent Communication:** Ongoing and clear. Share success stories. Communicate the business case around the decision. Include the compelling case as to why the decision was made. Every meeting begins with a reminder.

5. **Check in Regularly:** Scheduled check-ups are conducted to ensure alignment is precise. The tsunami of everything coming at us daily can knock us off track. Like tires on a car, even being out of alignment a few degrees hobbles performance.

Troy's five-step process reinforced laser-like precision and alignment among his team. Not only was the team working in concert now, but new opportunities to improve performance were identified as well.

Troy's team soon overtook Allison's as the top-performing region. Other regional vice presidents adopted Troy's five-step process and it became part of every leader's toolbox.

## Accountability Tip: Do Not Shoulder Their Accountabilities

Things do not always go according to plan. Even highly trained, competent employees who are impeccably clear on what is expected of them will run into complications or challenges on the journey to your must-achieve desired results. At times, rather than figuring out how to overcome the obstacle on their own they may turn to their manager or leader instead.

If that leader or manager chooses to offer a solution, who now has taken accountability for the issue?

The leader or manager has taken accountability from their employee.

Whose problem was it to solve?

Remember, in this scenario the employee is highly skilled, trained, competent, and clear on expectations.

Why is it that many managers and leaders fall into the trap of solving problems for employees that are skilled, trained, competent, and clear on expectations?

Often it is because doing so is the easiest and quickest option.

However, it is a trap. The next time that same highly skilled, well-trained, competent employee runs into even the smallest snag in the future, what is he or she most likely to do? You bet: that employee will come running back to you for help.

Do you have time to solve problems for every single one of your employees?

Are you building bench strength when you take on the problems of others?

To fortify and cement a culture of accountability, once employees are skilled up, trained, and clear on expectations it becomes their accountability to find a way forward.

Ellen, an effective leader within Peachtree Corporation, understands this key principle.

With her teenagers out the door and on the way to school, Ellen's attention now was on getting her ten-year-old son, Brock, off to school.

"Mom, I need you to do these math problems for me. I cannot do them. I am stuck."

Ellen asked Brock to point out the problems that were challenging for him. Brock pointed to the problems and claimed, "They are just too hard. I cannot do them."

Although in a rush to attend an important strategy meeting, Ellen took a deep breath and remained patient. She calmly coached Brock to think through past lessons and to draw upon the skills she knew he possessed.

Rather than solve the problems for him, she asked Brock leading questions that forced him to think through and solve the problems on his own.

Finally at work, Ellen entered the strategy meeting to discover the team was off topic and caught up in a heated debate about a supply-chain issue. Ellen could easily have solved the dilemma. After all, she had twenty-three years of experience and was widely regarded as "the" supply-chain expert in the entire organization.

Instead, she thought about Brock and his math homework that morning. With that in mind, Ellen decided to continue with the strategy that had proven to help her build effective, capable, and accountable teams throughout her career. Rather than solve the issue for them, she asked the team open-ended questions.

"What is preventing us from our desired result? What else can we do? What do we have control over? Of all we could do, what is the one thing that would help most? If we had to do something in the next day to make even a little progress, or we would all lose our jobs, what would we do?"

Ellen shared with me, "I can honestly say that I believe that is one of the principles that has helped us foster an organization with high levels of personal accountability. Sure, managers and leaders can often provide solutions and solve the problems. But that is a trap. If we solve their problems, we unload accountability from their shoulders and take it on our own. Even worse, we almost guarantee they will run to us whenever they encounter even a hint of a dilemma, hurdle, or problem.

"I do not have time for that. They are well trained, have the necessary resources available and are clear on what we must achieve. Not to sound harsh, but their problems are their problems. I have too much on my plate to solve everyone's problems.

"Early in my career, I allowed myself to get sucked into that ruse. I found it quicker and easier to solve problems for others. Then I realized that I was not building bench strength by doing so. I learned over time that by having the patience to ask the right questions my folks have discovered that they must take accountability to figure things out. We are a more resilient culture as a result of it.

"I need to be certain that expectations are explicit, that they have the appropriate training, skills, and competencies, and that they have the necessary resources available. With those pieces in place, it is then incumbent upon them to take accountability and figure things out."

## Do Not Dwell on Why: Keep Moving Forward

Learning from mistakes is an important part of growth. But dwelling on mistakes rather than developing a plan to move forward is unproductive.

Kelly spent a couple of weeks at her new job absorbing as much as she could about her new team and the culture. Her track record was unparalleled and the CEO of Trinity Corporation was expecting big things from his new superstar. He thought if anyone could resurrect this unit it was Kelly.

During her first few days, Kelly observed that team members spent an inordinate amount of time falling into behaviors that scuttle accountability. Initially thinking these were isolated or coincidental incidents, she became alarmed as she discovered these were the norm.

Kelly was dismayed to learn how much time her team members spent playing the blame-game. The amount of time devoted to wallowing in self-pity, dissecting and analyzing mistakes, complaining about other people and departments, and discussing why things will never change was astonishing.

Okay, I need to come up with a plan to eradicate that behavior, she thought as she drove home that night.

As she glanced at her mail once home, she noticed a letter from the Youth Center. It was from the director, thanking Kelly for her work there. Kelly volunteered with at-risk youths.

That's it, she thought. The kids! The solution is with the kids.

Kelly noticed unmistakable behaviors in the children who excelled versus those who did not. With guidance from their counselors, the group that did well had developed specific skills for holding one another accountable for their actions.

This group did not fall into the traps of blaming others, finger-pointing, denying reality, confusion, or solving others' problems. The counselors had nurtured a unique response and skill within this group that reinforced and developed each member's personal accountability.

When a classmate failed to follow through on a commitment or fulfill an agreement, these kids would first clarify the commitment or agreement that was made. They did not focus or ask why it was not kept. For the most part, time spent rehashing why things did not happen was not viewed as productive or, for that matter, helpful.

What was important was the acknowledgment that a commitment or agreement was made and not fulfilled. That reinforced ownership. They did not waste time assessing why—that was water under the bridge—but instead focused on the here and now. The current reality was what mattered.

The kids would then help their friends solve their problems by asking open-ended questions. Once the kids agreed on a solution and way forward, they would gain commitment and make a plan to follow up.

Kelly experienced firsthand how this simple approach helped instill a sense of personal accountability within at-risk youths. It helped move them from a mindset of victimization to one of creating solutions.

Kelly enlisted key members of her team to help her lead this approach within her unit. After some coaching and planning, the team of volunteers committed to modeling the behavior of the children from the Youth Center.

Following several weeks of determined effort, Kelly was able to sense a change in the demeanor of her unit. For the most part, employees found it refreshing and engaging to understand they had more control than they had historically convinced themselves they had.

The approach used by Kelly and her team sent a subtle yet unmistakable message that staying stuck playing the blame-game would not be accepted.

## Accountability in Times of Change: Overcome Fear and Denial

When organizations move through change efforts, about 80 percent of employees will find refuge in one of two places. Those two places are denial and discomfort/fear/pain. There will be a large group that will deny the change is needed or, for that matter, that it will even occur. Another large group will experience some degree of pain or discomfort (anxiety, stress, unease, distress, doubt, etc.) with the change. The approach to move these groups to take accountability differs.

It had been several years since Tom had last seen or talked with Eddie. He was not sure what to expect.

For more than twenty years Eddie had blamed other people or things for his problems. When asked why he could not improve his situation, without fail he responded with a catalog of reasons, excuses, and events to explain why the world was against him. Tom found it exhausting to spend time with Eddie.

In Eddie's mind, *he* had nothing to do with his difficulties. His situation was the fault of other people and events and he would attempt to convince everyone he met that such was the case. He was oblivious to the reality that he was in his situation as a direct result of how he chose to respond to his challenges.

Eddie was stuck in denial. He was unable and unwilling to make the connection that he was in his current situation—personally and professionally—as a direct result of every choice he had made. Family and friends had given up on attempts to counsel and encourage Eddie.

As Tom walked into the restaurant, he spotted Eddie seated at a table near the fireplace. Tom's gut churned a bit as he attempted to convince himself, I can get through this.

As Tom approached the table he noticed that Eddie was beaming. Tom did not expect this at all.

Within minutes Eddie was telling Tom about all the great things that were happening in his life and how happy he was. Tom was thinking to himself, Am I dreaming? This is not the person I have known all my life.

After a few minutes of Eddie's sharing everything that was going well in his life, Tom stopped him mid-sentence and asked, "Eddie, the last time we talked you had pretty much given up on life. What happened?"

"Simple," Eddie said. "After years of putting up with me denying I played a role in where I was in life, my wallowing in self-pity, complaining, and making excuses for my circumstances, my daughter pulled me aside one day, looked me in the eyes and calmly said, 'Dad, is this the life you want for yourself?'

"She sat there completely silent waiting for me to respond. I was ashamed to even look her in the eye, knowing deep in my heart that for years I had been avoiding and running from my problems.

"She then asked me, 'What is going to happen if you continue to live in denial?'

"For some reason, those simple questions coming from someone that cared so much about me had profound impact.

"At that moment, I admitted to what I had known but denied for years. My daughter forced me to face the truth and recognize reality. My reality was that I was living a life of blame. I found it easier to ignore my realities and blame other people or things when times got tough. Instead of reflecting on how my decisions and choices were the root of my problems, it was easier for me to hide from them. I was in denial and numb to reality.

"That day was a wake-up call for me. I accepted that if I continued denying my realities the consequences would not be desirable. To understand and believe I have complete control over my life has been pivotal in helping me improve my situation. It has been invigorating to realize I can choose to take accountability to achieve what I want in life."

Eddie's daughter was not educated, skilled, or trained in building personal and organizational accountability, but her approach to helping her father was spot-on.

When someone is in denial, and until they recognize their own realities, any attempts to encourage and build desire and willingness to move from it will drive them deeper into denial. To move someone out of denial, you must reflect the truth of the situation. They must draw their own conclusion about the consequences of their choices and their current state.

To move folks out of denial they must first recognize reality. Applying accountability is the prescription for these folks. They need help visualizing the consequences of their choices. It is important for this group that they draw their own conclusion about why the change is necessary.

They must understand the external or internal drivers (i.e., increased competition, new regulations, globalization, new technologies, opportunity to grab new markets, shrinking margins, commoditization, shorter product life cycles, economic conditions, etc.) that instigated the need for change.

Most often, when you emerge from denial and finally accept the truth, or recognize your realities, you typically move into the pain or discomfort arena. This is where you may now understand the reason driving the change, but still find it uncomfortable.

To move those experiencing some degree of pain into a place of personal accountability requires a different tactic. Those in the pain domain must be inspired, motivated, and energized. It is important to nourish this group by demonstrating and providing support and guidance.

The approach for those residing in pain is to offer encouragement, show belief, and urge them to keep moving. Using this approach with someone in denial is fruitless, as they will see no reason to take any action.

Organizational change, as with most significant personal transformations, often involves some degree of pain or discomfort. Consider your own moments of personal growth or rewarding transformational discoveries. My guess is that some form of discomfort and pain preceded them. Think back to Tito in chapter 2. His most rewarding, life-changing discovery was preceded by unimaginable emotional and physical pain and discomfort. As he said himself, it made him a better person.

Change spawns feelings of ambivalence, apprehension, unease, discomfort, or pain among the rank and file. There are two common mistakes made during times of change that must be avoided.

The first mistake is to slow down the change effort in an attempt to make it comfortable for everyone. Often the thinking here is that this allows folks to slowly adapt and feel better about it. Providing comfort is not the solution. In fact, comfort gives rise to complacency and incubates inaction, which are the primary culprits that sabotage change efforts. Providing comfort is not the solution.

It is vital to create and maintain alignment around the desired future state (the must-achieve desired results) and to keep moving. As current realities and external drivers evolve, course correction may be needed, but it is vital to never stop moving. Slowing or pausing in the midst of an important change effort sends the wrong message and is usually the death knell to change initiatives.

The combination of acknowledging the discomfort, infusing a sense of safety, and inspiring folks to keep moving is the appropriate formula. Consider the following.

The CEO of Sun Belt Technology was fielding questions from a group of employees during a town hall meeting. External drivers precipitated dramatic changes be made or the company would suffer crushing consequences. One of the folks attending raised a hand to comment on the newly announced direction and objectives that had been communicated to all employees.

"Mr. Murray, it is not going to be easy to do this. What you are asking us to do seems very difficult and will be tough."

Manny, the CEO, thanked the employee for the comment and commandingly replied, "Nothing important or significant is ever easily achieved. I have faith and confidence that we have the resources, talent, and technology to succeed. Lastly, we do not pay our employees to be comfortable. We will keep moving forward and support one another as we do."

His message was received loud and clear. Comfort—complacency and the status quo—was not an option. He did, however, acknowledge and accept the discomfort that some were feeling. More important, he provided encouragement, assurance, and belief in his employees.

The second mistake is to spend too much precious time attempting to prod and convince folks to buy in based on the benefits of the change. Face it, you will never convince everybody to buy in nor will you achieve complete consensus. Striving to gain buy-in from the majority of your employees prior to launch is nearly impossible. Doing so often creates more resistance among the "nevers" and the naysayers, and casts doubt in the minds of those on the fence ("maybes").

In a world that is moving in nanoseconds there is no time for incremental nudging and coaxing that often cause more damage than good. Waste time consensus building or slowing down change and the resistance will build. When your models and top performers notice the attention being paid to the naysayers, cynics, and nevers, it destroys their enthusiasm, morale and passion.

During times of change, consider the three segments of employees in the bell-shaped curve—the models, the maybes, and the nevers—that we discussed in chapter 2.

The first group, the models, will hear the message, understand it, believe it, buy-in, and choose to become involved. Moving ahead boldly and leveraging your models and top performers is the formula to success. Involvement is more essential than buy-in.

The second group, the maybes, will only "get it" and choose to become involved when they see or experience some success or a small win. This group looks for reasons to select in or select out. With every step taken closer to the desired outcome, more buy-in and involvement is gained with this group. As a team or organization creeps closer and closer to the desired state, buy-in, ownership, and accountability increase as employees are able to see the desired outcome more clearly.

The third group, the nevers, will never get it and will only see it and buy in once the desired result is achieved, as covered in chapter 2.

Once you have determined bold change is needed, strike while the iron is hot. Understand that change is uncomfortable for many and be prepared to engage them using the appropriate strategies.

## The Pinnacle: The Accountable Organization

No matter how powerful one person or one team is, a group that is united and aligned is exponentially more powerful.

"I love going there. All of the people there are great," commented Conner about his favorite restaurant.

Phyllis smiled at her son, thinking, I wish our customers were saying that about us.

Phyllis, the president of Synergetic Technologies, had been trying to resolve internal conflicts between functional units for years, and nothing seemed to work. There were brief periods where various "tourniquet" approaches slowed the bleeding, but it never solved their problems for the long term.

The lack of cross-functional accountability eventually would rear its crippling head and demolish any progress made and wreak havoc on efforts to unify functions.

Functioning in silos was stifling growth, suffocating innovation, devastating employee engagement, and most troubling—customer experience scores were plummeting.

Brent, Phyllis's husband, burst into the kitchen where Conner and Phyllis were discussing Conner's favorite restaurant. Brent was returning from his visit to the electronics store.

His anger and disgust were audibly and visually on display as he barked, "I will never spend another dime with them. They may be the worst company on the planet."

Brent went on to share with Phyllis details of the abysmal experience at the store. "Not one person was willing to take accountability to resolve the issue. I am finished with them," he seethed.

"All they did was lament how another department created the problem and that it happens all the time. As if sharing about their dysfunctional interdepartmental relationships would make things better.

"I do not care what department dropped the ball. Why would they waste their breath telling me who in their company was to blame? All I care about is the experience their company has created for me.

"Not one person was willing to take accountability and find a solution for me. What a horrible experience."

In her den later that evening, with her mind almost fully focused back on business, Phyllis reflected on Brent's experience earlier that day. Something Brent had said really struck and stuck with her. She thought to herself, What detrimental experiences are we creating for our customers because of our lack of cross-functional accountability?

Our customers, thought Phyllis, do not have interest in learning who was to blame, why we may have breakdowns between units, or what causes our lapses in operational performance. They see us as simply a single company, Synergetic Technologies. To our customers, it is a seamless operation. They don't give a darn about functional issues, how well we work together, or what department dropped the ball.

Operating in silos where each unit has defined their own objectives and metrics has created separation, Phyllis's thoughts continued. Operating in isolation has resulted in territorialism, self-preservation, hoarding resources, poor communication, lack of collaboration, and conflict. My functional leaders have been effective at creating high levels of accountability and ownership within their units. We need that same degree of accountability cross-functionally.

Phyllis committed the rest of that evening to reflecting and then deciding on her next step.

The next morning, Phyllis kicked off her staff meeting with a quotation from Albert Einstein: "Problems cannot be solved at the same level of awareness that created them."

"He's right, you know. We operate as four distinct silos, each with its own understanding of what makes it successful. This has led to communication breakdowns, isolation, and hoarding resources, ideas, and talent.

"Let me borrow again from Albert Einstein and ask this team, 'If we continue to operate and function as we always have, can we expect to realize a different result?'"

Phyllis did not have to wait for a response. Everyone in the room got the message loud and clear.

"The current reality is that your individual teams are highly accountable to your functions. The commitment, ownership, engagement, and accountability you have created within your functional areas have regularly produced peak results for your teams. Although your team members are dedicated to the success of your functions, oftentimes you make strategic decisions without considering the impact on other functions. At times this has sparked unintended consequences and generated internal competition, mistrust, conflict, dissension, and lack of collaboration and shared accountability.

"As I think about how our functions perform, The 'Human Body Metaphor,' primarily attributed to Edwards Deming is apropos.

"Imagine if every system (function) in the human body performed in isolation and at maximum ability at all times: Breathing in as much air as possible, circulating blood at maximum capacity, absorbing as much information as the mind is capable, reproductive system and digestive system at full tilt. What would result from that?

"The fact is we are interdependent and interconnected. Decisions made in one function directly affect the functioning of the others. Cooperation, synergy, and systems-thinking are what will thrust us toward optimum results.

"There have been times in the past where this team, those of you seated here, have rallied together and collaboratively achieved some minor miracles.

"Last night I reflected on those times in hopes of understanding what it was that kindled those glorious team efforts. Take a moment on your own and think back to those occasions. What was it that triggered all of that passion, energy, alignment, sense-of-urgency, collaboration, camaraderie, creativity, sharing, openness, and can-do mindset organizationally?

"I believe I have been able to identify the instigator." Phyllis then stayed silent and allowed her team to think.

Harold, smiling ear to ear, slowly stood from his chair, gazed out at his colleagues, and asked Phyllis if he could take a stab at answering her question. Of course, she obliged.

"We have been at our best in times of crisis," Harold declared. "When this company was on the verge of calamity or distinction.

"Think about it, guys. When times have gotten tough and things have looked dire, we have consistently pulled together to form a united front and have overcome huge challenges. Those have been the most memorable times here for me. In fact, my people often ask me why we do not operate that way all of the time."

Phyllis was beaming. Harold had landed upon the same answer she had the previous night.

Phyllis asked, "What is it, Harold, or any of you for that matter, that becomes crystal clear in times of crisis?"

"What is at stake for everybody, how we must work together, and not a number or metric," Julie boldly announced to her colleagues.

"Exactly," Phyllis gushed. "During times of crisis, what must be achieved and how we will execute becomes crystal clear very quickly. During those times, this team sitting here has banded together to accomplish some incredible results. In each instance collaboration escalated, resources were shared, there was openness and transparency around decision making, conflicts were resolved quickly, information was shared, and, most important, accountability soared.

"Unfortunately, after any crisis had been averted in the past, we slowly drifted back into our silos where the focus reverted back to our own desired outcomes and our own metrics. That is what creates separation, deficient cross-functional accountability, along with the other operational breakdowns we experience too often.

"As Einstein put it, we are not be able to solve existing conflicts if we stay at the same level of awareness (consciousness) that created them. My recommendation is that this team crafts a Vision of Achievement for Synergetic Technologies that we all mutually own.

"A Vision of Achievement that not only pinpoints the business results that, once achieved, ensure our mutual ongoing success, but also describes how we will execute and work together. This is not a rehash of our Vision Statement connected to our Mission Statement. Those are important, but are simply platitudes and have little meaning for anyone not involved in their creation. Our Vision of Achievement will specify in detail the behaviors, beliefs, and actions of how we will work together to achieve what matters most.

"As we have demonstrated in the past during trying times, we know what it takes and how to execute flawlessly and cross-functionally to achieve what matters most. Instead of waiting for crises to be thrust upon us to unify we will take control.

"This team will identify our desired future state—where we want to go—as well as our expectations of how we will interact together. We will all own this, commit to it, and take personal accountability to lead our functions in a manner to manifest our Vision of Achievement. We will define what success looks like and how we will execute together to achieve it. We will win or fail as a team."

With contributions from everyone the team devoted time during their next five staff meetings developing a Vision of Achievement. The final version read as if one person wrote it. The Vision of Achievement was unambiguous and specifically detailed how these teams would execute and work together. There was unequivocal alignment, commitment, ownership, and personal accountability among the entire team to embrace, embody, and manifest the Vision of Achievement.

Fourteen months after adopting and operationalizing their Vision of Achievement, Phyllis and her team were recognized by their board of directors for achieving performance levels never before reached.

## Key Learnings to Internalize

### One:

When leaders exude the principles Gencova leader Alex uncovered, they set the stage for their teams or organizations to maximize potential. A leader's willingness and ability to maintain focus and awareness of these proven concepts helps to ignite personal and organizational accountability. Leadership is primarily about visioning, encouraging, motivating, aligning, inspiring, modeling, challenging, and strategizing. Accountability flows when exemplary leadership practices are in play.

### Two:

Human beings need to feel appreciated. When you recognize and reward desired behavior with appreciative feedback, not only do you feed a genuine need for your employees, but you also open the door to tangible benefits to your organization.

Openly and visibly recognizing these desired behaviors by calling attention to them creates organizational alignment around those behaviors. When given genuine appreciative feedback, employees will demonstrate those specific behaviors more often and to a greater degree. Recognizing and rewarding desired behaviors—or catching people doing the right thing—is a simple pragmatic tool effective leaders use to increase accountability.

### Three:

Whether a team of fourteen or an organization of thousands, employees watch their colleagues closely to determine what behaviors are appropriate, acceptable, and tolerated within a culture. Employees take cues by observing those around them—peers, supervisors, managers, and top leaders. Being cognizant of the messages sent through their words, body language, tone, and actions is the most important practice leaders can internalize. Leadership, no matter your positional authority, has a lot to do with the congruency between what you say and what you do. When you act inconsistently with your messaging, credibility is lost.

**Four:**
The words you speak and, more important, the actions you choose over time, develop your personal brand, or the way people view you. Are you crystal clear on how you show up in the eyes of your colleagues? Do you know what your boss thinks about you? Are you absolutely certain what your family and friends think about you? The way others perceive you may or may not be aligned with your desired brand. The only way to know for sure is to ask for feedback. The feedback others have for you is simply the beliefs they hold about you—and you are the one that created the experiences that resulted in those beliefs.

**Five:**
Alignment around must-achieve desired results and objectives erodes naturally over time. We are all bombarded daily with new information, fires to put out, tasks to complete, e-mails to read, and issues in our personal lives. The daily tsunami of all of that stuff coming at us weakens and decays alignment. Just a little deviation in understanding, perception, or interpretation can result in employees slightly veering off course. Over time they'll move further and further from your charted course. Devoting time to maintaining alignment is essential to ensure you achieve what matters most.

**Six:**
When employees are stuck, asking questions that require more than a simple yes or no forces individuals to think through the

issues and take ownership and accountability. Socratic questioning often leads to moments of self-discovery. Most often, employees discover that there is always some small step they can take to move closer to a must-achieve desired result if they truly possess ownership for that result.

**Seven:**
When we hold others accountable in a positive and principled manner, it accelerates personal growth, enhances creativity, and increases leadership capacity. Most important, it can quickly shift a culture stuck in the quagmire of the blame-game and excuse-making into one of engagement, ownership, and accountability.

**Eight:**
In organizations, especially during times of change, a large segment of employees congregate in a place we call denial. Another sizable batch muster in a place we call pain/fear/discomfort. To thrust these two groups of employees into an accountable mindset requires different tactics.

When you're dealing with someone in the pain/fear/discomfort arena, the effective approach is to offer encouragement and support and demonstrate belief. Encourage these folks to move and take action. If you attempt to apply accountability to folks in pain/fear/discomfort, they often are driven into denial.

Those in denial require a different approach. For this group it is vital to apply accountability and reflect the truth of the current realities so they can draw their own conclusion that they must take accountability. They must be shown the mirror of reality and be made to understand the consequences of not taking action. If you offer the encouragement and belief that works so well for people in pain/fear/discomfort to someone in denial, they will not accept it and most often become more entrenched in denial.

**Nine:**

Personal accountability and team accountability are helpful, but top-performing organizations go a step further. They strive for organizational or cross-functional accountability. This is achieved when there is a mutually created and owned "Vision of Achievement" that drives agreed-on organizational behaviors and agreements of interplay between functions. The focus of the Vision of Achievement must be on desired organizational outcomes, rather than desired functional outcomes.

Cross-functional accountability is realized when you know you can count on every component of the organization to execute from a higher level of consciousness that is derived from the mutually created Vision of Achievement. This amplifies trust, which in turn balloons morale, engagement, collaboration, and commitment. Ultimately this results in increased performance and optimal experiences for all internal and external clients.

A Vision of Achievement is your blueprint for how you will work together. It should detail the behaviors, beliefs, and attitudes that exist in your desired future state. It provides an understanding of what is expected of me and what I expect of others. It moves us away from analyzing why things may not be working perfectly in our current state and instead creates focus on how we will work together in the future to achieve the Vision of Achievement that is mutually developed.

## Take Accountability Now

**One:**

Reflect on the eight principles Alex identified. All six are important ingredients to forge a culture of accountability. As you consider your must-achieve desired results, which of these six principles can you embrace to improve organizational accountability? Develop your plan to implement the practices of exemplary leadership to bolster organizational accountability.

**Two:**
What are the specific behaviors employees would exhibit in your optimal culture? What actions would be on display that you know would accelerate achievement of your must-achieve desired results? Make a list of five to seven behaviors you want your employees to demonstrate. Then go and offer appreciative feedback daily to employees you observe demonstrating one of those behaviors.

**Three:**
As you think about your must-achieve desired results, what visible actions can you display daily to reinforce your messaging to those watching you? Remember, words are cheap. People will tolerate what you say, but they will act on what they see you do. What is your plan to build capability in others and to energize and engage others?

**Four:**
Ask five colleagues at work and five friends or family members to provide you with five words that enter their minds when they think about you. Collect all the words (fifty total) and look for common words. Those common words, how people see you, is a snapshot of your personal brand. If it is not the brand you want, plan to create new experiences, to foster and generate new beliefs others will hold, and to develop the brand you wish to have.

**Five:**
Maintain alignment: regularly realign your team using the five-step process developed by Troy at Haliboo Dynamics. Investing a few minutes every week can keep your team a smoothly running, highly accountable machine.

**Six:**
Identify one colleague who may be stuck in the blame-game. Put together a list of six to eight open-ended questions that you

can use to help coach this person to take personal accountability for his or her situation. Here are five on the house:

(1) What is preventing you from making the progress you need to make? (2) Of everything that is holding you back, what are the things you can control? (3) What else can you do? (4) What are the consequences of not achieving this? (5) What actions are you going to take and why?

## Seven:

Identify today someone at work, or in your personal life, that has not followed through on a commitment or agreement. Utilize the process that Kelly of Trinity Corporation uses with her team to hold others accountable in a positive principled manner: (1) identify the commitment; (2) acknowledge without blame that it was not fulfilled; (3) problem-solve by asking open-ended questions; and (4) agree on a solution and plan to follow up.

## Eight:

Identify one colleague not currently operating with an accountable mindset and determine if the colleague is in denial, or experiencing some degree of pain/fear/discomfort. Then apply the appropriate approach to move this person into an accountable mindset. Remember: use accountability for people in denial by reflecting the truth of the realities; use encouragement and demonstrate belief to keep those in discomfort/fear/pain moving forward.

## Nine:

Identify silos that exist within your organization at a macro or micro level. Have these teams work together to develop a mutually owned Vision of Achievement that will eliminate separation between silos. The Vision of Achievement should precisely detail expectations of how functions commit to work together as well as agreed-on behaviors of interplay and execution.

## Live and Lead Accountably

Thank you for taking this excursion with me. You have begun to build the foundation to take accountability to achieve what matters most, both personally and professionally.

Accountability is not a destination. It is an ongoing journey that requires focus, self-awareness, and internal fortitude.

Return often to the ideas, concepts, and models included in these pages to reinforce and master the principles of accountability. Take accountability for your future.

Live and lead accountably.

# Index

Page references followed by *fig* indicate an illustrated figure.

**A**

Accountability
  as being about control, 83–86
  as the catalyst for organizational change,
      117–120
  as a choice, 35–64
  comparing the positive vs. negative
      perception of, 2–3
  cross-functionally, 151–155, 158
  defining the four categories of, 67–73
  do not should your employees's,
      142–144
  as everyone's job, 46–49, 63
  for how you are perceived by others,
      137–139
  live and lead with, 161
  the magic and rewards of taking, 6–11
  ownership matrix of, 79–80*fig*
  perception shift required for achieving,
      97–99
  shared, 56–61, 64–65
  as trumping responsibility, 109–111
  understanding the high stakes of blaming
      vs., 5–6
  working to heighten employees's,
      1–5
  *See also* Key Learnings to Internalize; Take
      Accountability Now
Accountability accelerator strategies
  changing how leadership approaches
      accountability, 4–5
  communicating outcome priorities, 4,
      12–15
  engaging employees on the front end vs.
      after-the-fact, 3–4
  helping employees understand outcome
      benefits to them, 4
  improving clarity using the "rules of the
      game," 21–28, 32, 33
  *See also* Take Accountability Now
Accountability categories
  accept and embrace ownership for your
      circumstances, 72, 73*fig*, 78–81,
      93–94

  determine solutions and devise a path
      forward, 73*fig*
  recognize entirely your current realities,
      70–72, 73*fig*, 74–76, 91–92, 93
  take action and relentlessly follow
      through, 71, 72, 73*fig*, 90–91
Accountability plan
  begin to develop a, 33–34
  to focus on the "maybes" employees,
      64–65
  to implement your solution, 91
  on overcoming barriers that impede your
      desired outcomes, 65
  shifting employee mindset toward shared
      accountability, 65
  write down three accountable behaviors to
      begin modeling, 65
Accountability scenarios
  Café Delight, 103–106
  Cambria Ventures, 78–81, 89
  Cameron Medical, 15–18
  Clearwater Technologies, 46–49
  CVP International, 21–27, 102
  the Deadhorse project, 129–133
  DigiTech, 106–109
  Gencova, 120–122
  Graff Engineering, 123–127
  Haliboo Dynamics, 140–141, 159
  Harton Industries, 43–46
  Harvest Financial, 98–99
  Hearst Newspaper Group, 128–129
  by Jack in the Box *E. coli* in
      food-poisoning crisis, 5–6
  Johnson & Johnson Tylenol crisis, 6
  Lakeshore Bionics, 19–21
  Mountainview Enterprises, 67–73*fig*, 91
  Peachtree Corporation, 142–144
  Sea Tech, 137–139
  Squire Medical, 1–5
  Sun Belt Technology, 149
  Superior Electronics, 111–113
  Synergetic Technologies, 151–155
  Trinity Corporation, 144–145, 160
  Tundra Oil, 56–61

Accountability scenarios (continued)
Wells Fargo, 52–54
*See also* Experiences; Organizations;
Stories
Accountable behaviors
being aware that all behaviors send a
message, 49–52, 63–64, 119, 120
Clearwater Technologies on identifying,
46–49
collaborating, rewarding, and learning
from mistakes, 49
creating change by igniting desired,
122–127
focusing on solutions instead of blame as,
49
Graff Engineering on transformation
toward, 123–127
modeling, 48–52, 63–64, 65, 119, 120–122
the power of genuine appreciation of,
125–126, 127–129
precise, open, and consistent
communication as, 49
spotlighting identifiable, 48–49
write down three that you can model, 65
*See also* Actions; Behaviors
Accountable culture
the accountable organization as pinnacle
of changing to an, 150–155
assessing if it is nurturing accountability or
not, 133–135
creating lasting change by shaping an,
120–122
CVP International leader's creating a,
21–28
do not shoulder your employees's
accountabilities within an,
142–144
helping employees to "see how things are
done around here" in, 129–131
how the power of genuine appreciation
supports an, 125–126, 127–129
how the relationship triad (experiences,
beliefs, and actions) in, 135–136
it is everyone's job to build a, 46–49, 63
keep moving forward and do not dwell on
mistakes made, 144–145
leadership alignment with the, 122,
139–141, 156
overcoming fear and denial in times of
change in an, 146–150, 157, 160
tactics for instilling shared accountability
into, 60–61, 64
Take Accountability Now to create a,
158–160

tune-ups required for sustaining an,
139–141
understanding that little things matter in
an, 131–133
*See also* Organizational culture
Accountable organizations
accountability as the catalyst for
transformational change to,
117–120
ignite desired accountable behaviors to
create, 122–127
Key Learnings to Internalize for
transforming to an, 155–158
as pinnacle of change, 150–155
the relationship triad (experiences, beliefs,
and actions) in, 135–136
Take Accountability Now for change
toward an, 158–160
Achievement
accountability trumps responsibility for,
109–111
cement the focus for, 100–103
Key Learnings to Internalize for mindset
shift for, 113–114
LAG versus LEAD measures of, 111–113,
114, 115
lessons from the Marauders on, 106–109
the magical mindset shift required for,
103–106
the shift in perception required for,
97–99
Take Accountability Now for mindset shift
to, 114–115
Actions
being aware of the messages sent by our,
49–52, 63–64, 119, 120
to change your current reality, 71, 72,
73*fig*, 90–91
how emotions drive our, 41–42
identifying accountable behaviors and,
46–49
understanding that they matter, 52–54
*See also* Accountable behaviors; Take
Accountability Now
Adrzejewsky, Adam, 136
"Agree to disagree," 74, 75
*Albany Times Union* (newspaper), 128–129
Alex's story, 120–122, 156, 158
*Alone* (History Channel TV show), 30–31
Appreciation and recognition
Bob Danzig's story as a testimonial to
power of, 128–129
as key attribute of leaders, 81, 94
the power of genuine, 125–126, 127–129

Arctic flight's line-forming phenomenon, 129–131
Audrey and Russell's story, 103–106

**B**

Babcock, W. B., 76
Badge story, 134, 135
Baldwin, Doug, 59–60
Behaviors
being aware of the messages sent by our, 49–52, 63–64, 119, 120
understanding that they matter, 52–54
*See also* Accountable behaviors
Beliefs
badge story example of, 134, 135
how emotions fuel our, 41–42
as part of the accountable relationship triad, 135–136
pizza story as example of, 132–133, 135
*See also* Perspectives of others
"Below the Water" behavior
craft five to seven open-ended questions on, 64
description of the, 39–42
as the epitome of the blame-game, 39–42, 62
open-ended questions used to move people from, 42, 62
reflect on how much time is spent on, 64
self-awareness of when you engaging in, 41, 51
Vinnie's story on, 42
Beltway Rail, 86
Blame-game
accountability versus playing the, 2–4
"Below the Water" behavior as the epitome of the, 39–42, 62, 132–133, 134–136
by Jack in the Box *E. coli* in food-poisoning case, 5–6
keep moving forward and do not dwell on mistakes playing the, 144–145
key learning on how lack of clarity feeds the, 32
open-ended questions to move past the, 145, 156–157, 159–160
reasons vs. excuses mindsets of the, 54–56
recognizing your own tendency to play the, 29
reflect on how much time is spent on, 64
understanding the high stakes of accountability vs., 5–6
understand that people are naturally drawn to the, 28
as waste of valuable time, 28–30

*Bloomberg Businessweek,* 43
Brainstorming, 89
Brianna's story, 134
Bruce's story (on Rudy Ruettiger), 86–89

**C**

Cabrera, Miguel, 43
Café Delight, 103–106
Cambria Ventures, 78–81, 89
Cameron Medical, 15–18
Celebrate wins, 81, 93
Challenging the status quo, 89, 95
Change. *See* Transformational change
Child helping his father exercise, 76–78
Choices
"Below the Water" behavior that avoids making, 39–42
destiny is determined by your decisions and, 35–39
Key Learnings to Internalize on accountability as a, 61–64
leading accountability as everyone's job and a, 46–49, 63
open-ended questions to move people to the accountability, 42, 62
parable of the two shoe salesman on, 39
to take personal accountability to help your team, 121
Tito's story on making a, 35–38
Tom Peter's "Tower Death" phrase on lack of, 38–39
understanding that accountability is a, 35
*See also* Decision making
Chris's story, 10–11
Clarity
Cameron Medical on refocusing communication for, 15–18
in communicating outcome priorities, 4, 12–15, 31–33, 121
consequences and sampling of root causes of lack of, 25*fig*
CVP International on creating, 21–27
impact on employee engagement, 21–28, 32
Key Learnings to Internalize on importance of, 31–32
Take Accountability Now with desired results, 32–33
Clearwater Technologies, 46–49
Collaboration
invite everyone to identify a solution, 89, 94
to learn from mistakes, 49

Commitment
  to the agreed upon solution, 145
  articulation of your, 141
  positive principles manner for holding
      people to their, 160
Communication
  being aware of the messages sent by our
      behaviors, 49–52, 63–64, 119, 120
  Cameron Medical on refocusing, 15–18
  clearly state outcome priorities, 4, 12–15,
      31–33, 121
  key leader attribute of practicing excellent,
      121–122
  Lakeshore Bionics on leadership, desired
      outcomes, and, 19–21
  leadership alignment through consistent,
      141
  of leadership's commitment to
      accountability, 4–5
  precise, open, and consistent as
      accountable behavior, 49
Communication plan, 33
Control
  key lessons on accountability and, 92
  lessons from Hyrum on taking
      accountability, 83–86, 94
  Take Accountability Now by taking, 93
*Conversations with Bobby* (Danzig), 129
Covey, Stephen R., 75
Craig's story, 98–99
Creativity and innovation, 81, 94
Crosby, Sidney, 139
Cross-functionally accountability, 151–155,
      158
Culture. *See* Accountable culture;
      Organizational culture
Current realities
  accepting ownership for your
      circumstances and, 72, 73*fig*,
      78–81, 93–94
  child helping his father exercise on, 76–78
  determining solutions and devise a path
      forward, 71, 72, 73*fig*, 86–90, 92,
      94–95
  lessons from Hyrum on taking control to
      change, 83–86, 92, 94
  recognizing your, 70–72, 73*fig*, 74–76,
      91–92, 93
  scheduling time to develop and execute
      plan to change, 95
  seek the perspectives of others on, 71,
      76–78, 93
  take action and follow through to change
      your, 71, 72, 73*fig*, 90–91

Yoda as accountability coach on changing
      your, 81–83
CVP International, 21–27, 102

**D**
Dan's story, 43–46
Danzig, Bob, 128–129
Deadhorse project
  the flight's line-forming phenomenon,
      129–131
  the "little things" forging an accountable
      culture during, 131–133
  the pizza story, 133, 135
Decision making
  destiny is determined by your, 35–39
  perseverance and determination elements
      of, 41
  *See also* Choices
Degal, Willie, 75
Deming, Edwards, 152
Denial
  Eddie and Tom's story on getting past,
      146–147
  as first reaction to change by employees,
      146
  Key Learnings to Internalize about, 157
  overcoming in times of change, 146–150,
      157, 160
Department of Justice, 53
Desired results (outcomes)
  accelerate accountability by stating
      priorities for, 4
  as "The Big Opportunities," 14
  Cameron Medical on resetting metric for,
      15–18
  emotions as fueling beliefs, actions, and,
      41–42
  ensure that employees internalize the
      desired, 12–15, 33
  how shared accountability can help create
      the, 58–60
  importance of clear communication on, 4,
      12–15, 31–33, 121
  Lakeshore Bionics on leadership,
      communication, and, 19–21
  leadership alignment around, 122,
      139–141, 156
  overcoming barriers that impede
      your, 65
  perception shift required for achieving,
      97–99
  Vision of Achievement on your, 154–155,
      160
  *See also* Work performance

Devine, Dan, 88
"Digging deeper," 30–31
DigiTech, 106–109
Do not assume, 89, 94
Dufresne, Andy (*Shawshank Redemption* character), 38

**E**

Eddie and Tom's story, 146–157
Einstein, Albert, 154
Ellen's story, 142–144
Emotions
   decision making and role of, 41
   fear and denial responses to change, 146–150, 157, 160
   as fueling beliefs, actions, and results, 41–42
Employee engagement
   Alex's story on accountable leadership for, 120–122
   on the front end vs. after-the-fact accountability, 3–4
   how clarity and knowing "rules of the game" impacts, 21–28, 32, 33
Employee mindset shifts
   to accountability trumps responsibility, 109–111
   to LEAD measures from LAG measures, 111–113
   the magical mindset shift required for achievement, 103–106
   reasons vs. excuses, 54–56
   to shared accountability, 65
Employees
   avoid playing the blame-game with your, 2–6
   being aware of both others's and their own behaviors, 49–52, 63–64, 119, 120
   clearly communicating outcome priorities to, 4, 12–15
   do not shoulder the accountabilities of your, 142–144
   ensure they internalize the desired results, 12–15, 33
   helping them "see how things are done around here," 129–131
   "maybes," 44, 45–46, 62–63, 65, 149–150
   "naysayers," 44–45, 150
   the "nevers" and three types of to fire immediately, 43–46, 64–65, 149
   shifting their mindset to shared accountability, 65

   the shift in perception required for accountability of, 97–99
   *See also* Jobs; Teams; Work performance
Envision success, 89, 95
Excuses vs. reasons mindsets, 54–56
Experiences
   to assess if your culture is nurturing accountability, 133–135
   how exemplary leaders take control and influence, 135
   as part of the accountable relationship triad, 135–136
   *See also* Accountability scenarios; Stories

**F**

Fear
   as first reaction to change by employees, 146
   Key Learnings to Internalize about, 157
   overcoming in times of change, 146–150, 157, 160
*Forbes* (magazine), 136
*The 4 Disciplines of Execution* (McChesney), 14
Franklin International Institute (now FranklinCovey), 83

**G**

Gail's story, 75
Gencova, 120–122
Genuine appreciation
   Bob Danzig's story as a testimonial to power of, 128–129
   Lauren's story on the power of, 125–126
   shaping an accountable culture, 127–128
Georgia Tech, 88
Graff Engineering, 123–127

**H**

Haliboo Dynamics, 140–141, 159
Harley's story, 54–55, 129
Harrington, Elizabeth, 135–136
Hartman, Captain, 39–42, 51, 54, 62
Harton Industries, 43–46
Harvest Financial, 98–99
Head and heart leadership, 122
Hearst Newspaper Group, 128–129
"The Helmick Sculpture," 136
Holy Cross College, 87
"Human Body Metaphor" (Edwards Deming), 152–153
Hyrum's lessons, 83–86

**I**

Innovation and creativity, 81, 94
Insanity definition argument, 119
*In Search of Excellence* (Peters), 19
Invite everyone, 89, 94

**J**

Jack in the Box *E. coli* food-poisoning case,
    5–6
Janet's story, 1–3
Janitor at NASA story, 99
Jobs
    accountability trumps responsibilities,
        109–111
    cementing the focus of work and,
        100–103
    employee engagement in their, 3–4, 21–28,
        32–33
    lessons learned from the Marauders on
        performance, 106–109
    the magical mindset shift for desired
        results of, 103–106
    powerful shift in perception for desired
        result of, 97–99
    shared accountability versus "my
        responsibilities" of, 57–58
    *See also* Employees
Johnson & Johnson Tylenol case, 6

**K**

Kate's story, 78–81
Kelly's story, 144–145, 160
Kennedy, John, 99
Kevin's story, 111–113
Key Learnings to Internalize
    accountability is a journey, 93
    for change toward an accountable
        organization, 155–158
    "digging deeper" for what you want to
        achieve, 30–31
    on four categories of accountability,
        91
    how clarity motivates and engages
        employees, 32
    making the choice for accountability,
        61–64
    mindset shifts to accountability for
        achievement, 113–114
    people are naturally drawn to the
        blame-game, 28
    playing the blame-game wastes valuable
        time, 28–30
    providing employees with clear outcome
        priorities is crucial, 31–32

on recognizing realities and taking
        ownership, 91–92
    *See also* Accountability
King Ray Studio, 136
"The Know-It-Alls" employees, 43
Kotter, John, 14, 44
Kouzes, Jim, 48

**L**

LAG measures, 111, 112, 113, 114, 115
Lakeshore Bionics, 19–21
Lauren's story, 124–127
Leader key attributes
    be a role model, 48, 49, 122, 155–156
    demonstrate leadership alignment, 122,
        139–141, 156
    focus on the few, 122
    head and heart approach to leadership,
        122
    include and involve, 122
    nurture silo-free mindset, 122, 160
    practice excellent communication,
        121–122
    provide clarity, 121
Leaders
    changing how they approach
        accountability, 4–5
    do not shoulder accountabilities of your
        employees, 142–144
    employee awareness of both their own
        and the behaviors of, 49–52, 63–64,
        119, 120
    Lakeshore Bionics on accountability role
        of, 19–21
    leading accountability as a choice by,
        46–49
    lie and lead accountably, 161
    modeling accountable behaviors, 48, 49,
        122, 155–156
    taking control and influencing experiences
        approach by exemplary, 135
    transformation change role of, 119–120
Leadership
    Alex's story on her accountable approach
        to, 120–122
    alignment by, 122, 139–141, 156
    transformation change role of, 119–120
Leadership alignment
    development and implementation of, 156
    five-step process for achieving, 141, 159
    as key attribute of an ideal leader, 122
    periodic check-ins to support, 141
    Troy's story on implementing, 139–141
*The Leadership Challenge* (Kouzes and
        Posner), 48

LEAD measures, 111–113, 114, 115
Live and lead accountably, 161
Lucas's story, 137–139

**M**
Maddock, G. Michael, 43
Magical mindset shift, 103–106
Malone, Dudley Field, 74
Manny's story, 149
Mary's story, 109–110
Max's Marauders story, 106–109
The "maybes" employees
 description of the, 44
 getting a buy-in to change by, 149–150
 increasing focus on the, 62–63, 65
 understanding the potential of, 45–46
McChesney, Chris, 14
Mindset shifts
 to accountability trumps responsibility, 109–111
 for always getting better, 139
 first ask "What is your job?" to begin, 97–98, 114
 Key Learnings to Internalize for accountability, 113–114
 to LEAD measures from LAG measures, 111–113, 114, 115
 the magical mindset shift required for achievement, 103–106
 nurture a silo-free, 122, 160
 reasons vs. excuses, 54–56
 to shared accountability, 65
 story about the janitor at NASA example of, 99
 Take Accountability Now for accountability, 114–115
 toward alignment, ownership, and personal accountability, 98–99
 *See also* Perspectives of others
Miracles for Kids, 2
Mistakes
 keep moving forward and do not dwell on, 144–145
 not blaming and learning from, 49
 rewarding effort instead of chiding, 49
Modeling
 Alex's story on her approach to accountability, 120–122
 being aware that all behaviors send a message, 49–52, 63–64, 119, 120
 key leader attribute of accountability, 48, 49, 122, 155–156
 write down three accountable behaviors to begin, 65

Morse, Mae, 128, 129
Motivation. *See* Employee engagement
Mountainview Enterprises, 67–73*fig*, 91

**N**
NASA janitor story, 99
"Natural laws of successful time and life management" (Hyrum Smith), 129
"Naysayers" employees
 description of the, 44–45
 resistance to change by, 150
The "nevers" employees
 don't waste time on the, 64–65
 as "naysayers" who actively impede, 44–45
 resistance to change by, 149–150
 as the 20 percent that will never contribute, 43–44
 what to do with the, 45–46
"The Nonbelievers" employees, 43
Notre Dame football team
 Rudy Ruettiger's story on the, 86–89
 "walk-ons" allowed by the, 87

**O**
Open-ended questions
 encourage accountability by crafting five to seven, 64
 helping people to move past blame-game and solve problems using, 145, 156–157, 159–160
 to move people to the accountability choice, 42, 62
Openness and transparency, 91, 94
Open the Books, 136
Organizational culture
 accountability is the catalyst for transformational change, 117–120
 a "want-to" culture trumps a "have-to," 118
 *See also* Accountable culture
Organizations
 accountable, 117–161
 cross-functionally accountability by, 151–155, 158
 examples of failures of once successful, 139
 *See also* Accountability scenarios
Outcomes. *See* Desired results (outcomes)
Ownership matrix of accountability, 79–80*fig*
Ownership taking
 apathy of those who do not engage in, 9
 "Below the Water" behavior versus, 39–42
 Chris's story on the rewards of, 10–11
 creating employee energy for, 4

Ownership taking (continued)
  CVP International leader's creating a
      culture of, 21–28
  for getting things done, 30
  Johnson & Johnson's Tylenol crisis and, 6
  making the choice for degree of, 28
  matrix of accountability and, 79–80*fig*
  taking the opportunity for, 2
  for your current realities, 72, 73*fig*, 78–81,
      93–94, 93–99
Ownership taking tactics
  appreciation and recognition, 81, 94
  celebrate wins, 81, 93
  encourage creativity and innovation, 81,
      94
  openness and transparency, 91, 94
  storytelling, 81, 94
  WIIFM (What is in it for me), 81, 94

**P**

Pac-North Building Supplies, 75
Pain/fear/discomfort, 146–150, 157, 160
Palo Alto Polytrauma and Blind
      Rehabilitation Center, 136
Parable of the two shoe salesman, 39
Parseghian, Ara, 87
Patton, George S., 19
Peachtree Corporation, 142–144
Performance. *See* Work performance
Perseverance, 41
Perspectives of others
  "agree to disagree" attitude toward,
      74, 75
  child helping his father exercise on,
      76–78
  how they can help clarify your current
      realities, 71, 76–78, 93
  leadership alignment by seeking the, 141
  the "young girl vs. old woman" example
      of, 75
  your accountability for the way you are
      perceived by, 137–139
  *See also* Beliefs; Mindset shifts
Peter's story, 76–78
Peters, Tom
  on being able to achieve what you want
      badly enough, 30
  on how senior leaders can contribute
      value, 19
  *In Search of Excellence* co-authored by, 19
  on the "Tower Death" concept, 38–39
  on when to fire an employee immediately,
      45
Phyllis's story, 150–155

Picture success, 89, 95
Pizza story, 132–133, 135
Posner, Barry, 48
Problem solving
  destiny is determined by decision making
      and, 35–39
  open-ended questions to help people
      with, 145, 156–157, 159–160
  perseverance and determination elements
      of decision making and, 41
  *See also* Solutions

**R**

Randy's story, 56–61, 64
Realities. *See* Current realities
Reasons vs. excuses mindsets, 54–56
Recognition. *See* Appreciation and
      recognition
Redding, Ellis "Red" (*Shawshank Redemption*
      character), 38
Relationship triad
  made up of experiences, beliefs, and
      action, 135–136
  VA scandal (2012–2015) example of failed,
      135–136
Responsibilities
  how accountability trumps, 109–111
  shared accountability versus "my," 57–58
Results. *See* Desired results (outcomes)
Reverse thinking, 89, 95
Rio 2016 Summer Olympics, 55–56
Robert's story, 22–24, 102
Ruettiger, Rudy, 86–89
"Rules of the game," 21–28, 32, 33
Russell and Audrey's story, 103–106

**S**

Sea Tech, 137–139
Securities and Exchange Commission, 53
Self-awareness
  craft five to seven open-ended questions
      to increase, 64
  of the messages our behaviors send,
      49–51, 63–64
  for stopping "Below the Water" behavior,
      41, 51
*The 7 Habits of Highly Effective People*
      (Covey), 75
Shared accountability
  a "my responsibilities" focus versus, 57–58
  Randy's story on, 56–61, 64
  shifting employee mindset toward, 65
  tactics for transforming culture into, 60–61,
      64

understanding the importance and benefits of, 58–60

*Shawshank Redemption* (movie), 38

Shipp, Josh, 38

Silo-free mindset, 122, 160

Smith, Hyrum
  his "natural laws of successful time and life management," 129
  life lessons learned from, 83–86, 129
  "We become what we think about" law of, 129

Solution guidelines
  1: rapid-fire brainstorming without analysis, 89, 94
  2: invite everyone, 89, 94
  3: do not assume, 89, 94
  4: challenge the status quo, 89, 95
  5: picture success, 89, 95
  6: reverse thinking, 89, 95

Solutions
  Cambria Ventures's guidelines for creative, 89
  developing plan to implement your, 91
  Key Learnings to Internalize on, 92
  learning to recognize and identify, 86–90
  the magical mindset shift for desired, 103–106
  making commitment to agreed upon, 145
  as one of the categories of accountability, 71, 72, 73*fig*
  open-ended questions to help people find, 145, 156–157, 159–160
  Take Accountability Now by pursuing, 94–95
  *See also* Problem solving

Squire Medical
  need to turn things around at, 1
  working to heighten accountability at, 1–5

Status quo
  challenging the, 89, 95
  Manny's story on rejecting the, 149
  *See also* Transformational change

Stories
  Alex, 120–122, 155, 158
  Audrey and Russell's, 103–106
  badge, 134, 135
  Bob Danzig, 128–129
  Brianna's, 134
  Bruce on Rudy Ruettiger, 86–89
  Chris, 10–11
  Craig, 98–99
  Dan, 43–46
  Ellen, 142–144
  Gail, 75

Harley, 54–55, 129
Janet, 1–3
janitor at NASA, 99
Kate, 78–81
Kelly, 144–145, 160
Kevin, 111–113
Lauren, 124–127
Lucas, 137–139
Mary, 109–110
Max's Marauders, 106–109
parable of the two shoe salesman, 39
Peter, 76–78
Phyllis, 150–155
pizza, 132–133, 135
Randy, 56–61, 64
Robert, 22–24, 102
Ted and Rhonda, 46–49
Tito, 35–38
Tom and Eddie, 146–147
"Tower Death," 38–39
Troy, 140–141, 159
Vinnie, 42
*See also* Accountability scenarios; Experiences

Storytelling for ownership tactic, 81, 94

Stumpf, John, 52–54

Sun Belt Technology, 149

Superior Electronics, 111–113

Synergetic Technologies, 151–155

**T**

Take Accountability Now
  assess how your team is doing on accountability, 93
  for change toward an accountable organization, 158–160
  clearly identify your desired results, 32–33
  develop a communication plan, 33
  develop a plan for increasing accountability, 33–34
  increasing ownership of your current realities, 93–94
  mindset shifts to accountability for achievement, 113–114
  reflect on time spend on "Below the Water" and blame-game, 64
  seek the perspectives of other people, 93
  take control and deploy tactics to meet desired results, 94–95
  *See also* Accountability; Accountability accelerator strategies; Actions

Teams
  Alex's story on leading accountable, 120–122

Teams (continued)
  choosing to take personal accountability to help your, 121
  take an accountability assessment of your, 93
  *See also* Employees
Ted and Rhonda's story, 46–49
"Three Types of People to Fire Immediately" (Maddock and Viton), 43
Tito's story, 35–38
Tom and Eddie's story, 146–147
Toomey, Patrick, 53
"Tower Death" story, 38–39
Transformational change
  accountability as the catalyst for lasting, 117–120
  the accountable organization as pinnacle of, 150–155
  assessing if culture is nurturing accountability or not and need for, 133–135
  do not shoulder your employees's accountabilities for lasting, 142–144
  helping employees "see how things are done around here" to support, 129–131
  how the definition of insanity is a good argument for making, 119
  how the power of genuine appreciation drives, 125–126, 127–129
  how the relationship triad impacts lasting, 135–136
  ignite desired accountable behaviors for, 122–127
  keep moving forward and do not dwell on mistakes made, 144–145
  leadership role in, 119–120
  overcoming fear and denial in times of, 146–150, 157, 160
  by shaping an accountable culture, 120–122
  tune-ups required for sustaining an accountable culture, 139–141
  understanding that little things matter when creating, 131–133
  *See also* Status quo
Transparency and openness, 91, 94
Trinity Corporation, 144–145, 160
Troy's story, 140–141, 159
Tundra Oil, 56–61
Twain, Mark, 70, 74

**V**
Veterans Affairs Administration (VA) scandal (2012– 2015), 135–136
"The Victims" employees, 43
Vinnie's story, 42
Vision
  accelerate accountability by stating priorities for achieving, 4
  ensure that employees understand the desired results and, 12–15
  Vision of Achievement to share your, 154–155, 160
Vision of Achievement, 154–155, 160
Viton, Raphael Louis, 43
Vons Companies, Inc., 5

**W**
Warren, Elizabeth, 52–53
*Washington Free Beacon* (newspaper), 135
"We become what we think about" law, 129
Wells Fargo
  Senate panel (2016) questioning of, 52–54
  Vision and Values Statement conflict with cross-selling by, 52, 53
"What is your job?" question, 97–98, 114
WIIFM (What is in it for me), 81, 94
Work performance
  cementing the focus of, 100–103
  how accountability trumps responsibility for, 109–111
  Key Learnings to Internalize for mindset shift for, 113–114
  LAG versus LEAD measures of, 111–113, 114, 115
  lessons learned from the Marauders on, 106–109
  the magical mindset shift for desired results of, 103–106
  powerful shift in perception for desired result of, 97–99
  shared accountability versus "my responsibilities" for, 57–58
  Take Accountability Now for mindset shift to improve, 114–115
  *See also* Desired results (outcomes); Employees

**Y**
Yoda, the accountability coach, 81–83
The "young girl vs. old woman" perspectives, 75